Wireless Communications and Networking for Unmanned Aerial Vehicles

A thorough treatment of UAV wireless communications and networking research challenges and opportunities. Detailed, step-by-step development of carefully selected research problems that pertain to UAV network performance analysis and optimization, physical layer design, trajectory and path planning, resource management, multiple access, cooperative communications, standardization, control, and security is provided. Featuring discussion of practical applications including drone delivery systems, public safety, IoT, virtual reality, and smart cities, this is an essential tool for researchers, students, and engineers interested in broadening their knowledge of the deployment and operation of communication systems that integrate or rely on unmanned aerial vehicles.

Walid Saad is a professor of Electrical and Computer Engineering at Virginia Tech and an IEEE Fellow.

Mehdi Bennis is an associate professor at the University of Oulu, Finland.

Mohammad Mozaffari is an experienced researcher at Ericsson, USA.

Xingqin Lin is a senior researcher at Ericsson, USA.

"This book, written by the most prominent experts in the field, provides a complete in-depth analysis of UAV wireless communications. It should become a reference material for all the students, engineers, and researchers who are building our next-generation wireless communication networks."

Merouane Debbah, CentraleSupélec

"This is the most comprehensive book on the rapidly evolving field of wireless communications and networking for UAVs. The authors are among the researchers who have made the most profound contributions to this emerging field. Their impressive command of the subject matter results in a thorough presentation taking theory, practice, and industrial standards into account. A must-read for researchers and engineers working in this field."

Halim Yanikomeroglu, Carleton University

Wireless Communications and Networking for Unmanned Aerial Vehicles

WALID SAAD

Virginia Polytechnic Institute and State University

MEHDI BENNIS

University of Oulu, Finland

MOHAMMAD MOZAFFARI

Ericsson Research Silicon Valley

XINGQIN LIN

Ericsson Research Silicon Valley

CAMBRIDGE
UNIVERSITY PRESS

University Printing House, Cambridge CB2 8BS, United Kingdom

One Liberty Plaza, 20th Floor, New York, NY 10006, USA

477 Williamstown Road, Port Melbourne, VIC 3207, Australia

314–321, 3rd Floor, Plot 3, Splendor Forum, Jasola District Centre, New Delhi – 110025, India

79 Anson Road, #06–04/06, Singapore 079906

Cambridge University Press is part of the University of Cambridge.

It furthers the University's mission by disseminating knowledge in the pursuit of education, learning, and research at the highest international levels of excellence.

www.cambridge.org
Information on this title: www.cambridge.org/9781108480741
DOI: 10.1017/9781108691017

© Cambridge University Press 2020

This publication is in copyright. Subject to statutory exception and to the provisions of relevant collective licensing agreements, no reproduction of any part may take place without the written permission of Cambridge University Press.

First published 2020

Printed in the United Kingdom by TJ International Ltd, Padstow Cornwall

A catalogue record for this publication is available from the British Library.

Library of Congress Cataloging-in-Publication Data
Names: Saad, Walid, author. | Bennis, Mehdi, author. | Mozaffari, Mohammad, 1989– author. | Lin, Xingqin, 1987– author.
Title: Wireless communications and networking for unmanned aerial vehicles / Walid Saad, Mehdi Bennis, Mohammad Mozaffari, and Xingqin Lin.
Description: First edition. | New York : Cambridge University Press, 2020. | Includes bibliographical references and index.
Identifiers: LCCN 2019043875 (print) | LCCN 2019043876 (ebook) | ISBN 9781108480741 (hardback) | ISBN 9781108691017 (ebook)
Subjects: LCSH: Drone aircraft – Control systems. | Drone aircraft – Computer networks. | Wireless communication systems.
Classification: LCC TL685.35 .S233 2020 (print) | LCC TL685.35 (ebook) | DDC 629.135–dc23
LC record available at https://lccn.loc.gov/2019043875
LC ebook record available at https://lccn.loc.gov/2019043876

ISBN 978-1-108-48074-1 Hardback

Cambridge University Press has no responsibility for the persistence or accuracy of URLs for external or third-party internet websites referred to in this publication and does not guarantee that any content on such websites is, or will remain, accurate or appropriate.

Walid Saad:
To Mary, Karim, Raphael, and everyone who believes the sky is not the limit

Mehdi Bennis:
To my beloved family

Mohammad Mozaffari:
To my family

Xingqin Lin:
This book is dedicated to my grandfather and father, Zuojiao Lin and Yuluan Lin, with love

Contents

	Acknowledgments	*page* xii
1	**Wireless Communications and Networking with Unmanned Aerial Vehicles: An Introduction**	1
	1.1 Brief Evolution of UAV Technology	1
	1.2 UAV Types and Regulations	2
	1.2.1 Classification of UAVs	3
	1.2.2 UAV Regulations	4
	1.3 Wireless Communications and Networking with UAVs	5
	1.3.1 UAVs as Flying Wireless Base Stations	6
	1.3.2 UAVs as Wireless Network User Equipment	8
	1.3.3 UAVs as Relays	9
	1.4 Summary and Book Overview	10
2	**UAV Applications and Use Cases**	12
	2.1 UAVs for Public Safety Scenarios	12
	2.2 UAV-Assisted Ground Wireless Networks for Information Dissemination	13
	2.3 Three-Dimensional MIMO and Millimeter-Wave Communication with UAVs	14
	2.4 Drones in Internet of Things Systems	16
	2.5 UAVs for Virtual Reality Applications	16
	2.6 Drones in Wireless Backhauling for Ground Networks	18
	2.7 Cellular-Connected UAV UEs	19
	2.8 UAVs in a Smart City	20
	2.9 Chapter Summary	21
3	**Aerial Channel Modeling and Waveform Design**	22
	3.1 Fundamentals of Radio Wave Propagation and Modeling	23
	3.2 Overview of Aerial Wireless Channel Characteristics	27
	3.3 Large-Scale Propagation Channel Effects	30
	3.3.1 Free-Space Path Loss	30
	3.3.2 Ray Tracing	31
	3.3.3 Log-Distance Path Loss Models	37
	3.3.4 Empirical Path Loss Models	40

		3.3.5	Shadowing	42
		3.3.6	Line-of-Sight Probability	44
		3.3.7	Atmospheric and Weather Effects	50
	3.4	Small-Scale Propagation Effects		51
		3.4.1	Time Selectivity and Doppler Spread	52
		3.4.2	Frequency Selectivity and Delay Spread	54
		3.4.3	Spatial Selectivity and Angular Spread	56
		3.4.4	Envelope and Power Distributions	58
	3.5	Waveform Design		60
		3.5.1	Waveform Basics	60
		3.5.2	Orthogonal Frequency Division Multiplexing	62
		3.5.3	Direct Sequence Spread Spectrum	64
		3.5.4	Continuous Phase Modulation	65
	3.6	Chapter Summary		67
4	**Performance Analysis and Tradeoffs**			**68**
	4.1	UAV Network Modeling: Challenges and Tools		68
	4.2	Downlink Performance Analysis for UAV BS		70
		4.2.1	System Model	70
		4.2.2	Network with a Static UAV	73
		4.2.3	Mobile UAV BS Scenario	79
		4.2.4	Representative Simulation Results	83
	4.3	Chapter Summary		89
5	**Deployment of UAVs for Wireless Communications**			**90**
	5.1	Analytical Tools for UAV Deployment		91
		5.1.1	Centralized Optimization Theory	91
	5.2	Deployment of UAV BSs for Optimized Coverage		94
		5.2.1	Deployment Model	94
		5.2.2	Deployment Analysis	96
		5.2.3	Representative Simulation Results	99
		5.2.4	Summary	100
	5.3	Deployment of UAV BSs for Energy-Efficient Uplink Data Collection		100
		5.3.1	System Model and Problem Formulation	101
		5.3.2	Ground-to-Air Channel Model	102
		5.3.3	Activation Model of IoT devices	102
		5.3.4	UAV BS Placement and Device Association with Power Control	103
		5.3.5	Update Time Analysis	106
		5.3.6	Representative Simulation Results	107
		5.3.7	Summary	111
	5.4	Proactive Deployment with Caching		112
		5.4.1	Model	112
		5.4.2	Optimal Deployment and Content Caching for UAV BSs	116

		5.4.3 Representative Simulation Results	118
		5.4.4 Summary	122
	5.5	Chapter Summary	122

6 Wireless-Aware Path Planning for UAV Networks — 123

- 6.1 Need for Wireless-Aware Path Planning — 123
- 6.2 Wireless-Aware Path Planning for UAV UEs: Model and Problem Formulation — 124
  - 6.2.1 Problem Formulation — 126
- 6.3 Self-Organizing Wireless-Aware Path Planning for UAV UEs — 128
  - 6.3.1 Path Planning as a Game — 128
  - 6.3.2 Equilibrium of the UAV UE Path Planning Game — 130
- 6.4 Deep Reinforcement Learning for Online Path Planning and Resource Management — 131
  - 6.4.1 Deep ESN Architecture — 131
  - 6.4.2 Deep ESN-Based UAV UE Update Rule — 133
  - 6.4.3 Deep RL for Wireless-Aware Path Planning — 134
- 6.5 Representative Simulation Results — 136
- 6.6 Chapter Summary — 144

7 Resource Management for UAV Networks — 145

- 7.1 Cell Association in UAV-Assisted Wireless Networks under Hover Times Constraints — 145
  - 7.1.1 System Model — 146
  - 7.1.2 Optimal and Fair Cell Partitioning for Data Service Maximization under Hover Time Constraints — 149
  - 7.1.3 Extensive Simulations and Numerical Results — 153
  - 7.1.4 Summary — 158
- 7.2 Resource Planning and Cell Association for 3D Wireless Cellular Networks — 159
  - 7.2.1 A Rigorous Model for 3D Cellular Networks — 159
  - 7.2.2 3D Deployment of a Cellular Network with UAV BSs: A Truncated Octahedron Structure — 161
  - 7.2.3 Latency-Minimal 3D Cell Association — 164
  - 7.2.4 Representative Simulation Results — 166
  - 7.2.5 Summary — 168
- 7.3 Managing Licensed and Unlicensed Spectrum Resources in Wireless Networks with UAVs — 169
  - 7.3.1 Model of an LTE-U UAV BS Network — 170
  - 7.3.2 Models for Data Rates and Queuing — 172
  - 7.3.3 Resource Management Problem Formulation and Solution — 174
  - 7.3.4 Representative Simulation Results — 176
  - 7.3.5 Summary — 179
- 7.4 Chapter Summary — 180

8 Cooperative Communications in UAV Networks — 181

- 8.1 CoMP Transmission in Wireless Systems with Cellular-Connected UAV UEs — 183
  - 8.1.1 A Model for CoMP in Networks with Aerial UAV UEs — 183
  - 8.1.2 Probabilistic Caching Placement and Serving Distance Distributions — 183
  - 8.1.3 Channel Model — 185
  - 8.1.4 Analysis of Coverage Probability — 186
  - 8.1.5 Representative Simulation Results — 189
  - 8.1.6 Summary — 191
- 8.2 Reconfigurable Antenna Arrays of UAVs: UAV BS Scenario — 192
  - 8.2.1 UAV-Based Antenna Array in the Sky: A Basic Model — 193
  - 8.2.2 Transmission Time Minimization: Optimizing UAV Positions within the Array — 195
  - 8.2.3 Control Time Minimization: Time-Optimal Control of UAVs — 199
  - 8.2.4 Representative Simulation Results — 202
  - 8.2.5 Summary — 204
- 8.3 Chapter Summary — 205

9 From LTE to 5G NR-Enabled UAV Networks — 207

- 9.1 Mobile Technologies-Enabled UAVs — 208
  - 9.1.1 Connectivity Aspects — 208
  - 9.1.2 Services beyond Connectivity — 209
- 9.2 Introduction to LTE — 210
  - 9.2.1 Design Principles — 211
  - 9.2.2 System Architecture — 212
  - 9.2.3 Radio Interface Protocols — 213
  - 9.2.4 Physical Layer Time-Frequency Structure — 215
- 9.3 UAV as LTE UE — 216
  - 9.3.1 Coverage — 216
  - 9.3.2 Interference — 217
  - 9.3.3 Mobility Support — 220
  - 9.3.4 Latency and Reliability — 223
- 9.4 UAV as LTE BS — 226
- 9.5 3GPP Standardization on Connected UAV — 227
  - 9.5.1 3GPP Release-15 Study Item on LTE-Connected UAV — 228
  - 9.5.2 3GPP Release-15 Work Item on LTE-Connected UAV — 231
  - 9.5.3 3GPP Release-16 Study Item on Remote UAV Identification — 232
- 9.6 Towards 5G NR-Enabled UAVs — 234
  - 9.6.1 A Primer on 5G NR — 234
  - 9.6.2 Superior Connectivity Performance — 236
  - 9.6.3 Service Differentiation with Network Slicing — 237
  - 9.6.4 Network Intelligence — 238
- 9.7 Chapter Summary — 238

10	**Security of UAV Networks**	240
	10.1 Overview on UAV Security Problems	240
	10.2 Security of UAV UEs in Delivery Systems	243
	10.2.1 Modeling the Security of a UAV Delivery System	244
	10.2.2 UAV Security as a Network Interdiction Game	246
	10.2.3 Security of UAV Delivery Systems in Presence of Human Decision Makers	250
	10.2.4 Representative Simulation Results	253
	10.2.5 Summary	256
	10.3 Concluding Remarks on UAV Security	257
	References	258
	Index	279

Acknowledgments

This work was supported by the U.S. National Science Foundation under Grants CNS-1446621, IIS-1633363, CNS-1836802, CNS-1739642, OAC-1541105, and OAC-1638283 and, in part by the Academy of Finland project CARMA, in part by the INFOTECH project NOOR, and in part by the Kvantum Institute strategic project SAFAR.

1 Wireless Communications and Networking with Unmanned Aerial Vehicles: An Introduction

The past few years witnessed a major revolution in the area of unmanned aerial vehicles (UAVs), commonly known as drones, due to significant technological advances across various drone-related fields ranging from embedded systems to autonomy, control, security, and communications. These unprecedented recent advances in UAV technology have made it possible to widely deploy drones across a plethora of application domains, ranging from delivery of goods to surveillance, environmental monitoring, traffic control, remote sensing, and search and rescue. In fact, recent reports from the Federal Aviation Administration (FAA) anticipate that sales of UAVs may exceed seven million in 2020, and many industries are currently investing in innovative drone-centric applications and research. To enable all such applications, it is imperative to address a plethora of research challenges pertaining to drone systems, ranging from navigation to autonomy, control, sensing, navigation, and communications. In particular, the deployment of UAVs in tomorrow's smart cities is largely contingent upon equipping them with effective means for communications and networking. To this end, in this book, we provide a comprehensive treatment of the wireless communications and networking research challenges and opportunities associated with UAV technology. This treatment begins in this chapter, which provides an introduction to UAV technology and an in-depth discussion on the wireless communication and networking challenges associated with the introduction of UAVs.

1.1 Brief Evolution of UAV Technology

A UAV is, in essence, an unmanned aircraft or robot that can fly in nearly unconstrained locations either autonomously or while being remotely controlled by an operator. In the early twentieth century, UAV technology was mostly restricted to military environment. For instance, many references [1–4] trace back the origin of drones to the nineteenth century when unmanned balloons were used to bomb the city of Venice in Italy. Then, after some failed or unused UAV-like experiments (such as the US Army's Kettering Bug [5]) in the early 1900s, military UAV technology started to improve and evolve during the Second World War and throughout the Cold War. These early attempts at providing unmanned aircrafts were mostly restricted to well-defined and

This work is supported by the U.S. National Science Foundation under Grants CNS-1446621, IIS-1633363, CNS-1836802, CNS-1739642, OAC-1541105, and OAC-1638283.

very confined military missions, such as reconnaissance or combat surveillance. Despite their restricted application space, these early developments in UAV technology provided an important foundation for the modern-day commercial drone revolution, which really started in the mid-2000s when new applications of UAVs, such as disaster relief, search and rescue, and infrastructure inspection, began to take shape, driven by a number of governments. Meanwhile, the first commercial UAV permit was issued in 2006. Following this event, the French company Parrot produced their Parrot AR Drone in 2010, which was arguably one of the first UAVs ready to be operated by end-users using a WiFi connection and a smartphone. The Parrot AR Drone was an important first step toward popularizing the idea of consumer-operated drones that can be employed for recreational as well as commercial purposes.

However, the true catalyzer for the UAV technology was Jeff Bezos' 2013 announcement about his intentions to deploy a UAV-based delivery system for Amazon. This announcement was also followed by similar ideas from other major companies such as Google, who debuted their drone-delivery Wing project in 2014. Since then, the interest and investment in UAV technology for commercial applications began an exponential growth both in terms of applications and technology. Most recently, UAVs have become inherently equipped with important communications, computer vision, and machine learning techniques that turned them into truly autonomous and multipurpose devices. This, in turn, led to a surge of new startups, research, and standardization efforts focused on the multifaceted technological and social challenges of UAV technology. These research efforts are rapidly culminating in major breakthroughs across multiple application domains. It is, therefore, inevitable to envision that the next five years or so will witness some of the first real-world deployments of drones across various sectors in the global economy. Such deployments will range from the initial introduction of drone-delivery systems in the near term to a wide-scale deployment of UAV-based autonomous, flying taxis in the long term.

These rapid recent developments in UAV technology have naturally led to many research problems that cut across multiple fields, including navigation, control, machine learning, and communications. In particular, the ability of UAVs to fly in nearly unconstrained locations, coupled with their flexibility and agility, makes them particularly appealing for wireless communications applications. Indeed, communications and networking provide one of the most important emerging applications for UAVs; thus, it is essential to investigate the challenges and opportunities brought forward by UAVs in this domain. The wireless communications and networking applications and challenges of UAVs naturally depend on the type of UAV and associated government regulations. As a result, next, we first provide a classification of UAVs depending on their types and then delve into the wireless communications and networking challenges and opportunities.

1.2 UAV Types and Regulations

Prior to delving into the wireless communications challenges of UAVs, we first provide an overview on the different types of UAVs available as well as recent regulatory progress regarding their deployment.

1.2.1 Classification of UAVs

In general, the terms "UAV" and "drone" can be used to refer to any type of flying, unmanned robot that can be remotely controlled and has multipurpose functions. However, depending on the application, one can choose different types of UAVs, while taking into account their capabilities (e.g., sensors, size, weight, battery life, etc.) and their flight abilities (e.g., altitude, ability to hover, etc.). Although one can provide different ways to classify UAVs, one initial classification can be done based on the flight altitude of the UAVs and their size. In particular, UAVs can be generally grouped into two key categories: low-altitude platforms (LAPs) and high-altitude platforms (HAPs). LAPs are usually small-sized UAVs that can fly at low altitudes that range from tens of meters up to a few kilometers. LAPs are able to move rapidly and are very flexible in their deployment. For instance, most UAVs that have been recently considered in end-user and commercial applications are essentially LAPs. Examples of LAPs include the previously mentioned Parrot AR Drone as well as the popular DJI Phantom drone series. According to the FAA, LAPs will be allowed to fly without permit for a maximum altitude of 400 ft. To exceed this altitude, LAP operators must seek special permissions from the FAA [6].

In contrast to the small and flexible LAPs, HAPs are larger and more capable UAVs that are used to fly at high altitudes (typically above 17 km). HAPs are often quasi-stationary and used for long-term mission purposes. Prominent examples of HAPs include Airbus' Zephyr [7], which is a stratospheric UAV that can operate as a pseudo-satellite while harnessing solar energy, and Google's Project Loon [8], a HAP balloon that can be placed at an altitude of 18 km to provide long-term wireless connectivity to rural areas. HAPs are generally much larger and much more enduring than LAPs; thus, they can be deployed for longer-term, satellite-like missions. Meanwhile, LAPs are more appropriate for time-sensitive missions due to their ability to quickly deploy and move. In general, HAPs can be operated continuously for up to a few months of continuous operations (and even longer if energy limitations are overcome). In contrast, current LAP technology limits their continuous operation to a few hours (depending on battery capability and ability to recharge if needed). Naturally, HAPs are also more costly than LAPs.

Both HAPs and LAPs can be further categorized depending on the type of robot/drone being used, as shown in Figure 1.1. For instance, LAPs can be further grouped into fixed-wing, rotary-wing, and balloon UAVs. Compared to rotary-wing UAVs, fixed-wing LAPs, such as small aircrafts, have a higher weight and speed, and they are able to remain aloft by moving forward. In contrast, rotary-wing UAVs have the ability to hover over a specific geographical area while remaining stationary if needed. Meanwhile, HAPs can be further grouped into airships, aircrafts, and balloons. Airships are the largest type of HAPs, and they have significant power and load capabilities. They are often deployed in a quasi-stationary manner for long-term continuous missions (up to a few years). HAP balloons, on the other hand, are relatively lightweight HAPs that can operate for a few months continuously. They are deployed primarily for stationary missions. Moreover, aircraft HAPs are also lightweight; however, in contrast to balloons, they can fly and move around an area (typically flying in a circle and in a less

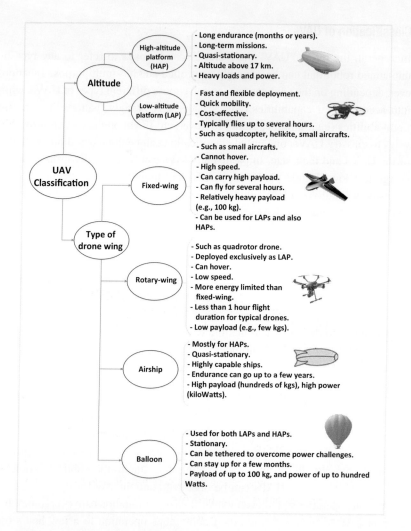

Figure 1.1 Classification of UAVs.

flexible manner than HAPs). HAP aircrafts are also suitable for missions of up to a few months.

As will be evident from the subsequent chapters of this book, both HAPs and LAPs, in their various categories, will play important roles in wireless communication scenarios. Indeed, the various features of LAPs and HAPs exposed in Figure 1.1 will naturally impact the role they will take from a wireless and networking perspective.

1.2.2 UAV Regulations

The application domains of UAVs is not only limited by their types, but it is also constrained by potential regulations that various governmental agencies may impose. For instance, although the application domain of UAVs includes countless use cases, these use cases are accompanied by various privacy, public safety, security, collision

Table 1.1 Initial regulations for the deployment of UAVs without any specific permit.

Country	Maximum altitude	Minimum distance to people	Minimum distance to airport
United States	122 m	N/A	8 km
Australia	120 m	30 m	5.5 km
South Africa	46 m	50 m	10 km
United Kingdom	122 m	50 m	N/A
Chile	130 m	36 m	N/A

avoidance, and data restrictions and concerns. To handle these concerns, numerous efforts have recently emerged to provide regulations to control the use and operation of UAVs while taking into account their types and capabilities. For regulatory purposes, five key criteria are often considered [9, 10]:

1 *Applicability*: applicability involves specifying the scope (considering type, weight, and role of UAVs) within which certain regulatory rules will be applied.
2 *Operational limitations*: these include restrictions on the locations where UAVs can fly or operate. For instance, many European cities, such as Helsinki in Finland, have recently designated various areas as no-fly zones for UAVs. Such location constraints naturally impact all sorts of applications in which UAVs will be used.
3 *Administrative procedures*: these include precise, legal processes that must be put in place in order to deploy and use a UAV.
4 *Technical requirements*: these include constraints on the communications, control, and mechanical capabilities of drones.
5 *Ethical constraints*: in order to operate UAVs (and any other autonomous systems), it is imperative to introduce ethical considerations that must be followed by UAV operators. Such considerations include ways to protect the privacy of generated data and the way in which a UAV can be used in commercial and military scenarios.

UAV regulations vary between different countries and types of geographical areas (e.g., urban or rural). In the United States, regulations for UAV operations are issued by the FAA and National Aeronautics and Space Administration (NASA). For instance, NASA is planning to develop UAV control frameworks in collaboration with the Federal Communications Commission (FCC) and the FAA. The FCC is currently investigating if a new spectrum policy needs to be established when operating drones for communication purposes. In Table 1.1, we list a number of UAV regulations in various countries [9]. All such regulations must be accounted for when treating UAV-related research problems, particularly communication problems.

1.3 Wireless Communications and Networking with UAVs

UAVs, in all of their types and sizes, provide ample opportunities for wireless communication applications. In general, all types of UAVs can be equipped with wireless interfaces. Such interfaces can operate at either unlicensed, WiFi frequencies or

Table 1.2 UAV networks versus terrestrial networks.

UAV Wireless Networks	Terrestrial Wireless Networks
• Spectrum is scarce.	• Spectrum is scarce.
• Three-dimensional network models.	• Mostly two-dimensional network models.
• Inherent ability for line-of-sight communication due to altitude.	• Difficulty to maintain line-of-sight.
• Elaborate and stringent energy constraints and models.	• Well-defined energy constraints and models.
• High dynamics due to high mobility.	• Mobility confined to a few models (e.g., pedestrians, cars, etc.).
• Hover and flight time constraints.	• No inherent timing constraints.

licensed, cellular frequencies. Naturally, equipping UAVs with wireless communications capabilities will pave the way for a plethora of new application domains for UAV technologies. Across these application domains, we can see three primary communication roles for UAVs: (a) UAVs as aerial base stations (or access points) that can be deployed to provide wireless networking and communications capabilities to various geographical areas, (b) UAVs can leverage existing infrastructure (e.g., cellular or WiFi) to communicate with one another or with ground devices, and (c) UAVs can be deployed as aerial relays that can provide an extension to the coverage and connectivity of existing wireless infrastructure. Across all those three roles, as summarized in Table 1.2, one can identify a number of key technical differences between conventional networks, terrestrial wireless networks, and wireless networks that must support UAVs.

For each one of the three use cases, various research challenges must be overcome, as discussed next.

1.3.1 UAVs as Flying Wireless Base Stations

The first natural use case for UAVs in communication applications is the flying base station (BS) case. In this use case, the UAV itself is used as a provider of wireless communication services. For instance, LAPs can be used to provide on-demand wireless networking capabilities to areas that lack coverage or that are currently congested, such as hotspot areas. Indeed, the flexibility and agility of LAPs allows network operators to use them for providing rapid and on-demand connectivity whenever needed. Meanwhile, HAPs can be deployed for longer-term wireless coverage purposes. In fact, HAPs are a central component of most recent proposals for providing connectivity to rural areas (e.g., Google's Loon project). This is due to the fact that HAPs can remain flying for long periods of time and, thus, can provide continuous broadband services to rural or remote areas in which ground wireless infrastructure is sparse or hard to deploy. Moreover, by jointly using LAPs and HAPs as flying base stations, one can construct a multitier three-dimensional (3D) wireless network that incorporates both short-term and long-term coverage solutions. Such a fully fledged UAV-based wireless network is

Table 1.3 UAV base station versus terrestrial base station.

UAV Base Stations	Terrestrial Base Stations
• Deployment is naturally three-dimensional.	• Deployment is typically two-dimensional.
• Unique propagation environment with scarcely available models.	• Well-established models for the propagation environment.
• Short-term, frequently changing deployments.	• Mostly long-term, permanent deployments.
• Mostly unrestricted locations.	• Few, selected locations.
• Mobility dimension.	• Fixed and static.

envisioned to be an important stepping stone toward delivering global wireless connectivity. A summary of the key differences between terrestrial BSs and UAV-based BSs is shown in Table 1.3.

Naturally, designing a wireless network that relies on flying UAV BSs (LAPs or HAPs), brings forward a number of unique research challenges and opportunities that stem from the unique features of UAV BSs shown in Table 1.3:

- The deployment of flying BSs is, by nature, done in 3D space. Indeed, the altitude dimension provides a new degree of freedom that a network operator can exploit to enhance connectivity, such as by establishing line-of-sight (LOS) links between flying BSs and ground users. However, the flying nature of UAV BSs also brings in new research challenges, such as the need for dynamically optimizing their deployment locations as well as managing their mobility.
- The air-to-ground wireless channel presents a new propagation environment whose characteristics can significantly differ from conventional terrestrial channel models (e.g., Rayleigh models). Indeed, propagation modeling and measurements are an important research challenge for UAV BSs. Along those same lines, there is a need for realistic air-to-air channel models (e.g., for communication between multiple UAVs of possibly different types) in order to deploy a fully fledged wireless cellular network that leverages UAVs. We do note that propagation challenges are not restricted to the UAV BS role, but they are pervasive across all wireless communication roles of UAVs.
- When dealing with UAV BSs, it is imperative to explicitly take into account the dynamics (e.g., control), mobility, and flight constraints of the UAVs. For instance, depending on their class (HAP or LAP) and type, UAVs can have different battery and power capabilities. These capabilities will directly impact the quality-of-service (QoS) that these UAVs can provide when servicing wireless users. For example, the hover time constraints of rotary-wing LAPs will impose a maximum wireless service time that such UAVs can deliver for a given geographical area. As such, characterizing the performance of a wireless network that relies on UAV BSs must explicitly factor in these UAV-specific constraints.
- Resource management in a network with UAV BSs differs substantially from resource management in classical cellular networks. On the one hand, the aforementioned flight constraints provide new resources (e.g., flight time, on-board energy) that must

be managed along with conventional wireless resources (e.g., spectrum). On the other hand, the ability of UAVs to fly and hover brings forward a unique opportunity to leverage high-frequency bands (e.g., millimeter wave) that can benefit from the ease with which UAVs can establish LOS connections. As a result, the design of new resource management schemes that are cognizant of these unique features of UAV BSs is a very important research challenge.

1.3.2 UAVs as Wireless Network User Equipment

To enable the various UAV applications previously mentioned, UAVs must be able to communicate with existing wireless networks, such as cellular or WiFi networks. In such scenarios, UAVs act as user equipment (UE) of the wireless network. When UAVs are used as UAV UEs of a ground wireless cellular network, they are often referred to as cellular-connected UAVs. Cellular-connected UAV UEs will enable a myriad of new application domains in which communications between UAVs and a ground cellular infrastructure is necessary for the UAVs to deliver application-specific data, to acquire control information, and to achieve the objective of their mission. Examples of such applications include delivery drones, real-time surveillance and multimedia transmission, and UAV-assisted transportation networks [11]. As discussed in the UAV BS case, the introduction of aerial UAV UEs that fly in unrestricted locations and communicate in 3D space, will lead to unique wireless networking challenges that are not dealt with in a ground network. In particular, deploying cellular-connected UAV UEs requires overcoming some of the following key challenges:

- Managing network interference becomes much more challenging when UAV UEs are deployed. This is due to the fact that flying UAV UEs will now generate LOS interference on ground BSs and ground UEs, which can potentially lead to significant performance degradation. As such, it is necessary to introduce new interference management solutions that are cognizant of the unique, 3D properties of UAV UEs and their capabilities.
- Current wireless infrastructure has been designed to maximize the performance of ground users. As a result, many design choices have been made without accounting for the possibility of having flying users. For example, current cellular network BSs have been developed in a way to maximize antenna coverage to the ground. As a result, current BSs will have their antennas tilted downward toward the ground. Consequently, these BSs cannot serve flying UAV UEs using their main antenna lobe and will have to rely on their side or back lobes. Hence, optimizing antenna usage for coexisting aerial and ground UEs is a key challenge for wireless communication with cellular-connected UAV UEs.
- For mission-critical applications such as delivery drones, the UAVs will need to use the cellular infrastructure to receive status information and control data. Such data will be very time sensitive and critical, and, thus, there is a need to develop new

techniques to guarantee low latency, reliable communications among UAV UEs, and ground cellular infrastructure.
- Given the difference in the propagation environment between ground users and UAV UEs, a network operator must design new techniques to identify ground and aerial users. Identification becomes particularly challenging when a terrestrial device (e.g., a smartphone) is attached to a UAV to act as a UAV UE. In such a case, the network cannot rely on traditional authentication or reporting mechanisms, and, thus, new identification techniques are needed. Performing device identification is a necessary step toward properly integrating UAV UEs into cellular systems, since it will allow the system to better map aerial and ground interference and then perform proper resource optimization and management.
- Most UAV-based systems plan the trajectory of their UAVs based on the specific mission objectives. In fact, it is very common to optimize the trajectory of UAVs in a way to minimize the mission time. However, when UAVs are deployed over a wireless network as UAV UEs, their trajectory will not only affect the mission objectives, but it will also impact the performance of the wireless network. For example, if the trajectory of a given UAV UE passes through many ground BSs, it may cause substantial LOS interference to those BSs and degrade the QoS of the wireless system. Hence, it is necessary to develop new wireless-aware trajectory optimization solutions that can balance the various objectives of a UAV system, including mission objectives and wireless network performance.
- Along with trajectory optimization, handover and mobility management are also two prominent technical challenges for cellular networks with UAV UEs. These challenges will be significantly exacerbated by the fact that the mobility of UAV UEs is much more dynamic than that of ground devices. In particular, the diversity of paths and locations that UAV UEs can visit, along with their 3D nature, will bring forward new mobility management challenges that are not dealt with in ground cellular systems.
- As is the case for the UAV BS scenario, UAV UEs will also face challenges pertaining to the aerial propagation environment as well as the need for dynamic resource management.

1.3.3 UAVs as Relays

The third use case scenario for UAVs in a wireless environment is one in which the UAVs act as relay stations that provide a relaying link between a transmitter and a receiver. In particular, the use of UAV relays is suitable for enhancing the coverage of a ground network or for overcoming obstacles (e.g., high hills or high-rise buildings) that can prevent the possibility of LOS communication between a transmitter and a receiver. The use of UAV relays has also been particularly popular for providing connectivity among the ground users of mobile ad hoc networks. Another important application for UAV relays is the use of a flying ad hoc network to provide backhaul connectivity to a ground wireless or cellular users. In UAV relay use case scenarios, the UAV will act as a transceiver that receives data from a ground device and then relays this data (via one or more hops) to other devices. While deploying UAV relays shares many challenges with the UAV BS and UAV UE cases, it also has its own unique challenges:

- To perform proper relaying, UAVs must rely on well-designed cooperative communication mechanisms. For instance, UAVs can potentially adopt classical cooperative relaying schemes, such as amplify-and-forward or decode-and-forward. However, the fundamental performance limits of such mechanisms were mostly studied for ground networks, and, hence, there is a need for a more comprehensive analysis on the relaying performance of flying, UAV-based networks. In addition, more advanced relaying mechanisms will also be needed to cope with unique features of UAVs, such as their mobility and dynamics.
- For proper relaying, UAVs will need to coordinate their positioning and potential transmission. To do so, the UAVs must rely on their control system. As a result, there is a need for new communication and control codesign mechanisms that can take into account, jointly, the performance of the control and communication systems. Such mechanisms will also be able to account for exogenous factors, such as wind, which can affect the performance of UAV relays. Here, it is noteworthy to mention that joint communications and control problems are also relevant for the UAV BS use case.
- The use of relaying will require UAVs to establish multi-hop communication links in the air. The formation and optimization of such multi-hop, airborne networks is a major research challenge when UAVs act as relays. For instance, given that the air-to-air link is not yet well understood, it is challenging to design dynamic routing and multi-hop communication algorithms that can adapt to this link's propagation environment. Moreover, the development of scaling laws tailored toward the flying nature of multi-hop UAV relays will also be needed to understand the performance limits of a flying multi-hop UAV network.
- The use of HAPs for relaying can also be an interesting research challenge. HAPs provide stable connections and, hence, can potentially help in relaying data from ground users and from LAPs. However, given the long distances over which HAPs, LAPs, and users will communicate, the design of power-efficient and reliable communication methods will be needed.

1.4 Summary and Book Overview

Clearly, deploying UAVs for wireless networking purposes brings in a plethora of challenges, use cases, and opportunities. In the rest of this book, we will explore those challenges and associated problems, while focusing on the following themes:

- In Chapter 2, we provide an in-depth overview of the various applications in which UAVs can be used for communication purposes. This overview will then drive the different research questions that follow in subsequent chapters.
- In Chapter 3, we focus on the physical layer aspects of UAV communications, particularly on radio propagation and waveform designs for aerial wireless users.

1.4 Summary and Book Overview

- In Chapter 4, we provide a rigorous performance analysis of wireless networks with UAVs, while focusing on the achievable network performance in terms of coverage, rate, and other related QoS metrics.
- In Chapter 5, we focus on the deployment of UAVs (particularly UAV BSs), and we study a number of problems for optimally deploying UAVs while optimizing wireless networking metrics.
- In Chapter 6, we turn our attention to issues of mobility management and, particularly, wireless-aware path planning for communication networks with UAV UEs.
- In Chapter 7, we introduce comprehensive frameworks that enable the optimization of wireless network resources (e.g., spatial, spectral, or temporal resources) while taking into account the unique features of UAV BSs and UAV UEs.
- In Chapter 8, we study the problem of cooperation among UAVs, and we also investigate how coordinated transmissions can be leveraged to improve wireless communication performance with UAV UEs.
- In Chapter 9, we provide a panoramic and practical overview on how mobile technologies, such as long-term evolution (LTE) wireless cellular systems and the emerging fifth-generation (5G) new radio networks, can support UAVs.
- In Chapter 10, we conclude this book by delving into the security of UAV networks. In particular, we discuss a number of frameworks to mitigate prominent cyber attacks that can target UAV systems, particularly UAV systems that are equipped with communication capabilities.

Notations: In the rest of this book, given that each chapter is self-contained and develops comprehensive analytical models for the treated research problems, the notations used in each chapter are specific to that chapter and do not extend to other chapters.

2 UAV Applications and Use Cases

Chapter 1 provided a broad motivation for wireless communications and networking with UAVs. In this chapter, we expand on the motivation of Chapter 1 by providing a holistic overview of several key applications and use cases of UAVs in various wireless networking scenarios. These scenarios are relevant to all of the UAV roles discussed in Chapter 1, including UAV BS and UAV UE. For the role of UAV BS, we focus on the use of UAVs in a variety of applications, including public safety, the Internet of Things, caching, edge computing, and smart cities. Then, we discuss a handful of important applications for UAV UEs. These applications will then drive much of the analysis and technical discussions in subsequent chapters.

2.1 UAVs for Public Safety Scenarios

Natural and man-made disasters such as floods, fires, and earthquakes have devastating impacts on the economy and human life. In the aftermath of large-scale disasters, wireless ground infrastructure, such as BSs and cell towers, are often damaged and telecommunication services become unavailable. Such services are essential for connecting victims, first responders, and individuals who are located in the disaster zone. In these situations, it is necessary to establish robust, flexible, and reliable wireless communication systems that can provide support for public safety and disaster management tasks. Indeed, having reliable wireless connectivity can significantly reduce economic losses and fatalities in the aftermath of disasters. Clearly, in such scenarios, relying solely on the preexisting ground infrastructure is not apropos.

To this end, one promising approach to overcome the lack of disaster-resilient wireless infrastructure is to deploy flying UAV BSs that can provide wireless services to a desired geographical area during or after a disaster (see Figure 2.1 as an example). Considering the fact that a UAV BS is not physically attached to ground infrastructure, it can freely and effectively move in three-dimensional space and in nearly unconstrained locations. Hence, a UAV BS can provide a rapid approach to enable necessary wireless connectivity in disaster-affected environments. In fact, with flexibility, reconfigurability, and mobility, a drone-based aerial wireless network can enable reliable connectivity among the various human actors present in a disaster-affected area, such as first responders, residents, and potential victims. Furthermore, aerial UAV BSs can autonomously

2.2 UAV-Assisted Ground Wireless Networks for Information Dissemination

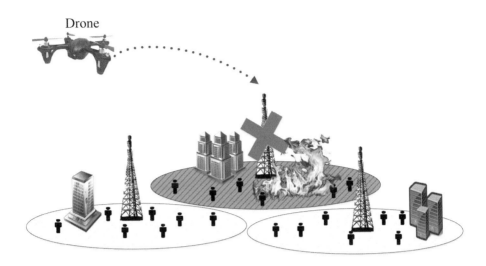

Figure 2.1 The use of drones with wireless communication capabilities in public safety scenarios.

update and optimize their location so as to ensure that users located in a large geographical area are completely covered and are able to quickly receive their time-critical communication services. Indeed, UAV BSs are one of the most promising technologies to provide connectivity for public safety scenarios. Beyond serving disaster-affected areas, UAV BSs can provide communication support to individuals stranded in remote areas (e.g., lost in a mountainous area). Moreover, in public safety cases, one can also use UAV UEs to provide delivery of medicine and other pressing needs to remote or poorly connected areas. Clearly, UAV BSs and UAV UEs with wireless communication capabilities will play a prominent role in disaster management, public safety, and emergency situations. Such a role has already been explored in recent years, such as during Hurricane Harvey, where drone-based communication was used. We will explore the various fundamental and theoretical challenges of UAV communications in such scenarios, across Chapters 3–7, and we will revisit some of these challenges, from a practical cellular networking perspective, in Chapter 9.

2.2 UAV-Assisted Ground Wireless Networks for Information Dissemination

Given the flexibility, maneuverability, and high chance of establishing LOS drone-to-ground communications, UAVs equipped with communication capabilities can be leveraged for information dissemination as well as coverage expansion of ground wireless and cellular networks [12, 13]. For example, as we can observe from Figure 2.2, UAV BSs can facilitate fast and efficient information dissemination in device-to-device (D2D) communication networks as well as mobile ad hoc networks. Naturally, wireless D2D devices suffer from a limited communication range (due to energy constraints) and significant interference in large networks. In such scenarios, mobile UAVs (including both UAV BSs and UAV relays) can assist the D2D network by quickly disseminating

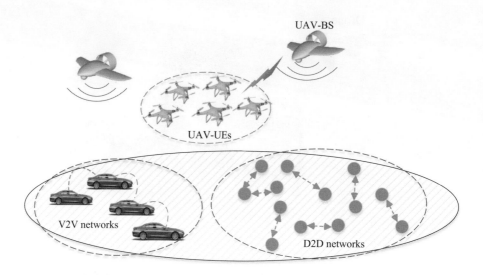

Figure 2.2 Drone-assisted ground networks with D2D capabilities.

important information and multi-casting any desired data to D2D devices thus reducing the number of D2D transmissions, which, in turn, translates into reduced interference for the network and reduced energy consumption for the devices. One key application of a drone-D2D integrated network is in emergency scenarios where critical information needs to be disseminated within a short period of time. In fact, UAVs can cooperate with D2D networks to improve energy efficiency, connectivity, and spectral efficiency of cellular networks. We also note that, in UAV-assisted ground networks, the joint design of drones' mobility and device clustering can also bring substantial performance gains.

Meanwhile, UAVs can also assist vehicle-to-vehicle (V2V) communications by disseminating safety and traffic messages among vehicles within a vehicular network. In this case, a ground vehicular network can leverage either dedicated UAV BSs to send this information or it can also rely on existing, flying UAV UEs that can take on the role of a UAV relay to disseminate information. Clearly, by leveraging the three key roles of UAVs discussed in Chapter 1, one can significantly improve the coverage, reliability, latency, and efficiency of information dissemination in various terrestrial networks that support D2D and V2V communication links.

2.3 Three-Dimensional MIMO and Millimeter-Wave Communication with UAVs

Another promising use case of UAVs is in enhancing communication at high-frequency millimeter-wave bands as well as leveraging notions pertaining to full-dimensional MIMO, massive MIMO, and reconfigurable antenna array systems. In cellular networks, 3D MIMO (in horizontal vertical directions) has recently received significant attention [14–20]. As we can observe from Figure 2.3, 3D beamforming allows the control of a beam toward any location in 3D, which can be useful for interference mitigation [21].

2.3 Three-Dimensional MIMO and Millimeter-Wave Communication with UAVs

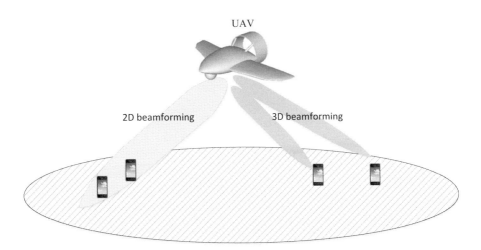

Figure 2.3 3D beamforming using a drone.

In comparison with traditional 2D MIMO systems, the use of 3D MIMO can potentially yield a superior performance across various metrics, such as data rate, and it can also provide simultaneous support for a larger number of users. 3D MIMO is particularly effective when there is a need for serving a significantly large number of users located at various elevation angles (e.g., in high-rise buildings and streets) [12, 20]. Considering the fact that UAV BSs can be naturally deployed at high altitudes relative to ground users (as well as relative to conventional ground BSs, as discussed in Chapter 1), the elevation angles between UAVs and users can be simply identified. In addition, the ability to establish LOS links between flying UAVs and ground users allows the employment of robust 3D beamforming when using UAV BSs.

Meanwhile, multiple UAVs can be used in coordination with one another so as to form a single, flexible, reconfigurable, and wireless antenna array system in the sky [22]. For such a UAV-based antenna array, each antenna element is a UAV that can adjust its position. Such UAV-based antenna array has a number of benefits over a traditional fixed antenna array system: (1) variable number of antenna elements, (2) maximizing beamforming gain by optimizing the locations of drones, (3) mechanical beamforming with moving drones, and (4) creating any arbitrary array geometry in 2D and 3D. In Chapter 8, we will provide an in-depth study on the benefits of such a cooperative UAV antenna array.

Another promising application of UAVs within the wireless domain is that of drone-based millimeter-wave communication [12, 23–25]. Naturally, due to the high propagation loss in high frequencies, millimeter-wave communications will need LOS links between transmitters and receivers. In this regard, drones, which are capable of establishing LOS links, can be a key enabler for providing high-capacity wireless services via millimeter-wave communications. As we will also briefly discuss in Chapter 9, this wireless application of UAVs will be particularly apropos for 5G systems and beyond [26] in which millimeter-wave communication will be central to the cellular network architecture. We can envision UAV BSs that provide millimeter-wave connectivity as well as UAV UEs that use such connectivity to transmit surveillance or virtual reality data to ground BSs. The blend of UAVs, across all their roles, and high-frequency, millimeter-wave bands will be a promising wireless domain for drone communication technologies.

2.4 Drones in Internet of Things Systems

The massive deployment of a diverse set of Internet of Things (IoT) devices that range from smart meters to wearables and wireless implants introduces new wireless networking challenges that can potentially benefit from the presence of UAVs. In particular, small, power-limited IoT devices, such as those used in many key IoT application domains such as healthcare, transportation, and smart cities, among others [27–30], require a robust wireless networking infrastructure that can provide them with the necessary long-term coverage needed to deliver their services. In such scenarios, a number of fundamental challenges such as connectivity, coverage, reliability, latency, and energy efficiency must be carefully addressed. Specifically, the battery limitations of IoT devices significantly impacts coverage range and reliability of IoT communications. Therefore, it is necessary to have a wireless networking solution that can meet IoT requirements for deep coverage, reliability, and energy efficiency. Moreover, the scattered and massive nature of IoT devices requires a flexible wireless system that can provide pervasive coverage across very large geographical areas.

Clearly, for addressing such IoT challenges, one can exploit UAVs with communication capabilities. Indeed, flying drones are seen as a major enabler for meeting the wireless networking requirements of IoT devices. For example, UAV BSs can effectively support IoT services in both uplink and downlink scenarios across scattered geographical areas. For uplink, UAV BSs can dynamically move based on the locations of IoT devices to collect IoT data in an energy-efficient and reliable way. This dynamic use of UAVs for IoT data collection will be explored in detail in Chapter 5. For downlink, drones can be optimally deployed close to IoT devices and provide deep coverage for the IoT system [31]. Clearly, the use of UAVs and the deployment of the IoT will have many intertwined wireless communication challenges and opportunities.

2.5 UAVs for Virtual Reality Applications

Another use case of drones is in virtual reality (VR) applications, as shown in Figure 2.4. In this case, UAVs can be used for delivering VR applications, for collecting tracking data on VR users, for monitoring VR user movement, and for providing general wireless connectivity to VR applications. For instance, UAV UEs can be equipped with cameras to collect the 360° contents that will be requested by the ground VR users [32, 33]. For example, a given ground VR user can use UAV UEs with cameras to engage in VR

2.5 UAVs for Virtual Reality Applications

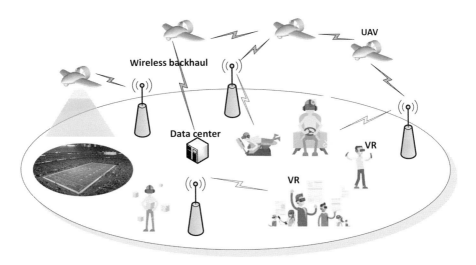

Figure 2.4 Drones in VR applications and millimeter-wave communications.

environments that encompass the landscape of a mountain, an entire city, or a remote stadium. In these scenarios, the UAVs, acting as UAV UEs, will first collect the VR images that are requested by the ground VR users using their onboard cameras, and then they will transmit these VR images to the ground BSs. Finally, the BSs will send the VR images to the ground VR users. Deploying UAVs for such VR applications requires overcoming a number of challenges, such as:

- *High rate, low delay*: The data size of VR content, such as a full, immersive image, will generally be very large. Meanwhile, the delay requirement of each VR user is typically less than 20 ms. In consequence, for transmitting remote, VR content via the use of UAV UEs equipped with VR apparatus, it is necessary to design new wireless communication techniques that can ensure high data rates as well as low-latency communication.
- *Limited energy*: The use of cameras to capture VR images can consume significant energy from the onboard energy source of the UAV. The power of the drones (particularly LAPs) is typically provided by a limited-capacity battery. Consequently, there is a need for new techniques that can guarantee the required VR application QoS while minimizing the energy consumption needed for the VR content collection, processing, and transmission by the UAVs.

Beyond data collection and VR content transmission, one can leverage sensor-equipped UAV UEs for tracking VR users. For instance, in conventional VR systems, user body movement (e.g., hand movement or orientation change) can be readily detected by the sensors that are installed on walls or ceilings. However, such a static deployment of sensors limits the application range of VR, since it is clearly not practical to continuously change the locations of the sensors as VR users move from one location to another (indoor or outdoor). For example, a VR user might install its VR sensors in the living room. In such a case, even if wireless VR support is available for the user's VR apparatus, it will be cumbersome to change the locations of VR sensors if the user decides to use the VR device in a different room. Indeed, to provide more seamless support for VR applications, particularly for wireless VR applications, it can be desirable to use sensors that can dynamically change their location. In such scenarios, one can readily equip those VR sensors on UAV UEs and exploit the mobility of the drones to track each VR user's movement. Naturally, such scenarios will also bring forth many wireless challenges ranging from the need for reliable sensor data collection to the need for low-latency transmissions.

Last but not least, in a VR use case, UAV BSs can also be used to provide wireless VR connectivity, particularly for outdoor VR applications. Moreover, for an outdoor VR user, one can also employ UAV relays to transmit VR contents that cannot be directly transmitted from ground BSs to ground VR users. Clearly, the VR domain will admit numerous UAV wireless communication and networking use cases.

2.6 Drones in Wireless Backhauling for Ground Networks

In conventional cellular networks, the most common method to connect wireless base stations to a core network is through a wired backhaul. Nevertheless, wired connections are prohibitively expensive, and their deployment could be infeasible due to geographical limitations and restrictions [34–36]. To address the shortcomings of wired backhauling, wireless backhauling has been introduced as a reliable and cost-effective solution. Meanwhile, wireless backhauling can become inefficient due to the presence of obstacles and due to the potential increase in wireless interference. Such factors can impair the performance of wireless backhauling and limit its use cases [37]. To cope with this limitation, it is very natural to exploit UAVs that can complement and support ground wireless and wired backhaul networks. Indeed, UAVs can be deployed to provide a cost-effective, reliable, and high-speed wireless backhauling support for terrestrial networks [38]. Particularly, it is possible to achieve an optimal placement of UAVs (for backhauling purposes) in order to avoid blockage and set up LOS and reliable communication links. Furthermore, high-speed wireless backhaul connections can be established to deal with high traffic demands in congested areas by equipping UAVs with millimeter-wave communication capabilities. Another advantage of using UAVs for backhauling is their ability to form a reconfigurable aerial network and introduce a robust wireless backhauling solution via multi-hop LOS links. With a flexible UAV-based wireless backhauling technique, the capacity, coverage, and reliability of backhaul connections can be boosted. In addition, the deployment and maintenance expenses associated with traditional wired backhauling can be substantially reduced. For backhauling purposes, UAVs assume the role of UAV relays or UAV BSs. In addition, as we will see in Chapter 7, one can exploit the use of HAPs in order to further provide backhaul support to ground networks as well as to LAP-based UAV networks. Another promising area in this domain is the integration of HAPs and satellite networks to further enhance backhauling [39].

2.7 Cellular-Connected UAV UEs

As we discussed in Chapter 1 and in some of the previous sections, UAVs can also act as flying users (a.k.a., UAV UEs or drone UEs) within a cellular network. UAV UEs can be widely used in various scenarios, including surveillance, package/drug delivery, transportation, IoT, remote sensing, and VR. One evident example of cellular-connected UAVs applications is Amazon's Prime Air UAV delivery service [40]. Such diverse use cases of UAV UEs can be effectively enabled by exploiting the flexibility, maneuverability, and 3D mobility of UAVs. In general, beyond the few scenarios discussed in the previous sections, as illustrated in Figure 2.5, the main applications of cellular-connected UAVs can be categorized into three domains [11]: (1) UAV-based delivery systems (UAV-DSs), (2) UAV-based real-time multimedia streaming (UAV-RMS) networks, and (3) UAV-enabled intelligent transportation systems (UAV-ITSs).

UAV UE-based delivery systems allow efficient, low-cost, and quick transportation of packages, goods, and other items. Moreover, as discussed previously, UAV-DSs play an active and essential role in performing mission-critical applications by flying to remote

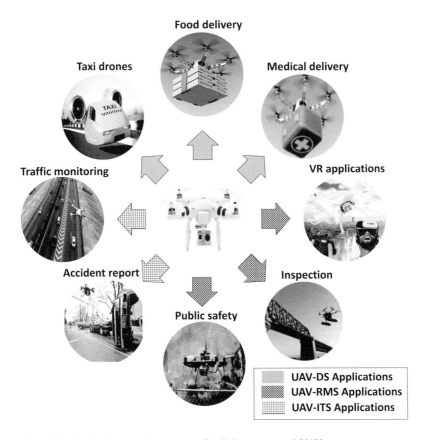

Figure 2.5 Applications and use cases of cellular-connected UAVs.

areas that are not readily accessible from the ground. In addition, employing flying taxis for public transportation is another application of UAV-DSs where, instead of delivering goods, UAV UEs are used to transport passengers. Meanwhile, in the UAV-based RMS scenario, UAV UEs can be used for high-speed online video streaming and broadcasting, virtual reality, and real-time tracking of mobile terminals. Finally, UAV UEs can be a key enabler of intelligent transportation systems for controlling traffic, reporting incidents, and ensuring the safety and security of roads. In addition, UAV-ITSs can significantly facilitate vehicular platoons by reducing network congestion in vehicle-to-vehicle communications as well as continuously tracking the status of platoon systems. All of these applications of UAV UEs bring forward critical wireless networking challenges and will be addressed, in detail, in the subsequent chapters, particularly Chapters 6, 7, 8, 9, and 10.

2.8 UAVs in a Smart City

Understanding the global vision set forth by smart and connected cities is prohibitively affected by practical technological challenges. The examples of which include integrating the services offered in smart cities with an IoT environment (as shown in Figure 2.6) and a reliable cellular infrastructure that is resilient to catastrophic situations and able to handle massive amount of data without sacrificing the quality of service. UAV-assisted wireless communication is a promising solution to overcome these challenges. UAVs can be used as data aggregators to effectively collect data across multiple devices in different geographical areas and relay the data to powerful clouds for analytics purposes. In addition, UAV BSs can be employed to enhance the coverage of wireless networks in case of an emergency or a disastrous situation. UAVs can also be used to sense the radio environment maps [41] across a city to help network operators optimize the frequency

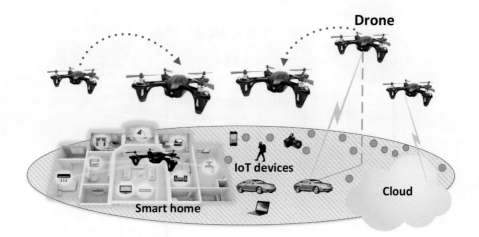

Figure 2.6 Drones in a smart city.

planning efforts. Moreover, UAVs that are used as delivery drones for other commercial purposes in smart cities can be viewed as an important UAV UE use case in which UAVs can be serviced by both ground and flying BSs. In addition to data collection capabilities, UAVs can be leveraged to act as mobile cloud computing platforms [42, 43] to facilitate fog computing and data offloading for devices that have limited computational and memory resources. Note that the drones operating within the smart city environment may require to be temporarily placed on designated buildings for purposes such as battery recharge. In this case, allocating on-demand resources to accommodate drone operation becomes a challenge. UAV UEs, of all application types, are expected to fill the skies within a smart city. Indeed, UAVs, in all their roles, will become an inseparable part of smart cities, from wireless connectivity and operational viewpoints.

2.9 Chapter Summary

In this chapter, we have provided an in-depth overview on the various applications in which UAVs can be used for wireless communication purposes. For UAV BSs, we have presented the use of UAVs in a variety of applications, such as public safety, coverage enhancement, multiple-antenna systems, IoT, caching, and smart cities. Moreover, we have discussed the key roles of UAV UEs within cellular-connected UAV systems for supporting wide range of use cases from VR applications and medical delivery to flying taxi scenarios. In the subsequent chapters, given the significant use of UAVs in many applications, we will analyze the design considerations, deployment optimization, fundamental limits, and implementation aspects of UAV-enabled wireless networks. In particular, in Chapter 3 we will focus on physical layer and channel modeling aspects of UAV communications, particularly in cellular-connected UAV systems. In Chapter 4, we will describe deployment optimization techniques for UAV BSs for coverage enhancement, IoT, and caching applications. In Chapter 5, the performance and fundamental tradeoffs of UAV BS-assisted wireless networks will be presented. Chapter 6 will focus on mobility management of UAV UEs. In Chapter 7, we will discuss how wireless resources can be optimized for UAV BSs and UAV UEs. Chapter 8 will introduce frameworks for enabling cooperative communications and reconfigurable antenna array in UAV systems. In Chapter 9, we will focus on implementation aspects of using UAVs in LTE and emerging 5G systems. Finally, Chapter 10 will present the security challenges in UAV UE-based delivery systems.

3 Aerial Channel Modeling and Waveform Design

The radio channel plays a fundamental role in wireless communications systems. It impacts transceiver design, link budget, interference levels, and network management and operation. Extensive empirical measurements have been carried out in the past few decades to develop radio channel models, especially for terrestrial wireless environments across a wide range of frequencies from sub-GHz to millimeter-wave frequencies. In order to properly deploy UAVs for wireless communication purposes and to enable all the UAV-related applications discussed in Chapter 2, it is imperative to develop comprehensive channel models tailored to the unique challenges of wireless networks with UAVs. Compared to terrestrial radio channels, aerial radio channels exhibit many different characteristics. For example, at a medium height or above, the direct LOS signal path between the transmitter and receiver is less likely to be obstructed by other objects in the propagation environment. Another example is airframe shadowing that occurs when the signal path is obstructed by the body of the UAV itself during the UAV movement.

Waveform design is another fundamental aspect of wireless communication systems that must also be studied for networks with UAVs. A waveform is the shape and form of a wireless signal that carries information bits. Important design considerations include spectral efficiency, power efficiency, robustness to interference, and implementation complexity. Due to the diverse UAV applications (as discussed in Chapter 2), different wireless communication systems possibly with different waveform choices may be required to serve different uses. For a UAV communication system requiring high data rates (e.g., in surveillance use cases or for hotspot coverage), a multicarrier waveform such as orthogonal frequency division multiplexing (OFDM), which is spectrally efficient and can be efficiently implemented digitally, could be a good candidate. For a UAV communications system that needs to be robust to interference and jamming, a spread spectrum waveform such as direct sequence spread spectrum (DSSS) can be considered due to its inherent property of "hiding" the signal below the noise floor and its resistance to narrowband jamming. For yet another UAV communications system designed particularly for a low-power Internet of Things system that is sensitive to power efficiency, a single carrier waveform with continuous phase modulation (CPM) may be used since they have constant envelope and the associated power amplifiers can work in a nonlinear regime to achieve high power efficiency.

This chapter focuses on aerial channel propagation modeling and waveform design. We begin by introducing the fundamentals of radio wave propagation and modeling in Section 3.1. In Section 3.2, we provide an overview of the salient characteristics of aerial wireless channels for UAVs, with a focus on how they differ from the more familiar and well-studied terrestrial wireless channels. Next we characterize large-scale propagation channel effects including path loss, shadowing, LOS probability, and atmospheric and weather effects in Section 3.3. Ray-tracing models for approximating wave propagation are also discussed. In Section 3.4, we look at the small-scale propagation effects due to the constructive and destructive combining of multipath signal components. Key physical phenomena, including time, frequency, and spatial selectivity, are explained. Statistics of the corresponding key parameters from representative measurement campaigns and simulation results for aerial wireless channels are surveyed. We also discuss several statistical models for the envelope and power distributions of aerial wireless channels. In Section 3.5, we turn our attention to waveform design, by reviewing "just enough" background in the basics and discussing some of the most widely used wireless waveforms. These include OFDM, DSSS, and CPM, which have been adopted in the fourth-generation (4G)/fifth-generation (5G), third-generation (3G), and second-generation (2G) mobile communications systems, respectively. We will be far from exhaustive here as waveform design is a rich subject for digital communications systems. The goal is to have a small set of exemplary waveforms in our repertoire to discuss the main design considerations for UAV wireless communications and networking. We conclude the chapter with a short summary.

3.1 Fundamentals of Radio Wave Propagation and Modeling

Radio propagation is the behavior of electromagnetic waves when they propagate in the environment. Basic radio wave propagation phenomena include reflection, diffraction, refraction, scattering, and absorption, in addition to the direct LOS propagation. Figure 3.1 illustrates the basic radio wave propagation phenomena in an air-to-ground propagation scenario with a UAV UE. Below we briefly describe these propagation phenomena.

- Reflection is the change of radio wave propagation direction when the radio wave impinges upon a different medium so that the radio wave is reflected back to the first medium. How much energy is reflected back to the first medium depends on the material properties, angle of incidence, radio wave frequency, and radio wave polarization.
- Diffraction is the bending of a radio wave around the corners of an obstacle, allowing the radio wave to propagate behind the obstacle. With diffraction, radio waves can overcome the earth's curvature and propagate beyond the horizon (e.g., the ground waves at frequencies below 300 kHz). The diffraction capability depends on the size of the obstacle relative to the wavelength. In general, when two radio waves encounter the same obstacle, the one with a larger wavelength diffracts more.
- Refraction is the change of radio wave propagation direction when the radio wave travels from one medium to another. An interesting refraction phenomenon is the one at 0.3–30 MHz frequencies, whereby radio waves transmitting into the sky can refract

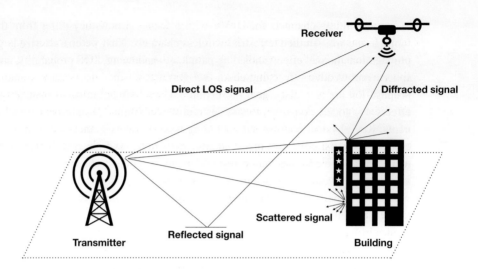

Figure 3.1 An illustration of basic radio wave propagation phenomena in an air-to-ground propagation scenario with a UAV UE.

from the ionosphere layer (a region of the atmosphere from about 60 km to 500 km) back to the earth beyond the horizon, allowing for long-range radio communications.
- Scattering is the reflection of a radio wave from irregularities on the surface of an obstacle into different reflected directions. Scattering is a weaker radio wave propagation phenomenon in sub-6 GHz wireless channels but may be substantial in millimeter-wave frequencies where diffraction becomes lossy and less reliable.
- Absorption is the loss of the energy of a radio wave when the wave hits an obstacle. The attenuation of the radio wave depends on the properties of the obstacle and the wavelength. For example, the penetration loss is usually low for glass materials but may be high for concrete materials. Low-frequency radio waves can penetrate through brick walls but millimeter waves in the 58–60 GHz band can be absorbed by water and oxygen significantly.

The details of radio wave propagation can be obtained by solving Maxwell's equations with boundary conditions. This requires knowledge of the characteristics of the physical objects in the environment where radio wave propagation occurs. It is, in general, difficult to obtain analytical solutions for the electromagnetic field in a realistic propagation environment. As an approximation, ray-tracing techniques can be used to yield accurate predictions on radio wave propagation for a given environment. In these techniques, rays are generated from source points, and each ray propagating in the environment may experience reflection, diffraction, refraction, scattering, and absorption. The models provided by ray-tracing techniques are deterministic. Depending on the ray-tracing environment and prediction accuracy requirement, the number of rays that need to be generated can be high, possibly leading to high computation burden and long processing time. Another disadvantage is that the results only apply to the specific scenario for which ray tracing is performed. We will discuss ray tracing in more detail in Section 3.3.2.

Large-scale propagation channel effects include path loss of radio signal as a function of propagation distance and shadowing due to signal path obstructions by large objects such as buildings. Small-scale propagation channel effects refer to the constructive and destructive combining of the multiple signal paths that occur at a small spatial scale of the order of the signal wavelength. These channel effects cause variations of the radio channel over time, frequency, and space. To facilitate the design and analysis of wireless communications systems with time-varying channels, analytical statistical models based on empirical measurements are often used. These statistical models are also more suitable to represent classes of channels. They have been used to characterize both large-scale and small-scale propagation channel effects.

Next, we introduce a mathematical channel model for wireless channels to describe the channel statistics in time, frequency, and space domains. Consider a single transceiver pair. Denote by $\vec{k}$ the wave-number vector that describes the phase variation of a plane radio wave emitted from the transmit antenna in a reference coordinate. The wave-number vector is given by $\vec{k} = \frac{2\pi f_c}{c} \vec{u}$, where f_c is the frequency of the radio wave, c is the speed of light, and $\vec{u} = \frac{\vec{k}}{\|\vec{k}\|}$ is the unit wave-number vector. Denote by $\vec{r}$ the original reference position of the receive antenna. Assuming that the receive antenna is moving with velocity $\vec{v}$, the position vector of the receive antenna at time t can be written as $\vec{x}(t) = \vec{r} + \vec{v}t$. The time-varying channel impulse response is then given by

$$h(t, \vec{r}) = e^{-j<\vec{k},\vec{x}(t)>} = e^{-j\frac{2\pi f_c}{c}<\vec{u},\vec{v}>t} e^{-j<\vec{k},\vec{r}>}, \qquad (3.1)$$

where $<\vec{p}, \vec{q}>$ denotes the inner product of the two vectors $\vec{p}$ and $\vec{q}$. We can see that the frequency f_c of the radio wave has been shifted by an amount Δf equal to

$$\Delta f = \frac{f_c}{c} <\vec{u}, \vec{v}> = \frac{v \cos(\varphi)}{c} f_c, \qquad (3.2)$$

where φ denotes the angle between the unit wave-number vector $\vec{u}$ and the velocity vector $\vec{v}$, and $v = \|\vec{v}\|$ is the moving speed of the receive antenna. This mathematical wireless channel model is illustrated in Figure 3.2.

The change of the frequency of a wave due to the movement of an observer relative to the wave source is well known as the effect of *Doppler shift*. When the observer is either directly moving toward or away from the source, i.e., $\varphi = 0$ or $\varphi = \pi$, the frequency is either increased or decreased with a maximum Doppler shift value $|\Delta f| = \frac{v}{c} f_c$. It should be noted that the amount of Doppler shift is frequency and velocity dependent.

When the electromagnetic wave propagates in the environment, the wave strength is attenuated due to path loss and shadowing and the wave is delayed when arriving at the receive antenna. The propagation delay and other propagation phenomena, such as reflection, diffraction, refraction, and scattering, lead to a phase change of the received wave. From a baseband perspective, these effects can be captured by a complex channel gain denoted by $a(t)$. Accordingly, the time-varying channel impulse response is given by

$$h(t, \tau, \vec{r}) = a(t) e^{-j2\pi \Delta f t} e^{-j<\vec{k},\vec{r}>} \delta(\tau - \tau_d), \qquad (3.3)$$

where $\delta(\tau)$ denotes the Dirac delta function and τ_d is the delay.

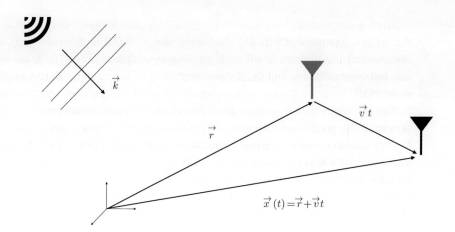

Figure 3.2 An illustration of the mathematical wireless channel model.

Thus far, we have implicitly assumed that there is only one radio wave arriving at the receive antenna. Due to, for example, scattering, the receive antenna may receive multiple radio waves that experience different levels of attenuation, Doppler shifts, and delays. By further incorporating the multipath effect, the time-varying channel impulse response is given by

$$h(t, \tau, \vec{r}) = \sum_{\ell=0}^{L-1} a_\ell(t) e^{-j2\pi \Delta f_\ell t} e^{-j<\vec{k_\ell}, \vec{r}>} \delta(\tau - \tau_{d,\ell}), \quad (3.4)$$

where we have added a subscript ℓ to our notation to distinguish the complex channel gain, Doppler shift, wave-number vector, and delay associated with each path ℓ, and L denotes the number of multipath components.

Note that a small change in path distance can lead to a large phase change of the received wave at the receive antenna. In particular, a path distance change of a quarter carrier wavelength, i.e., $\frac{c}{4f_c}$, leads to a path delay change of $\frac{1}{4f_c}$ and, accordingly, a significant phase change of $\frac{\pi}{2}$. Minor movements of the objects in the propagation environment can lead to rapid phase change in each path. If the path distance change is due to a movement at speed v, the time required for the $\frac{\pi}{2}$ phase change is $\frac{c}{4f_c v}$. For a 4G LTE system operating at a carrier frequency of 2 GHz, a quarter of the carrier wavelength is 3.75 cm. For a medium wireless device speed of 30 km/h, the time required for the $\frac{\pi}{2}$ phase change is 4.5 ms.

The receiver may not be able to distinguish two radio waves if their delay difference is small. Precisely, let W be the channel bandwidth, and let $\tau_{d,1}$ and $\tau_{d,2}$ be the delays of the radio waves 1 and 2, respectively. Assume that the received signal is sampled at the fundamental *Nyquist* sampling rate W. The receiver is not able to resolve radio waves 1 and 2 if their delay difference is much smaller than the sampling interval, i.e., $|\tau_{d,1} - \tau_{d,2}| \ll \frac{1}{W}$. For paths that are not resolvable, they smear together and appear in the same *tap* to the receiver. Therefore, it is more appropriate to think of each term ℓ in (3.4) as an aggregate multipath component (i.e., a tap) contributed by multiple paths instead of a single individual path. The different paths contributing to the same tap may have different complex channel gains and different Doppler shifts. As a result, the magnitude of the tap can change significantly in a short time scale due to the constructive and destructive combining of the multiple paths that fall in the same tap.

The constructive and destructive combining of the multiple paths causes rapid variation in the received signal strength. This phenomenon is called *fading* in the literature. For narrowband channels, the receiver cannot resolve signal paths at a fine scale, resulting in very few taps. Each tap in this case may be a sum of many paths. The received signal is more prone to rapid variation in the received signal strength due to the constructive and destructive combining of the many paths. By contrast, for a wideband channel, the receiver can resolve the signal paths at a fine scale. With more taps in the wideband channel, each tap may be contributed by fewer paths, and thus the tap's amplitude may not change as fast as in a corresponding narrowband channel.

Fading occurs at a fine time scale. Due to the movements of the transmitter, the receiver, or the scatters in the propagation environment, the scatters giving rise to multiple paths may change over time. As a result, the profile of the channel taps (propagation delays, Doppler shifts, and number of significant taps) may change over time. A model more general than (3.4) may consider incorporating time-varying delay $\tau_\ell(t)$, time-varying Doppler shift $\Delta f_\ell(t)$, and time-varying number $L(t)$ of significant taps. However, the variations of these in the aggregate multipath components typically occur at a larger time scale, and thus in (3.4), we assume that they do not change during the time interval of interest.

3.2 Overview of Aerial Wireless Channel Characteristics

The fundamentals of radio wave propagation and modeling described in the previous section apply not only to terrestrial wireless channels but also to the aerial wireless channels for networks with UAVs (in all of their use cases). Compared to terrestrial wireless channels, aerial wireless channels exhibit many different characteristics. This section provides an overview of the salient characteristics of aerial wireless channels. In later sections, we will delve into more details on the impact of the distinct characteristics of aerial wireless channels on the channel modeling.

To start with, we define air–ground (AG) channel, air–air (AA) channel, and ground–ground (GG) channel as follows.

- AG channel is the wireless channel between a transmitter up in the air and a receiver on the ground, or the other way around.
- AA channel is the wireless channel between a transmitter and a receiver that are both up in the air.
- GG channel is the wireless channel between a transmitter and a receiver that are both on the ground.

Height dependency. The characteristics of a wireless channel strongly depend on the heights of the transmitter and receiver. As discussed in Chapter 1 and shown in Figure 1.1, UAVs come in various sizes, shapes, and weights, and fly at different speeds and heights. A UAV can also fly at different heights during different flight phases. Take the path loss of the wireless channel between a UAV and a ground station (GS) as an example. When the UAV is flying below the antenna height of the GS, the propagation environment may be similar to that of terrestrial wireless channels, and thus existing terrestrial wireless channel models may be applicable to a large extent. When the UAV is flying well above the antenna height of the GS, the propagation environment may be close to free space since there are usually no surrounding objects in the sky. In this case, the path loss may be characterized by the free space path loss or a two-ray model. For the intermediate heights, new measurements may be needed to develop a corresponding path loss model. Due to height dependency, different aerial wireless channel models or parameters may need to be used for different phases of a flight. In [44], the author divided a flight into three main phases (parking and taxiing, takeoff and landing, and en-route) and proposed a class of aeronautical wideband channel models. Similarly, in [45], the authors characterized the channel characteristics in 5 GHz band for the flight phases including parking and taxiing, takeoff and landing, i.e., when the aircraft was near an airport. Height-dependent channel models have also been used by the 3G partnership project (3GPP) for UAV communication performance evaluation [21].

Airframe shadowing. One distinct characteristic of aerial wireless channels is airframe shadowing, which refers to the obstruction of the LOS path (often called the "specular path") between the transmitter and receiver by the aircraft body. For example, when a fixed-wing UAV is making a banking turn to change the heading direction, the obstruction of the LOS path by the UAV body may occur. The features of airframe shadowing depend on the structural, material, and flight characteristics of the aircraft. The pitch, roll, and yaw rates of change during the flight of a small UAV may be quite different from those of a large fixed-wing UAV, thereby leading to different airframe shadowing features. Airframe shadowing could be a severe channel impairment. In [46], the authors found that the airframe shadowing can be up to 28 dB in 5.7 GHz band during banking turns. In [47], the authors reported that the median measured airframe shadowing loss was on average 15.5 dB in 5 GHz band and that the duration of the airframe shadowing event was on average 35.2 s. Without proper measures, airframe shadowing may cause link outage and thus disrupt UAV communication sessions.

Higher likelihood of LOS propagation. Compared to terrestrial wireless channels, aerial wireless channels feature a higher likelihood of LOS propagation due to the absence of surrounding objects aloft in the sky, as shown in Table 1.2. This is especially true for the AA channel where both the transmitter and receiver are up in the air. For the AG channel, the likelihood of LOS propagation in general increases with the height of the antenna aloft in the air, as shown in [48]. Higher likelihood of LOS propagation is generally favorable for wireless communications as it leads to stronger received signal strength. However, without proper interference management schemes, higher likelihood of LOS propagation may also lead to stronger co-channel interference [49] (a topic that we also address from a path-planning perspective in Chapter 6).

Multipath components. For aerial wireless channels, multipaths may exist due to earth-face reflection and scattering from irregularities on the surfaces of ground objects. For a large UAV, scattering from the surface of its own body may lead to multipaths as well. Nevertheless, due to the absence of surrounding objects aloft in the sky, radio waves experience less scattering when they propagate. As a result, the number of multipath components tends to be less for aerial wireless channels, and in general it decreases with the height of the antenna aloft in the air. When the LOS path exists, the ratio of the energy in the LOS path to the energy in the non-LOS (NLOS) paths, known as the K factor, is usually larger for aerial wireless channels than for terrestrial wireless channels. The channel is more deterministic with a larger K factor. In [50], the authors reported that the mean values of the K factors were 14 dB in 1 GHz band and 28.5 dB in 5 GHz band in suburban environments. In the presence of earth-face reflection, the reflected path may lead to signal variation that may be described by the well-known two-ray model. This was observed, for example, in [51], which found that the aerial wireless channels could be modeled by an LOS path and earth-face reflection with scattered components based on measurements in 5 GHz band.

Antenna configuration. Antenna configuration has a major impact on how the wireless channels should be modeled. For example, if a UAV is served by the side lobe of a GS whose antenna is down-tilted for terrestrial devices, the earth-face reflection or scattered paths to the UAV may become strong as they are amplified by the main lobe of the GS's antenna while the LOS path is amplified by the side lobe of the GS's antenna [48]. In such a scenario, it might be important to properly model the earth-face reflection or scattered paths in the corresponding aerial wireless channel model. The antenna characteristics at the UAV such as type (omni-directional or directional), position (mounted at the bottom or on the top of the UAV body), orientation, and polarization may have implication on the modeling of the corresponding aerial wireless channels as well [52].

In this section, we have provided an overview of some salient characteristics of aerial wireless channels, with a focus on how they differ from the more familiar and extensively studied terrestrial wireless channels. There are other important characteristics of aerial wireless channels, such as frequency dependency and Doppler effects. For frequency dependency, this is a common phenomenon for both terrestrial wireless channels and aerial wireless channels. For example, the tropospheric attenuation is often negligible for sub-6 GHz bands but can be severe for millimeter-wave bands. Doppler effects, i.e., Doppler spread and Doppler shift, will be introduced due to the motions of the transmitter, the receiver, or the surrounding objects in the propagation environment. For aerial wireless channels, Doppler effects also exhibit height dependency. Large Doppler spread may occur when the UAV is close to the ground. The Doppler spread may become small at high altitudes and is concentrated around the Doppler shift typically caused by the UAV motion. Some other UAV specific electronic and mechanic characteristics may also have impact on aerial wireless channels. For more details, we refer to, for example, the works in [52, 53] and references therein.

3.3 Large-Scale Propagation Channel Effects

Large-scale propagation channel effects mainly include path loss of radio signal as a function of propagation distance and shadowing due to signal path obstructions by large objects such as buildings. The propagation loss is often expressed as a sum of distance-dependent path loss and shadowing. Several popular path loss models for aerial wireless channels are described in Sections 3.3.1–3.3.4, while shadowing, including the canonical log-normal shadowing, model and the relatively unique airframe shadowing in aerial wireless channels are discussed in Section 3.3.5. In addition to path loss and shadowing, this section also describes ray-tracing models, LOS probability models, and atmospheric and weather effects.

3.3.1 Free-Space Path Loss

Free-space path loss is a useful starting point for the characterization of wireless channel path loss. It is a particularly useful model for aerial wireless channels that feature higher likelihood of LOS propagation. At high altitudes without the presence of earth-face reflection, the propagation of electromagnetic waves in AA channels is close to free-space propagation [54]. Similarly, for an AG channel with either the transmitter or the receiver high in the air, free-space propagation (possibly with some modification) is a reasonably accurate approximation for characterizing the path loss of the AG channel in the presence of LOS, as confirmed by many measurement campaigns [51, 55–58].

The model characterizing the free-space electromagnetic wave propagation goes back to the work done by Harold T. Friis [59]. Assume that $u(t)$ is a complex baseband signal at the transmitter. The corresponding passband signal can be written as $s(t) = \Re(u(t)e^{j2\pi f_c t})$, where f_c is the carrier frequency and we have assumed that the initial phase is zero for simplicity. For a receiver located at a distance d away from the transmitter, the signal is scaled by a factor $\frac{\lambda\sqrt{G_t G_r}}{4\pi d}$ and experiences a propagation delay of $\frac{d}{c}$, where λ is the carrier wavelength, G_t and G_r are the antenna field radiation patterns of the transmit and receive antennas in the LOS direction, respectively, and c is the speed of light [60]. The received signal will be given by:

$$r(t) = \Re\left(\frac{\lambda\sqrt{G_t G_r}}{4\pi d} u\left(t - \frac{d}{c}\right) e^{j2\pi f_c\left(t - \frac{d}{c}\right)}\right) \tag{3.5}$$

$$= \Re\left(\frac{\lambda\sqrt{G_t G_r}}{4\pi d} e^{-j2\pi \frac{d}{\lambda}} u\left(t - \frac{d}{c}\right) e^{j2\pi f_c t}\right). \tag{3.6}$$

The received power P_r, in Watts, is equal to

$$P_r = P_t G_t G_r \left(\frac{\lambda}{4\pi d}\right)^2, \tag{3.7}$$

where P_t denotes the transmit power. The free-space propagation may be intuitively understood as follows. The effective isotropic radiated power (EIRP) is the product of the transmit power and the transmit antenna gain, i.e., $P_t G_t$. The power flux density (Watts/m^2) at the receiver is equal to the EIRP divided by the surface area of a sphere with radius d, i.e., $\frac{P_t G_t}{4\pi d^2}$. The received power P_r is then given by the product of the power flux density and the effective antenna area that captures useful energy. The gain of the receive antenna may be related to the effective area of the antenna and the operating carrier wavelength as follows [59]:

$$G_r = \frac{4\pi}{\lambda^2} A_e, \qquad (3.8)$$

where A_e denotes the effective area of the receive antenna. Multiplying the power flux density $\frac{P_t G_t}{4\pi d^2}$ by the effective antenna area $A_e = G_r \frac{\lambda^2}{4\pi}$ yields (3.7).

The free-space path loss denoted by PL is defined as $\frac{P_t}{P_r}$ and follows from (3.7):

$$PL = \frac{1}{G_t G_r} \left(\frac{4\pi d}{\lambda} \right)^2. \qquad (3.9)$$

We can see that the free-space path loss is proportional to the square of the distance between the transmit and receive antennas. As the distance increases, the received power decreases. Further, the free-space path loss is inversely proportional to the square of the signal wavelength. As an example, comparing the free-space path loss in 1 GHz band to the free-space path loss in 60 GHz millimeter-wave band, the latter experiences more than 35 dB higher path loss than the former. This simple example shows that using millimeter-wave frequencies for UAV communications needs to overcome this additional large path loss when compared with using sub-6 GHz frequencies.

Because of the small form factor in millimeter-wave frequencies, substantially more directional antennas may be permitted at the transmitter or the receiver. Adaptive arrays may be used for UAV communications at millimeter-wave frequencies. The narrow beams formed by the adaptive arrays may provide high antenna gains to reduce the path loss in millimeter-wave bands. Array processing, however, may impose computation burden at the UAV, and its feasibility depends on the type of the UAV.

3.3.2 Ray Tracing

Two-Ray Models

In an aerial wireless channel, in addition to the direct path, there may exist other paths between a transmitter and a receiver, particularly when a UAV is flying at a low height. In such cases, the free-space path loss may not be accurate when used alone. A two-ray model that considers both the direct path and a ground-reflected path between a transmitter and a receiver turns out to be a useful model for aerial wireless channel modeling. It was observed, for example, in [51] that aerial wireless channels can be modeled by an LOS path and earth-face reflection with scattered components based on measurements in 5 GHz band. The series of measurements carried out in [50, 61, 62] showed that the propagation path loss in several representative areas (over-water settings, hilly and mountainous areas, suburban and near-urban environments) mostly followed the two-ray model with adjustments. Another measurement study in [63] also reported that about 86% of the measured channel responses over sea surface in 5.7 GHz band can be represented by the two-ray model.

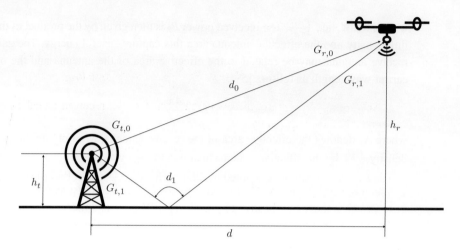

Figure 3.3 An illustration of a flat-earth two-ray model in an AG propagation scenario with a UAV.

We consider a flat-earth two-ray model, where the flatness is a valid assumption when the maximum distance between the transmitter and receiver is not larger than a few tens of kilometers. Earth curvature may need to be taken into account when considering scenarios with larger transceiver distances. We refer interested readers to [61, 64] for a curved-earth two-ray model.

The flat-earth two-ray model is illustrated in Figure 3.3. The propagation distances of the LOS path and the ground-reflected path are denoted by d_0 and d_1, respectively. The received signal is the superposition of the signal along the LOS path and the signal along the ground reflection path. Using similar notation as in the free-space path loss model, the received signal of the two-ray model, denoted by $r_{\text{2ray}}(t)$, can be written as

$$r_{\text{2ray}}(t) = \Re\left(\frac{\lambda}{4\pi}\left(r_{b,0}(t) + r_{b,1}(t)\right)e^{j2\pi f_c t}\right), \tag{3.10}$$

where

$$r_{b,0}(t) = \frac{\sqrt{G_{t,0}G_{r,0}}}{d_0}e^{-j2\pi\frac{d_0}{\lambda}}u\left(t - \frac{d_0}{c}\right), \tag{3.11}$$

$$r_{b,1}(t) = \frac{\Gamma\sqrt{G_{t,1}G_{r,1}}}{d_1}e^{-j2\pi\frac{d_1}{\lambda}}u\left(t - \frac{d_1}{c}\right) \tag{3.12}$$

where $G_{t,0}$ and $G_{r,0}$ are the antenna field radiation patterns of the transmit and receive antennas in the LOS direction, respectively, $G_{t,1}$ and $G_{r,1}$ are the antenna field radiation patterns of the transmit and receive antennas along the direction of the ground reflection path, respectively, and Γ is the reflection coefficient for the ground.

Denote by h_t and h_r the heights of the transmitter and the receiver, respectively. The propagation distance difference of the LOS path and the ground reflection path, denoted by Δd, is given by

$$\Delta d = d_1 - d_0 = \sqrt{(h_t + h_r)^2 + d^2} - \sqrt{(h_t - h_r)^2 + d^2}, \tag{3.13}$$

where d denotes the horizontal distance between the transmitter and the receiver.

3.3 Large-Scale Propagation Channel Effects

- When d is very large compared to $h_t + h_r$, we can use a Taylor series approximation for Δd to obtain that

$$\Delta d = d_1 - d_0 \approx \frac{2h_t h_r}{d}, \quad d \gg h_t + h_r. \tag{3.14}$$

Accordingly, the phase difference between the two received signal components, denoted by $\Delta \phi$, is given by

$$\Delta \phi = 2\pi \frac{\Delta d}{\lambda} \approx \frac{4\pi h_t h_r}{\lambda d}. \tag{3.15}$$

- When the delay spread $\frac{\Delta d}{c}$ is much smaller than the Nyquist sampling interval $\frac{1}{W}$, where W is the signal bandwidth, the direct LOS path and the ground reflection path are not resolvable and smear together in the same tap at the receiver. In this case, we have

$$u\left(t - \frac{d_0}{c}\right) \approx u\left(t - \frac{d_1}{c}\right). \tag{3.16}$$

In other words, the transmission is narrowband.
- For large d, the following approximations hold:

$$d_0 \approx d_1 \approx d, \tag{3.17}$$

$$G_{t,0} \approx G_{t,1}, \quad G_{r,1} \approx G_{r,1}, \tag{3.18}$$

The reflection coefficient for the ground is approximately $\Gamma \approx -1$ [60].

For large d, the received signal power can be approximately calculated as follows:

$$P_r \approx P_t \left(\frac{\lambda}{4\pi}\right)^2 \left\| \frac{\sqrt{G_{t,0} G_{r,0}}}{d_0} + \frac{\Gamma \sqrt{G_{t,1} G_{r,1}}}{d_1} e^{-j\Delta\phi} \right\|^2 \tag{3.19}$$

$$\approx P_t \left(\frac{\lambda}{4\pi}\right)^2 \frac{G_t G_r}{d^2} \left(\frac{4\pi h_t h_r}{\lambda d}\right)^2 \tag{3.20}$$

$$= P_t G_t G_r h_t^2 h_r^2 d^{-4}, \tag{3.21}$$

where, in the first line, we have used the narrowband approximation (3.16), and in the second line, we have used the approximations (3.15), (3.17), (3.18), and $\Gamma \approx -1$. Note that, starting from the second line, we have started using G_t to denote the approximate value for $G_{t,0}$ and $G_{t,1}$, and G_r to denote the approximate value for $G_{r,0}$ and $G_{r,1}$.

The two-ray path loss, denoted by $PL_{2\text{ray}}$, is defined as $\frac{P_t}{P_r}$ and follows from (3.21):

$$PL_{2\text{ray}} \approx \frac{d^4}{G_t G_r h_t^2 h_r^2}. \tag{3.22}$$

As seen from (3.22), at large values of d, the two-ray path loss increases with the distance raised to the fourth power. In contrast, the free-space path loss increases with the distance raised to the second power only. In the regime of large values of d, we can also observe from (3.22) that the two-ray path loss does not depend on the carrier frequency.

It is important to understand the series of assumptions behind the approximate path loss formula (3.22) for the considered flat-earth two-ray model. As a rule of thumb, the approximation is valid if the phase difference between the two received signal components satisfies that $\Delta\phi \approx \frac{4\pi h_t h_r}{\lambda d} < 0.6$ radian [65], which is equivalent to that $d > \frac{20\pi h_t h_r}{3\lambda}$. Consider an urban macro scenario where a GS transmits in 700 MHz band at the height of 25 m and a UAV receiver is up in the air at the height of h_r. Plugging these numbers into $d > \frac{20\pi h_t h_r}{3\lambda}$ yields that $d > 1222 h_r$. For a UAV altitude higher than 82 m, the horizontal distance between the GS and the UAV would have to be more than 100 km such that the path loss formula (3.22) holds. But such large transceiver horizontal separation distance would undermine the basic flatness assumption in the flat-earth two-ray model.

In the case that the horizontal distance d between the transmitter and receiver is not large enough compared to h_t and h_r, a more accurate characterization of the two-ray path loss is given by [64]

$$PL_{2\text{ray}} \approx \frac{1}{4G_t G_r}\left(\frac{4\pi d}{\lambda}\right)^2 \sin^{-2}\left(\frac{2\pi h_t h_r}{\lambda d}\right). \tag{3.23}$$

Note that (3.23) reduces to (3.22) when d is large compared to h_t and h_r. It can be seen from (3.23) that there exist alternate maxima and minima in the two-ray path loss.

- When $\frac{h_t h_r}{\lambda d} = \frac{n}{2}, n = 1, 2, ...$, $PL_{2\text{ray}}$ goes to infinity. In these cases, the two received signal components are out of phase and completely cancel each other, leading to zero received signal power.
- When $\frac{h_t h_r}{\lambda d} = \frac{(2n+1)}{4}, n = 0, 1, ...$, $PL_{2\text{ray}} \approx \frac{4\pi^2 d^2}{\lambda^2 G_t G_r}$. In these cases, the two received signal components are in phase and add constructively, leading to maximal received signal power.

We can also further observe that the path loss transits from a maximum to a minimum when $\frac{h_t h_r}{d}$ changes by a factor of $\frac{\lambda}{4}$. We define $\Delta > 0$ as the required distance increase such that

$$\frac{h_t h_r}{d} - \frac{h_t h_r}{d + \Delta} = \frac{\lambda}{4}. \tag{3.24}$$

With simple algebraic manipulation, we can obtain that

$$\Delta = \frac{\lambda d^2}{4 h_t h_r - \lambda d}. \tag{3.25}$$

Note that (3.25) is valid only if $d < \frac{4 h_t h_r}{\lambda}$, because Δ is negative when $d \geq \frac{4 h_t h_r}{\lambda}$, contradicting to the assumption that $\Delta > 0$. This result leads to an interesting observation: The two-ray path loss pattern with alternate maxima and minima occurs up to a critical distance of $\frac{4 h_t h_r}{\lambda}$. At the critical distance, the last minimum of the two-ray path loss is reached, after which the path loss increases sharply with the distance raised to the fourth power.

In the vertical domain, without loss of generality, let us fix h_t and consider the change of h_r needed for the path loss to transit from a maximum to a minimum. It is easy to see that the required height change is equal to $\frac{\lambda d}{4h_t}$. Unlike in the horizontal domain, a critical height value does not exist here. In other words, for an aerial channel modeled by the two-ray path loss model, the received signal power at the UAV experiences a sequence of maxima and minima when the UAV is moving vertically. In contrast, when the UAV is moving horizontally, the received signal power at the UAV experiences a sequence of maxima and minima only up to a critical distance.

The two-ray path loss pattern with alternate maxima and minima is an example of multipath fading, also known as small-scale fading. We will explore small-scale fading in detail in Section 3.4.

General Ray Tracing

The free-space and two-ray models are among the most simplified electromagnetic propagation models. General ray tracing can be applied to estimate wireless channel characteristics, such as path loss, angle of arrival (AoA), angle of departure (AoD), and tap delays [66] [67]. General ray tracing, based on geometric optics and diffraction theories, is a numerical method of solving Maxwell's equations in a high-frequency regime.

The fundamental concept in ray tracing is the *ray* concept, which we have used in the previous discussions of free-space and two-ray models without an explicit definition. For radio propagation modeling, one can assume that a ray travels in a straight line in a homogeneous medium or a tube in which the energy is contained and propagated, and the traveling of the ray obeys laws of reflection, refraction, and diffraction [68]. We can categorize rays emanated from a point source into four types: direct rays, reflected rays, diffracted rays, and scattered rays. The categorization, however, is not mutually exclusive: A ray may undergo a combination of reflection, diffraction, and scattering.

A direct ray is the ray from the source to the field point directly, for example, the one considered in the free-space path loss model. A reflected ray experiences reflection one or more times before reaching the field point. The propagation directions of reflected rays obey the law of reflection. The reflected fields can be determined by Fresnel's equations. The primary factors determining reflection mechanism include the frequency, angle of incidence, conductivity of the reflecting surface, and polarization of the incident wave. We have already considered one example of reflected rays in the two-ray path loss models.

Diffraction is the bending of a ray around the corners of an obstacle. Compared to reflected rays, diffracted rays are more difficult to characterize. An incident ray can spawn a continuum of diffracted rays. Calculating diffraction coefficients is more complicated, and different formulations may lead to different results. Diffraction can be characterized by using the geometrical theory of diffraction [69] or the improved uniform theory of diffraction [70], among others. Wedge diffraction can be applied to further simplify diffraction characterization by treating the diffracting object as a wedge instead of a more general shape [71, 72]. These methods are more pertinent to computer simulations rather than an analytical study of system performance.

Scattering is the reflection of a ray from irregularities on the surface of an object into different reflected directions. A scattered ray has a path loss proportional to the product of the length of incident segment and the length of reflected segment reaching the field point due to spreading loss after scattering. For example, scattering causes free-space path loss to be proportional to distance raised to the fourth power. Though scattering is a weaker propagation phenomenon, it may not be negligible [73], especially in millimeter-wave frequencies [74]. For scattering from buildings, the scattered rays may be divided into specular and nonspecular components [75] and modeled using the effective roughness concept [76].

Determining the rays from the source to the field point is key in ray tracing. A basic ray-tracing method may consist of three steps: ray launching, ray tracing, and ray reception [77]. In ray launching, a large number of rays are generated and emanated from the source point as uniformly as possible. In ray tracing, a ray is traced from the source point and is determined if it intersects any object in the propagation environment. This is usually the most computation-heavy step. A ray may be dropped in the tracing if its power drops below a certain threshold or it has undergone a given number of reflections, diffraction, and/or scatterings. In ray reception, a ray is considered reaching the field point if the ray tube illuminates the receiving field point, followed by calculating the respective field. We refer to [68, 78] for more in-depth overviews of ray-tracing methods.

Ray tracing is a popular deterministic radio channel modeling approach that reproduces the electromagnetic waves in a specific environment. Its main advantages are high accuracy [79] and enabling evaluation of wireless channels in situations where measurements are not sufficient or difficult. Its main disadvantage is that it only applies to the specific environment under investigation. It is computation heavy, but there are methods to achieve better computational efficiency [68].

As an effective method for propagation modeling, ray tracing has been applied to aerial wireless channel modeling. In [55], the authors used ray-tracing simulations to develop path-loss and shadowing models. In [80], ray-tracing simulations were used to verify the proposed theoretical LOS model for AG radio propagation in urban environments. To better understand the wireless channels for low-level maritime UAV operations, the work [81] developed a simulator to generate a random sea surface in a deep-water location and collected simulation results to characterize marine wireless channels as a function of frequency and observable sea surface height for fixed transmitter and receiver locations. A ray-tracing effort was also conducted in [82] for marine wireless channel modeling over the sea surface.

Ray-tracing analysis was conducted in [83] for urban terrain to validate the experimental results. A statistical propagation model based on urban environment properties for predicting the AG path loss was proposed in [57] and was validated with ray-tracing simulations. The work [84] also used ray-tracing simulations to study AG channels in millimeter-wave frequencies (28 GHz and 60 GHz).

To sum up, ray tracing has a sound physical basis and is a powerful method for propagation modeling. The ray-tracing method has been used for aerial wireless channel modeling. Combined with empirical and analytical methods, it is playing an increasingly important role in aerial wireless channel modeling for UAV wireless communications systems and networking.

3.3.3 Log-Distance Path Loss Models

A fundamental essence of the large-scale channel effects of radio propagation is that the path loss increases exponentially with distance. The rate at which the path loss increases with distance is called *path loss exponent*. For example, the path loss exponent equals 2 for the free-space path loss and 4 for the two-ray path loss (when the distance is large). Log-distance path loss models generalize the free-space path loss model and two-ray path loss models in the sense that the path loss exponent is regarded as a parameter of the models and is determined based on the propagation environment. Log-distance path loss models are still much simplified from real propagation environments, but they capture the essence of the large-scale channel effects of radio propagation. Thus, they have been widely used for various designs of wireless communications systems, not only in terrestrial but also in UAV wireless communications systems.

Basic Log-Distance Path Loss Model

The basic form of the log-distance path loss model can be expressed by

$$PL = K \left(\frac{d}{d_0}\right)^{\alpha}, \tag{3.26}$$

where K is a unit-less scaling factor, d_0 is a reference distance for the antenna far field, and α is the path loss exponent. The path loss in dB scale is given by

$$PL \text{ (dB)} = K \text{ (dB)} + 10\alpha \log_{10}\left(\frac{d}{d_0}\right). \tag{3.27}$$

The reference distance d_0 is determined from measurements close to the transmitter. Due to the effects of antenna near field, the model is generally only applicable to far field with $d > d_0$. The value of K can be determined through field measurements at distance d_0. The path loss exponent α depends on the propagation environment. The value of α is usually determined by minimizing the mean squared error between the model and the empirical measurements. The typical values of α for terrestrial radio environments range from 2 to 6 [65]. It was also proposed in the literature to model α as a Gaussian random variable [85], which is, however, a less common approach.

For aerial radio environments, several measurement campaigns have been carried out to obtain the corresponding path loss exponents. Table 3.1 lists α values for different environments reported in the literature. It can be seen from Table 3.1 that the existing measurement campaigns focus on LOS propagation conditions, while measurement results for NLOS propagation conditions are lacking. The reasons may be twofold: (1) LOS propagation conditions are common for many existing UAV applications, and (2) it may be more challenging to conduct aerial wireless channel measurements in NLOS scenarios, especially in low-altitude urban environments.

It can be further observed from Table 3.1 that the path loss exponents for aerial wireless channels tend to be smaller than for terrestrial wireless channels, agreeing with intuition. Most of the path loss exponent values in Table 3.1 range from 1.5 to 3.0, which is a consequence of the almost universal presence of LOS propagation conditions in the measurements. Interestingly, there are some uncommon small path loss exponent

Table 3.1 Path loss exponents measured under different aerial propagation environments.

Reference	Scenario	UAV type	LOS/NLOS	Heights[a] (m): (GS, aircraft)	Frequency (GHz)	Path Loss Exponent
[61]	Over water	S-3B Viking aircraft	LOS, NLOS	(4.9–235, 762–808) AMSL	0.996–0.977; 5.030–5.091	1.9 (fresh); 1.9 (sea); 1.9 (fresh); 1.5 (sea)
[63]	Over sea	Learjet 35A	LOS, NLOS	(2.1–7.65, 370–1830) AMSL	5.7	0.14–2.46
[62]	Hilly, mountainous	S-3B Viking aircraft	LOS, NLOS	(346.6–2760.6, 1089–4029) AMSL	0.968	1.3–1.8
[86]	Rural	Commercial UAV	LOS	(–, 15–120) AGL	5.060	1.0–1.8
[87]	Open field, suburban	Quadrocopter	LOS	(0.07–1.5, 4–16) AGL	0.8	2.0–2.9
[50]	Suburban, near-urban	S-3B Viking aircraft	LOS, NLOS	(171–776, 762–1745) AMSL	3.1–5.3	2.54–3.04
					0.968	1.7
					5.060	1.5–2.0
[88]	Cluttered environment	–	LOS, NLOS	(–, 457–975) AMSL	2	4.1
[89]	Open field	Quadrocopter	LOS	(3, 20–110) AGL	5.2	2.01
[45]	Near airport	BeechcraftB-99	LOS	(2, up to 914) AGL	5.8	2.0–2.25
[90]	Private airfield	Senior Telemaster	LOS	up to (4.3, 46) AGL	5	1.80
[91]	Private airfield	Self-designed	LOS	(–, Up to 125) AGL	2.4	1.92 (AA); 2.13 (AG)
[92]	–	ARES unmanned aircraft	LOS	(2.4, 200) AGL	2.4	2.34

[a] AMSL: above mean sea level; AGL: above ground level.

values (less than 1.5) observed in the measurements, especially for the over-sea scenarios. In particular, the work in [63] reported a path loss exponent value of 0.14 in one of the measured scenarios and pointed out that it was due to evaporation duct above the sea surface that resulted in enhanced radio wave propagation.

Modified Log-Distance Path Loss Models

There exist other modified forms of log-distance path loss models, beyond the basic form given in (3.27). One form closely similar to the basic form (3.27) is the floating intercept model given by

$$PL \text{ (dB)} = 10\alpha \log_{10}(d) + \beta. \tag{3.28}$$

Compared to the basic form (3.27), the floating intercept model eliminates the reference distance d_0 and the corresponding path loss value K at distance d_0. Instead, the floating intercept model depends on two parameters α and β, where α is the slope that still bears the meaning of the path loss exponent and β denotes the intercept. The values of α and β are usually jointly determined by minimizing the mean squared error between the model and the empirical measurements.

The floating intercept model was used in the work [86] for radio channel modeling for UAV communication over cellular networks. Noting the dependency of aerial wireless channels on UAV height, the authors proposed in [86] to extend the floating intercept model with height dependent α and β:

$$PL \text{ (dB)} = 10\alpha(h_u) \log_{10}(d) + \beta(h_u), \tag{3.29}$$

where $h_u \in [1.5, 120]$ m denotes the UAV height, and $\alpha(h_u)$ and $\beta(h_u)$ determined based on measurements are given by

$$\alpha(h_u) = \max(3.9 - 0.9 \log_{10}(h_u), 2), \tag{3.30}$$

$$\beta(h_u) = -8.5 + 20.5 \log_{10}(\min(h_u, h_{FSPL})), \tag{3.31}$$

where h_{FSPL} is the height where free-space propagation is assumed.

Another class of modified log-distance path loss models spells out the dependency of the factor K in the basic form (3.27) on, for example, the antenna characteristics, certain propagation environment factors, and frequency. For example, the authors in [87] gave the following modified log-distance path loss model:

$$PL \text{ (dB)} = PL_0 \text{ (dB)} + 10\alpha \log_{10}\left(\frac{d}{d_0}\right) - 10 \log_{10}\left(\frac{\Delta h}{h_{opt}}\right) + C_p + 10 \log_{10}\left(1 + \frac{\Delta f}{f_c}\right), \tag{3.32}$$

where PL_0 is the path loss at reference distance d_0, $\Delta h = |h_{gnd} - h_{opt}|$, h_{gnd} is the UAV height, h_{opt} is the minimum height of the UAV that gives the lowest path loss for a given environment, C_p is a constant loss factor capturing foliage loss and losses resulted from UAV antenna orientations, f_c is the carrier frequency, and Δf is the Doppler variation in the frequency.

3.3.4 Empirical Path Loss Models

In aerial wireless channels, especially when a UAV is flying at low altitudes and near ground clutters such as buildings and trees, the complicated reflection, diffraction, and scattering effects due to the presence of obstacles may not be accurately modeled by the free-space path loss, two-ray path loss, or log-distance path loss models. A number of empirical path loss models have been proposed in the literature to predict the path loss for aerial wireless channels. The empirical measurements for a given path loss model were usually conducted in a specific propagation environment for given frequency and distance ranges. These models may find applications beyond the measured environments, but validations are needed when applying the empirical models to general environments. In this section, we describe several empirical path loss models for aerial channel modeling.

Multi-Slope Log-Distance Path Loss Model

Multi-slope log-distance path loss model is an extension of the single slope log-distance path loss model. It is essentially a piecewise linear approximation of the empirical measurements. The model may specify $N-1$ breakpoints, $d_1, ..., d_{N-1}$ with each segment having a corresponding slope value. The multi-slope log-distance path loss model has been used to model path loss for terrestrial wireless channels (see e.g., [93]).

A special case of multi-slope log-distance path loss model is the dual-slope model with only one breakpoint:

$$PL \text{ (dB)} = \begin{cases} K \text{ (dB)} + 10\alpha_1 \log_{10}\left(\frac{d}{d_0}\right) & \text{if } d_0 \leq d \leq d_1 \\ K \text{ (dB)} + 10\alpha_1 \log_{10}\left(\frac{d_1}{d_0}\right) + 10\alpha_2 \log_{10}\left(\frac{d}{d_0}\right) & \text{if } d > d_1. \end{cases} \quad (3.33)$$

In this dual-slope model, the path loss increases exponentially with distance at the rate of α_1 up to the breakpoint distance d_1, after which the path loss increases exponentially with distance at the rate of α_2.

In the context of aerial wireless channel modeling, the dual-slope model has been used for over-water path loss modeling based on empirical measurements in L-band of 960–977 MHz and in C-band of 5.030–5.091 GHz [61]. The authors in [94] also used the dual-slope model to fit the empirical measurements collected in a suburban scenario in 5.76 GHz band.

The work in [48] reported path loss measurement data collected in a helicopter measurement campaign in the carrier frequency of 1.8 GHz. The results are reprinted in Figure 3.4. In the measurement, the GS height was about 50 m, the height of the surrounding clutters (trees, buildings, etc.) was about 25 m, and the UE height was about 50 m above the ground. From the measurement data, we can see the existence of breakpoint: The path loss increases exponentially with distance at the rate of close to 2 up to the breakpoint distance of about 10 km, after which the path loss increases exponentially with distance at a much higher rate. Figure 3.4 also shows the 3GPP rural macrocell LOS and NLOS path loss models for ground UE and the benchmark free space path loss. As shown in Figure 3.4, the 3GPP rural macrocell LOS and NLOS path loss models for

3.3 Large-Scale Propagation Channel Effects

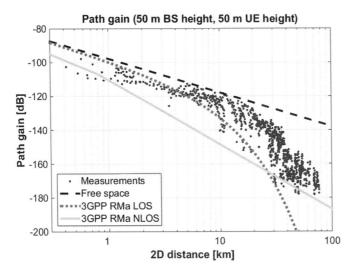

Figure 3.4 An illustration of breakpoint in the path loss of AG channel with measurement data. © IEEE. Reprinted, with permission, from [48].

ground UE may not be accurate enough for aerial UAV UEs, and, thus, 3GPP developed new channel models for aerial UAV UEs in Release 15 [21].

Height-Dependent Two-Ray Model

The two-ray path loss expressions (3.22) and (3.23) are obtained under a series of simplified assumptions. If we do not apply any common simplifications except the narrowband transmission assumption, the two-ray path loss will then be given by:

$$PL \text{ (dB)} = -20\log_{10}\left(\frac{\lambda}{4\pi}\right) - 10\log_{10}\left|\frac{\sqrt{G_0}}{d_0} + \frac{\Gamma\sqrt{G_1}}{d_1}e^{-j\phi}\right|^2, \quad (3.34)$$

where we recall that λ is the wavelength, G_0 and G_1 are the products of the transmit and receive antenna field radiation patterns in the LOS direction and in the ground reflection direction, respectively, d_0 and d_1 are the propagation distances of the LOS path and the ground-reflected path, respectively, $\phi = \frac{2\pi(d_1-d_0)}{\lambda}$ is the phase difference between the two received signal components, and Γ is the ground reflection coefficient.

Built upon the two-ray path loss (3.34) and motivated by the observation of the path loss dependency on the elevation angle [55], the authors proposed a height-dependent two-ray model for AG wireless channels in [83]:

$$PL \text{ (dB)} = -20\log_{10}\left(\frac{\lambda}{4\pi}\right) - 10\alpha(h)\log_{10}\left|\frac{\sqrt{G_0(h)}}{d_0} + \frac{\Gamma\sqrt{G_1(h)}}{d_1}e^{-j\phi}\right|, \quad (3.35)$$

where $\alpha(h)$, $G_0(h)$, and $G_1(h)$ depend on the height h of the UAV's antenna. Based on the empirical measurements, the authors introduced three different height zones to which different model parameters were assigned, as detailed in [83].

Excess Path Loss Model

An excess path loss model extends a reference path loss by adding an excess path loss component. For example, the free-space path loss is chosen as the reference path loss in [56], while the mean terrestrial path loss is chosen as the reference path loss in [95].

The empirical excess path loss model developed in [56] is of particular interest as it provides insights on AG wireless channel path loss in urban street environments. The propagation environments are complex and challenging due to the surrounding buildings around the GS, but urban street environments may become increasingly common for emerging UAV applications. The authors divided the excess loss, denoted as PL_{excess}, into two parts in [56]:

$$PL_{excess} \text{ (dB)} = PL_{lb} \text{ (dB)} + PL_{rt} \text{ (dB)}. \tag{3.36}$$

A similar approach has been used for terrestrial wireless channel path loss modeling (see e.g., [96]). The term PL_{lb} includes the diffraction loss due to the building closest to the GS, referred to as the last building. The last building is modeled by means of two knife edges on both outer walls. To approximately calculate the diffraction loss, Deygout's method can be used for the case where one edge dominates over the other [64]. In particular, the diffraction loss can be further decomposed into two terms: one term due to the main diffracting edge (i.e., the one closest to the GS) and the other due to the secondary knife edge. The term PL_{lb} may further include contributions from other rays besides the direct ray diffracted at the last building and reaching the GS, e.g., the ray reflected on the opposite side of the street [97]. The term PL_{rt} includes the diffraction loss caused by the multiple buildings between the UAV and the last building.

3.3.5 Shadowing

The deterministic path loss models that we previously discussed do not consider possible random variations due to blocking objects of different locations, sizes, and dielectric properties as well as changes in reflecting surfaces and scattering objects. These random effects give rise to random variation about the path loss known as *slow fading*, which calls for statistical models to characterize this random attenuation. It was first shown in [98] and later validated by many indoor and outdoor measurements that the random attenuation about the path loss can be characterized by the *log-normal shadowing* model [60, 65].

In the log-normal shadowing model, the random attenuation about path loss is modeled by a random variable χ_{dB} with a log-normal distribution given by (in dB):

$$f_{\chi_{dB}}(x) = \frac{1}{\sqrt{2\pi}\sigma_{\chi_{dB}}} \exp\left(-\frac{(x-\mu_{\chi_{dB}})^2}{2\sigma_{\chi_{dB}}^2}\right), \tag{3.37}$$

where $f_{\chi_{dB}}(x)$ denotes the probability density function of χ_{dB}, $\mu_{\chi_{dB}}$ is the mean, and $\sigma_{\chi_{dB}}$ is the standard deviation of the random variable χ_{dB}. The mean attenuation due to shadowing may be incorporated into the deterministic path loss model, in which case the mean $\mu_{\chi_{dB}} = 0$ in the log-normal shadowing model.

Table 3.2 Standard deviations of log-normal shadowing measured under different aerial propagation environments.

Reference[a]	Heights (m): (GS, aircraft)	Frequency (GHz)	$\sigma_{\chi_{dB}}$
[62]	(346.6–2760.6, 1089–4029) AMSL	0.968	3.2–3.9
		5.060	2.2–2.8
[50]	(171–776, 762–1745) AMSL	0.968	2.6–3.1
		5.060	2.9–3.2
[86]	(–, 15–120) AGL	0.8	3.4–6.2
[87]	(0.07–1.5, 4–16) AGL	3.1–5.3	2.8–5.3
[45]	(20, up to 914) AGL	5.8	1.2–9.8
[88]	(–, 457–975) AGL	2	5.24

[a] Refer to Table 3.1 for more information regarding the measurement setups in each reference.

Characterizing the spatial auto-correlation of shadowing is important since blocking objects for different communication links in close proximity are strongly correlated. The spatial auto-correlation of shadowing is commonly modeled by a first-order autoregressive process. In this model, for two locations separated by distance δ, the spatial auto-correlation of shadowing denoted by $R(\delta)$ is given by [99]:

$$R(\delta) = \mathbb{E}[(\chi_{dB}(d) - \mu_{\chi_{dB}})(\chi_{dB}(d+\delta) - \mu_{\chi_{dB}})] = \sigma_{\chi_{dB}}^2 \exp(-\delta/d_c), \quad (3.38)$$

where d_c is the decorrelation distance denoting the distance at which the auto-correlation equals the value $\sigma_{\chi_{dB}}^2/e$.

The log-normal shadowing model has also been widely assumed in aerial wireless channel modeling. Table 3.2 summarizes the $\sigma_{\chi_{dB}}$ values measured under different aerial propagation environments. The values of $\sigma_{\chi_{dB}}$ typically range from 5 to 12 dB in terrestrial macrocells and from 4 to 13 dB in terrestrial microcells [60]. In contrast, it can be seen from Table 3.2 that the values of $\sigma_{\chi_{dB}}$ tend to be smaller in aerial wireless channels, and the higher the antenna height, the smaller the value of $\sigma_{\chi_{dB}}$. These results agree with intuition: As the antenna height increases, there are fewer blocking objects and less randomness in reflected and scattered rays, leading to smaller shadowing variation.

A distinct shadowing phenomenon in aerial wireless channels is airframe shadowing, as highlighted in Section 3.2. Recall that airframe shadowing refers to the blockage of the LOS path by the aircraft body, which may occur during aircraft maneuvering. Airframe shadowing characteristics depend on the structural, material, and flight characteristics of the aircraft and are independent of the local ground site conditions and link distance [47]. Airframe shadowing has been mainly studied for manned aircraft at high altitudes, and not much work has been done to study airframe shadowing for UAV at low altitudes.

For an aircraft-to-satellite channel, the work [100] reported that the wing shadowing can yield up to 15 dB attenuation in 20 GHz band. The work [101] simulated airframe shadowing and reported up to 15 dB shadowing attenuation in 5.12 GHz band. In [46], the authors measured airframe shadowing during aircraft maneuvering including pitch, roll, and yaw with two different flight profiles: linear flight route and circular flight route.

The measured carrier frequency was in 5.7 GHz band and the aircraft altitude was kept at 3.2 km during the measurement. It was found that the shadowing can be up to 9.5 dB and 28 dB for the linear flight route and circular flight route, respectively. From their measurements, the authors recommended that the shadowing effect can be approximated by the log-normal distribution with a standard derivation in the range from 6.49 dB to 6.77 dB.

The work in [47] reported measurement results for airframe shadowing depth, duration, and multiple antenna diversity gain. The results were based on over 200 aircraft wing/engine shadowing events. The median shadowing loss was, on average, 15.5 dB in C-band (5060 MHz) and 10.8 dB in L-band (968 MHz), respectively. The shadowing event duration was on average 35.2 s in C-band and 25.5 s in L-band, respectively. The authors also proposed an algorithm to reproduce shadowing events and showed that the deployment of multiple aircraft antennas was useful to mitigate the airframe shadowing. The authors of [102] also observed airframe blockage during a large turn of the aircraft.

Simulations were preformed in [103] to study shadowing in wireless networks via HAPs in built-up areas. Four different types of environments (suburban, urban, dense urban, and urban high-rise) and three different frequency bands (2.0 GHz, 3.5 GHz, 5.5 GHz) were considered. The authors proposed an elevation angle-dependent shadowing model as follows:

$$\chi_{dB} = \begin{cases} \chi_{dB,LOS} & \text{LOS} \\ \chi_{dB,NLOS} + \chi_{dB,\theta} & \text{NLOS}, \end{cases} \quad (3.39)$$

where $\chi_{dB,LOS}$ is a zero mean log-normally distributed random variable with 3 to 5 dB standard deviation, $\chi_{dB,NLOS}$ is a zero mean log-normally distributed random variable with 8 to 12 dB standard deviation, and $\chi_{dB,\theta}$ represents the additional log-normal shadowing as a function of elevation angle denoted by θ. The following fractional rational function was derived as a best fit to simulation results to approximate the mean $\mu_{\chi_{dB},\theta}$ and standard deviation $\sigma_{\chi_{dB},\theta}$:

$$\mu_{\chi_{dB},\theta}, \sigma_{\chi_{dB},\theta} = \frac{g+\theta}{h+i\theta}, \quad (3.40)$$

where g, h, and i are empirical parameters (different for $\mu_{\chi_{dB},\theta}$ and $\sigma_{\chi_{dB},\theta}$).

3.3.6 Line-of-Sight Probability

Ensuring an unobstructed LOS between the transmit and receive antennas is an important design consideration for radio frequency link design. In this section, we first introduce some fundamental concepts such as *Fresnel zones* and *knife-edge diffraction* model to distinguish between the meaning of LOS path and the meaning of unobstructed LOS propagation. We then discuss LOS probability modeling in aerial wireless channels.

LOS Path and Unobstructed LOS Propagation

LOS path – a notion we have introduced and have been using in this chapter – has a straightforward meaning: the shortest direct path between the transmit and receive antennas. Unobstructed LOS propagation – a notion we have touched upon earlier in this chapter – is a bit more convoluted to define precisely. To be considered as unobstructed, the LOS path and some volume of adjacent space (defined via the concept of Fresnel zones) should be free of obstructions.

The concept of Fresnel zones is illustrated in Figure 3.5, where a transparent plane is placed between the transmitter and receiver. The plane is perpendicular to the LOS path. It is at a distance d_1 from the transmitter and d_2 from the receiver. Figure 3.5 shows a set of concentric circles with the center being the intersection of the LOS path and the plane. For a point on the n-th concentric circle, consider the path consisting of two segments: The first segment is the direct path from the transmitter to the point on the n-th concentric circle, and the second segment is from the point to the receiver. The length of this path is $\frac{n\lambda}{2}$ greater than the length of the LOS path, and the length difference of the two paths is referred to as *excessive path length*.

We define r_n as the radius of the n-th concentric circle. If $d_1, d_2 \gg r_n$, the radius can be approximated as [65]

$$r_n \approx \left(\frac{n\lambda d_1 d_2}{d_1 + d_2} \right)^{\frac{1}{2}}. \tag{3.41}$$

It can be seen that the radii of the concentric circles depend on the frequency and the location of the plane. The maximum radii are obtained when $d_1 = d_2$, i.e., when the plane is midway between the transmitter and receiver.

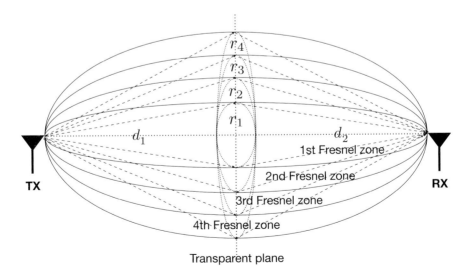

Figure 3.5 An illustration of Fresnel zones.

By joining all the points for which the excessive path length is $\frac{n\lambda}{2}$ and repeating this for all n, we can construct a series of concentric ellipsoids. The first Fresnel zone is the innermost ellipsoid space in which the LOS path passes through; the second Fresnel zone is the second innermost ellipsoid space, excluding the first Fresnel zone; and so on. For unobstructed LOS propagation, the first Fresnel zone should be free of obstructions to some extent. If there is an obstacle that introduces into the first Fresnel zone, the signals reflected or diffracted by the obstacle and the LOS signal sum at the receiver constructively or destructively depending on the total phase difference. When the obstacle blocks the direct LOS path, the signal loss could be significant. The level of diffracted signal reaching the receiver depends on the frequency, shape of the obstacle, height of the obstacle above the LOS path, and relative locations of the transmitter, receiver, and obstacle. A rule of thumb is to keep at least 60% of the first Fresnel zone free of obstructions, in which case the propagation may be considered as unobstructed LOS.

Estimating diffraction loss for a signal diffracted by a general diffracting object is complex. The simplest and most commonly used model is the Fresnel knife-edge diffraction model. The geometry of this model is shown in Figure 3.6. Let Δd be the excessive path length of the diffracted path relative to the LOS path. For the receiver located in the shadowed region, Δd is approximately given by:

$$\Delta d \approx \frac{(d+d')}{2dd'} h^2, \qquad (3.42)$$

where d is the distance between the transmitter and the point of the knife edge, d' is the distance between the point of the knife edge and the receiver, and h is the height of the knife edge relative to the LOS path.

The diffracted signal travels an extra distance Δd, leading to a phase shift relative to the LOS path given by

$$\Delta \phi = \frac{2\pi \Delta d}{\lambda} = \frac{\pi}{2} v^2, \qquad (3.43)$$

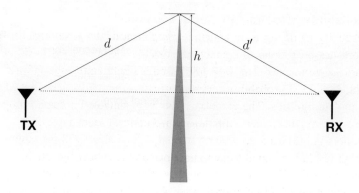

Figure 3.6 An illustration of the Fresnel knife-edge diffraction model.

where v is the Fresnel–Kirchoff diffraction parameter given by

$$v = \left(\frac{2(d+d')}{\lambda dd'}\right)^{\frac{1}{2}} h. \tag{3.44}$$

The diffraction gain compared to the free-space propagation is given by

$$G_d \text{ (dB)} = 20 \log_{10} \left| \int_v^\infty \left(\frac{1+j}{2}\right) e^{-j\frac{\pi t^2}{2}} dt \right|. \tag{3.45}$$

Approximation equations for (3.45) exist in the literature (see e.g., [60]). The Fresnel knife-edge diffraction model has also been used in aerial channel modeling. For example, the work [104] combined a two-ray reflection model with the knife-edge diffraction model to model the path loss for cellular connected UAV.

Scattering may lead to stronger received signal than predicted by reflection and diffraction models alone. Rough surfaces induce scattering, the effects of which are different from the specular reflection induced by flat surfaces. The surface roughness and *radar cross section* (RCS) of a scattering object are the two most important factors that impact outdoor radio propagation [65].

A surface may be considered rough for a given incidence angle θ_i if its minimum to maximum protuberance is greater than a critical height h_c given by

$$h_c = \frac{\lambda}{8 \sin \theta_i}. \tag{3.46}$$

RCS (units of m^2) is defined as the ratio of the scattered power density (in the direction of observation) to the incident power density, i.e.,

$$\sigma_{rcs} = \lim_{r \to \infty} \frac{4\pi r^2 S_s}{S_i}, \tag{3.47}$$

where σ_{rcs} denotes the RCS, S_s denotes the scattered power density, S_i denotes the incident power density, and r denotes the distance from the scattering object to the observation point. As the RCS increases, a higher received power from the scattered rays can be expected.

LOS Probability Modeling for Aerial Channels
LOS probability modeling is an essential component of the aerial channel modeling. Much work has been done to characterize the likelihood of LOS in different aerial propagation environments. The work [80] used a single knife-edge diffraction model to determine the LOS probability for AG channels. The model was verified with ray-tracing simulation results. The model may be more suitable for some European cities with dense and irregular streets since it assumed uniform street angles.

The authors in [48] used ray-tracing simulation to collect LOS probability statistics using a high-resolution digital terrain map of a rural area near Stockholm, Sweden. In the simulation, the GS height was 35 m above the terrain. The results are reprinted in Figure 3.7 that shows the LOS probabilities obtained from the rural area map data. For comparison, Figure 3.7 also shows the 3GPP LOS probability formula for ground UE.

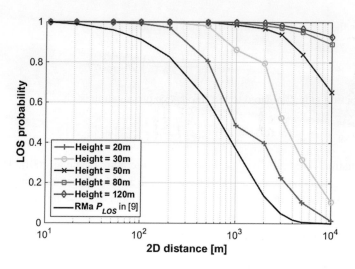

Figure 3.7 LOS probability statistics obtained from ray-tracing simulation using a terrain map of a rural area near Stockholm, Sweden. The "original reference [9]" in the legend for RMa P_{LOS} refers to the 3GPP technical report 38.901 that documents the 3GPP studies on channel models for frequencies from 0.5 to 100 GHz [105]. © IEEE. Reprinted, with permission, from [48].

A general observation can be made from Figure 3.4: The higher the antenna height, the higher the LOS probability of the aerial channel.

Built upon a statistical model recommended by the International Telecommunication Union (ITU) [106], works [57] and [107] proposed a modified sigmoid function to model the likelihood of LOS in AG channels. The model has been used by several other works for the analysis of UAV communications systems [108–110].

In the sequel, we describe the statistical model recommended by ITU [106]. The model characterizes the likelihood of the existence of an LOS path between a transmitter and a receiver in the presence of building blockage. The characterization of the buildings is simple. It is based on three statistical parameters related to an urban environment:

- Parameter α: the ratio of the land area covered by buildings to the total land area (dimensionless);
- Parameter β: the mean number of buildings per unit area (buildings/km^2); and
- Parameter γ: a scale parameter that describes the Rayleigh distributed building heights with probability density function $f_H(x)$ given by

$$f_H(x) = \frac{x}{\gamma^2} e^{-\frac{x^2}{2\gamma^2}}, \quad x \geq 0, \tag{3.48}$$

where H represents the random building height in meters.

Under this model, assuming that the buildings are arranged on a regular grid, a ray of 1 km length passes over $\sqrt{\alpha\beta}$ buildings. The expected number of buildings lying in the propagation path between a given transmitter and its receiver, denoted by N_b, is then given by

3.3 Large-Scale Propagation Channel Effects

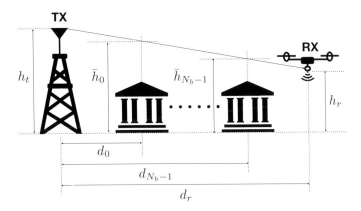

Figure 3.8 An illustration of the geometry of the LOS statistical model recommended by ITU in [106].

$$N_b = \lfloor d_r\sqrt{\alpha\beta}/10^3 \rfloor, \tag{3.49}$$

where d_r is the horizontal distance from the transmitter to the receiver in meters, and the floor operation $\lfloor \cdot \rfloor$ is introduced to get an integer number of buildings.

For a given transceiver pair, the LOS path exists if and only if each building lying in between the transmitter and receiver is below the height of the point at which the direct path crosses right above the building. Figure 3.8 shows the geometry, from which we can see that the model assumes the terrain to be flat or of constant slope over the area of interest.

We define $\bar{h}_j$ as the height of the point at which the direct path crosses right above the building $j, j = 0, ..., N_b - 1$. By simple geometric analysis, we can see that $\bar{h}_j$ is given by:

$$\bar{h}_j = h_t - \frac{d_j(h_t - h_r)}{d_r}, \tag{3.50}$$

where h_t and h_r are the heights of the transmitter and the receiver, respectively, and d_j is the horizontal distance from the transmitter to the building j.

We define $\{H_j\}_{j=0,...,N_b-1}$ as a sequence of random variables with H_j being the height of the building j. The model assumes that $\{H_j\}_{j=0,...,N_b-1}$ are independent and identically distributed (i.i.d.) with Rayleigh distribution given by (3.48). The probability p_{LOS} that an LOS path exists can be computed as follows:

$$p_{\text{LOS}} = \mathbb{P}(\cup_{j=0,...,N_b-1}\{H_j < \bar{h}_j\}) = \prod_{j=0}^{N_b-1} \mathbb{P}(H_j < \bar{h}_j), \tag{3.51}$$

where the second equality follows from the i.i.d. assumption of $\{H_j\}_{j=0,...,N_b-1}$. By the cumulative distribution function of the Rayleigh distribution given by $F_H(x) = 1 - e^{-\frac{x^2}{2\gamma^2}}, x \geq 0$, we can find that:

$$p_{\text{LOS}} = \prod_{j=0}^{N_b-1} \left(1 - e^{-\frac{\bar{h}_j^2}{2\gamma^2}}\right). \tag{3.52}$$

The model further assumes that the buildings are evenly spaced between the transmitter and the receiver. The distance from the transmitter to the building j is given by:

$$d_j = \left(j + \frac{1}{2}\right)\frac{d_r}{N_b}, \quad j = 0, ..., N_b - 1. \tag{3.53}$$

Plugging (3.49), (3.50), and (3.53) into (3.52) yields:

$$p_{\text{LOS}} = \prod_{j=0}^{\lfloor d_r\sqrt{\alpha\beta}/10^3 \rfloor - 1} \left(1 - \exp\left(-\frac{\left(h_t - \frac{(j+\frac{1}{2})(h_t - h_r)}{\lfloor d_r\sqrt{\alpha\beta}/10^3 \rfloor}\right)^2}{2\gamma^2}\right)\right). \tag{3.54}$$

We emphasize that the LOS probability (3.54) is independent of the frequency. This is because the model considers the probability of an LOS path (often known as geometrical or optical LOS), not the probability of unobstructed LOS propagation (often known as radio LOS). The latter depends on Fresnel zones whose radii depend on the frequency, as discussed in the first part of this section.

In [107], the authors proposed that the LOS probability (3.54) can be approximated by a modified sigmoid function:

$$p_{\text{LOS}}(\theta) = \frac{1}{1 + a\exp(-b(\theta - a))}, \tag{3.55}$$

where θ represents the elevation angle of the antenna in the sky with respect to the observation point on the ground, and a and b are model parameters that depend on the propagation environment. This approximation may be more friendly in terms of analytical studies.

3.3.7 Atmospheric and Weather Effects

The propagation of electromagnetic waves is affected by atmospheric effects. The gaseous molecules, such as oxygen and water vapor, contribute to some extent to signal attenuation. The attenuation may be modeled by additional path loss factors that are additive in decibel units. As the UAV height increases, the atmospheric effects on electromagnetic wave propagation vary due to the change of humidity, pressure, and temperature in the atmosphere.

Note that the atmospheric effects may not always attenuate the propagation of electromagnetic waves. In Section 3.3.3, we pointed out that some aerial channel measurements in the over-sea scenarios reported very small path loss exponent values (less than 1.5). The authors in [63] explained that it was due to an evaporation duct above the sea surface that resulted in enhanced radio wave propagation. Similar signal enhancements due to evaporation ducts were also reported in the works [111] and [112].

The evaporation duct is a result of the rapid decrease in vapor pressure from the sea surface (saturated with water vapor) to a height at which the water vapor pressure reaches an ambient value. This is reflected in a decrease in modified refractivity. An evaporation duct may be considered as a natural waveguide for the propagation of electromagnetic waves. Depending on the environment, the heights of evaporation ducts may vary from a few meters to a few tens of meters.

It is also known that weather impacts the propagation of electromagnetic waves. The weather attenuation is usually a function of link distance, rainfall rate, and the size and shape of the raindrops. The impact is more significant in the millimeter-wave bands than in the sub-6 Ghz frequencies, due to the larger electrical size of raindrops in the millimeter-wave bands. As the link distance increases, the weather attenuation may become dramatic in the millimeter-wave bands. Therefore, for the design of UAV systems with millimeter-wave communication, the impact of weather, especially the rain attenuation effect, should be taken into account [113].

There is a need to calculate the rain attenuation for use in prediction methods. One model recommended by the ITU can be found in [114]. In this model, the specific attenuation γ_{rain} (dB/km) is a function of the rainfall rate R (mm/h) given by the power-law relationship:

$$\gamma_{rain} = k(f) R^{\alpha(f)}, \tag{3.56}$$

where $k(f)$ is the multiplicative coefficient, $\alpha(f)$ is the power-law exponent, and both depend on the frequency f. The coefficients $k(f)$ and $\alpha(f)$ are given by

$$\log_{10} k(f) = \sum_{j=1}^{4} a_{k,j} \exp\left(-\left(\frac{\log_{10} f - b_{k,j}}{c_{k,j}}\right)^2\right) + m_k \log_{10} f + c_k \tag{3.57}$$

$$\alpha(f) = \sum_{j=1}^{5} a_{\alpha,j} \exp\left(-\left(\frac{\log_{10} f - b_{\alpha,j}}{c_{\alpha,j}}\right)^2\right) + m_\alpha \log_{10} f + c_\alpha, \tag{3.58}$$

where f (in GHz) is in the range from 1 GHz to 1000 GHz. Values for the constants $\{a_{k,j}, b_{k,j}, c_{k,j}, m_k, c_k\}$ for the coefficient $k(f)$ and values for the constants $\{a_{\alpha,j}, b_{\alpha,j}, c_{\alpha,j}, m_\alpha, c_\alpha\}$ for the coefficient $\alpha(f)$ can be found in [114].

3.4 Small-Scale Propagation Effects

Small-scale propagation channel effects describe the phenomena causing the rapid fluctuation of radio signals over time, frequency, and space. The fundamental cause of small-scale propagation channel effects is the multipath components introduced by wireless channels. A mobile receiver on the ground may receive radio waves from different directions with different propagation delays and Doppler shifts. These radio waves may have randomly distributed amplitudes, phases, and AoAs. In Section 3.3.2, the developed ray-tracing models capture the multipath effects for deterministic channels.

This section focuses on the statistical characterization of the multipath aerial wireless channels. Here, we will use the general channel model (3.4) introduced in Section 3.1

for analyzing small-scale propagation channel effects. Note that the model in (3.4) characterizes the channel impulse response (CIR), which is a function of time (t), delay (τ), and space ($\vec{r}$). To see the frequency dependency more explicitly, the delay domain in the CIR can be transformed into the frequency domain by applying a Fourier transform:

$$h(t, f, \vec{r}) = \int_{-\infty}^{\infty} h(t, \tau, \vec{r}) e^{-j2\pi f \tau} d\tau, \tag{3.59}$$

where $h(t, f, \vec{r})$ is known as the channel transfer function (CTF).

The channel (3.59) is generally modeled as a wide-sense stationary random process in statistical channel modeling. This implies that the channel auto-correlation is invariant in time, frequency, and space. Thus, it can be expressed by

$$R_h(\Delta t, \Delta f, \Delta \vec{r}) = \mathbb{E}\left[h(t, f, \vec{r}) h^*(t + \Delta t, f + \Delta f, \vec{r} + \Delta \vec{r})\right], \tag{3.60}$$

where $\Delta t, \Delta f,$ and $\Delta \vec{r}$ denote the time difference, frequency difference, and space difference, respectively.

For communication systems with UAVs in which the propagation environments may rapidly change (due to, e.g., a UAV moving at a high speed), the channel may be wide-sense stationary for only a small spatial interval [61]. This is not a unique phenomenon for aerial wireless channels: The stationary interval or time has been considered for high-speed terrestrial vehicular channel characterization [115, 116]. In the sequel, we analyze time, frequency, and spatial selectivity, and the corresponding Doppler spread, delay spread, and angular spread, separately. We will primarily assume that the channel is wide-sense stationary unless stated otherwise. It should be understood that the small-scale channel effects are analyzed in the stationary interval or time during which the assumption of wide-sense stationary channels is valid.

3.4.1 Time Selectivity and Doppler Spread

Time selectivity refers to the temporal variation of the wireless channel $h(t, f, \vec{r})$. For simplicity, we omit the arguments f and $\vec{r}$ and simply denote the channel as $h(t)$ in this section. One measure of time selectivity is the *coherence time*, denoted by T_c, which is the interval during which the channel at different time instants is correlated. With the channel auto-correlation function given by $R_h(\Delta t) = \mathbb{E}[h(t)h^*(t + \Delta t)]$, the coherence time can be formally defined as the value of the time difference satisfying that $R_h(T_c) = 0.5 R_h(0)$. The longer the coherence time, the less the time selectivity of the channel. We can categorize the channel as *fast fading* or *slow fading*, depending on its coherence time relative to the transmission duration of a transport block. If the coherence time is much shorter (resp. longer) than the transmission duration of the transport block, the channel is categorized as fast fading (resp. slow fading).

The time selectivity is caused by the motions of the transmitter, receiver, and surrounding objects that lead to possibly different time varying Doppler shifts on the multipath components. The different multipath components that may have different time varying phase shifts caused by Doppler effects sum constructively and destructively, affecting the amplitude of the resulting composite signal. Doppler spectrum, denoted by

$S_h(\mu)$, can be computed by applying a Fourier transform to the auto-correlation function $R_h(\Delta t)$ to transfer from the time domain to the dual Doppler domain. The so-called Doppler spread D_s is defined as the root mean square of the Doppler spectrum given by $D_s = \sqrt{\mathbb{E}[\mu^2] - (\mathbb{E}[\mu])^2}$, where $\mathbb{E}[\mu^n]$ is the n-th moment of the Doppler spectrum $S_h(\mu)$ and is given by $\mathbb{E}[\mu^n] = \int_{-\infty}^{\infty} \mu^n S_h(\mu) d\mu / \int_{-\infty}^{\infty} S_h(\mu) d\mu$. The relationship of the coherence time and Doppler spread is reciprocal: The smaller the Doppler spread, the longer the coherence time.

For aerial channels, there are three typical contributors: an LOS component, a specular ground reflection component, and diffuse scattering components [117]. The LOS component has a frequency shift as a result of the relative motion of the transmitter and receiver, while the ground reflection and scattering usually influence the Doppler spread most. A simplified stochastic model was proposed in [118] to characterize the Doppler spectrum for the diffuse components. The derived Doppler spectrum is given by

$$S_h(\mu) = \begin{cases} \frac{P_d f_\Theta(\theta)}{(v/\lambda)^2 - (\mu - f_c)^2} & |\mu - f_c| < \frac{v}{\lambda} \\ 0 & \text{otherwise,} \end{cases} \tag{3.61}$$

where P_d represents the total received power of the diffuse components and $f_\Theta(\theta)$ is the probability density function of the AoA of the diffuse components with respect to the direction of the aircraft velocity v. The work [119] extended this model by further considering the geometry of the scatters on the ground in a two-dimensional model. The derived theoretical results were shown to match their measurement data.

The two-dimensional model in [119] was further extended to a three-dimensional geometric stochastic channel model in [120]. The results reported in [120] showed that the specific flight trajectories indirectly impact the Doppler power spectrum. For example, assume homogeneous ground scattering and consider the scenario where two UAVs fly behind each other at the same height and with the same speed in the same direction. It was shown that the Doppler spectrum became wider for increasing delay. For the delay corresponding to the specular ground reflection path, the Doppler spectrum exhibited a "W" shape. Interestingly, the Doppler spectrum converged to the familiar Jakes's spectrum. For a different scenario with the same setup except that the two UAVs fly toward each other in opposite directions, a different shape of the Doppler spectrum was shown. For a given delay, the Doppler spectrum only exhibited positive frequencies. It was further observed that the position of the Doppler spectrum varied for different delays.

Doppler effects in AG channels were also measured or simulated in a few other works [45, 121–123]. The work [44] presented a class of Doppler power spectrum models for different scenarios, including parking, taxing, takeoff, landing, and en-route. The suggested parameter sets were based on published measurement results and empirical data. A more recent work [124] presented results on Doppler power profiles measured in a typical airport environment in 970 MHz frequency band. The GS antenna was placed at the rooftop of a building at the height of 23 m. Three flight scenarios were investigated.

- *En-route cruise:* The aircraft flew at an average speed of 216 m/s in the altitude range [8.5, 10.5] km AMSL. The transceiver distance ranged from 140 km to 350 km. It was found that all multipath components arrived with the same Doppler shift as the LOS path. This is because the multipath components mainly originated from the objects around the GS antenna and thus were seen under about the same AoA as the LOS path at a high height, agreeing with intuition.

- *Climb and descent:* The aircraft flew at an average speed of 170 m/s in the altitude range [3, 9] km AMSL. The transceiver distance ranged from 20 km to 50 km. It was found that almost all the multipath components arrived with the same Doppler shift as the LOS path. However, compared to the en-route cruise scenario, the Doppler spread increased.
- *Takeoff and landing:* The aircraft flew at an average speed of 90 m/s in the altitude range [30, 330] m AGL. The transceiver distance ranged from 0.5 km to 7.5 km. It was found that the multipath components contributed significantly to Doppler frequencies other than the one corresponding to the LOS path. This is because when an aircraft was flying close to the GS, e.g., during the phases of takeoff and landing, the aircraft saw the scattering objects under a variety of different AoAs.

3.4.2 Frequency Selectivity and Delay Spread

Frequency selectivity refers to the variation of the wireless channel $h(t, f, \vec{r})$ with respect to frequency. For simplicity, we omit the arguments t and $\vec{r}$ and simply denote the channel as $h(f)$ in this section. One measure of frequency selectivity is the *coherence bandwidth*, denoted by B_c, which is the range of frequencies over which the channel at different frequencies is correlated. With the channel auto-correlation function given by $R_h(\Delta f) = \mathbb{E}[h(f)h^*(f + \Delta f)]$, the coherence bandwidth can be formally defined as the value of the frequency difference satisfying that $R_h(B_c) = 0.5 R_h(0)$. The wider the coherence bandwidth, the less the frequency selectivity of the channel. We can categorize the channel as *frequency-selective fading* or *flat fading*, depending on its coherence bandwidth relative to the signal bandwidth. If the coherence bandwidth is much narrower (resp. wider) than the signal bandwidth, the channel is categorized as frequency-selective fading (resp. flat fading).

The frequency selectivity is caused by the multiple propagation paths arriving at the receiver with delays proportional to the corresponding path distances. By applying a Fourier transform to the auto-correlation function $R_h(\Delta f)$ to transfer from the frequency domain to the dual delay domain, we can compute the so-called *power delay profile*, denoted by $S_h(\tau)$, which models the distribution of the delays and amplitudes of the different multiple propagation paths. The root mean square of the power delay profile defines the *delay spread* as $T_s = \sqrt{\mathbb{E}[\tau^2] - (\mathbb{E}[\tau])^2}$, where $\mathbb{E}[\tau^n]$ is the n-th moment of the power delay profile $S_h(\tau)$ and is given by $\mathbb{E}[\tau^n] = \int_{-\infty}^{\infty} \tau^n S_h(\tau) d\tau / \int_{-\infty}^{\infty} S_h(\tau) d\tau$. The relationship of the coherence bandwidth and delay spread is reciprocal: The smaller the delay spread, the larger the coherence bandwidth.

A number of propagation studies have been carried out to measure the power delay profiles of aerial wireless channels in different environments and to estimate the corresponding delay spreads. Table 3.3 lists T_d values measured in different environments as reported in the literature. Not surprisingly, Table 3.3 shows that the delay spread value depends on the terrain and geometry. For example, the work [88] presented delay spread

Table 3.3 Delay spreads measured under different aerial propagation environments.

Reference[a]	Environment	Frequency (GHz)	T_d (ns)
[61]	Over water: Oxnard	5.030–5.091	9.8 (mean)
			9.8 (median)
			364.7 (max)
			2.0 (std deviation)
	Over water: Cleveland	5.030–5.091	9.8 (mean)
			9.8 (median)
			73.3 (max)
			1.1 (std deviation)
[63]	Over sea: 1.83 km AMSL	5.7	20–38 (median)
	Over sea: 0.91 km AMSL	5.7	30–35 (median)
	Over sea: 0.37 km AMSL	5.7	335–480 (median)
[62]	Mountainous	5.030–5.091	10.1 (mean)
			9.8 (median)
			177.4 (max)
			4.4 (std deviation)
	Hilly: Latrobe	5.030–5.091	17.8 (mean)
			11.3 (median)
			371.3 (max)
			12.5 (std deviation)
	Hilly: Palmdale	5.030–5.091	19.3 (mean)
			11.7 (median)
			1044.3 (max)
			51.1 (std deviation)
[87]	Open field, suburban	3.1–5.3	≤ 2
[50]	Near-urban: Cleveland	5.030–5.091	12.8 (mean)
			10.6 (median)
			217.5 (max)
			8.5 (std deviation)
	Near-urban: Latrobe	5.030–5.091	13.9 (mean)
			11.0 (median)
			1190.8 (max)
			13.6 (std deviation)
	Urban: Palmdale	5.030–5.091	59.6 (mean)
			11.0 (median)
			4242.9 (max)
			134.4 (std deviation)
	Urban: Cleveland	5.030–5.091	9.9 (mean)
			9.6 (median)
			2029.5 (max)
			17.4 (std deviation)
[88]	Cluttered environment	2	98.1 (for 7.5^o EA[b])
			54.9 (for 15° EA)
			24.3 (for 22.5^o EA)
			18.3 (for 30° EA)

[a] Refer to Table 3.1 for more information regarding the measurement setups in each reference.
[b] EA: elevation angle

statistics under different elevation angles. The results agree with intuition: The delay spread decreases as the elevation angle increases. The delay spread values in Table 3.3 are mostly about a few tens of nanoseconds and are seldom larger than a few hundreds of nanoseconds. This is mainly a consequence of the almost universal presence of LOS propagation conditions in the measurements.

Given these measurements, different models have been proposed to characterize the power delay profiles of aerial wireless channels. For example, the *Saleh-Valenzuela* model [125], which is a popular power delay profile model originally proposed for indoor environments, can be used to model the power delay profile of an aerial wireless channel when the multipath components have a "clustering" nature. According to the Saleh-Valenzuela model, the complex baseband channel response is given by:

$$h(t) = \sum_{k=0}^{\infty} \sum_{\ell=0}^{\infty} a_{k,\ell} \delta(t - \tau_k - \tau_{k,\ell}), \qquad (3.62)$$

where k is the cluster index, τ_k is the time of arrival of the first path associated with the cluster k, $\tau_{k,\ell}$ is the time of arrival of the ℓ-th path associated with the cluster k measured with respect to τ_k, and $a_{k,\ell}$ is the complex channel gain of the ℓ-th path associated with the cluster k.

The mean square values $\{\mathbb{E}[|a_{k,\ell}|^2]\}_{k,\ell}$ in the Saleh-Valenzuela model are doubly exponential: The powers of the first paths from the clusters decay exponentially, and the powers of the paths within each cluster also decay exponentially. Mathematically,

$$\mathbb{E}[|a_{k,\ell}|^2] = \mathbb{E}[|a_{0,0}|^2] \exp(-\tau_k/\Gamma) \exp(-\tau_{k,\ell}/\gamma), \qquad (3.63)$$

where Γ is the rate of decay of the powers of the first paths from the clusters and γ is the rate of decay of the powers of the paths within each cluster.

In [50, 61, 62], the authors conducted a series of measurements for different aerial environments and provided power delay profile measurement results. These results have been employed to build stochastic tapped delay line models, which can be used in link level simulations for UAV communications systems.

3.4.3 Spatial Selectivity and Angular Spread

Spatial selectivity refers to the variation of the wireless channel $h(t, f, \vec{r})$ with respect to space. For simplicity, we omit the arguments t and f and simply denote the channel as $h(\vec{r})$ in this section. Spatial selectivity is of particular interest in multi-antenna systems where signals received at different antennas may undergo different propagation environments and fade differently. One measure of spatial selectivity is the *coherence distance*, denoted by D_c, which is the maximum antenna spacing for which the received signals at different antennas are correlated. With the channel auto-correlation function given by $R_h(\Delta \vec{r}) = \mathbb{E}[h(\vec{r})h^*(\vec{r} + \Delta \vec{r})]$, the coherence distance can be formally defined as the largest position displacement such that $R_h(\Delta \vec{r}) \leq 0.5 R_h(\vec{0})$ holds for

3.4 Small-Scale Propagation Effects

all $\Delta\vec{r}$ satisfying $\|\Delta\vec{r}\| \leq D_c$. The larger the coherence distance, the less the spatial selectivity of the channel.

By applying a Fourier transform to the auto-correlation function $R_h(\Delta\vec{r})$ to transfer from the space domain to the dual wave-number domain, we can compute the wave-number spectrum denoted by $S_h(\vec{k})$. The root mean square of the wave-number spectrum defines *wave-number spread* as $W_s = \sqrt{\mathbb{E}[(\vec{k})^2] - (\mathbb{E}[\vec{k}])^2}$, where $\mathbb{E}[(\vec{k})^n]$ is the n-th moment of the wave-number spectrum $S_h(\vec{k})$ and is given by $\mathbb{E}[(\vec{k})^n] = \int_{-\infty}^{\infty} (\vec{k})^n S_h(\vec{k}) d\vec{k} / \int_{-\infty}^{\infty} S_h(\vec{k}) d\vec{k}$. The relationship of the coherence distance and wave-number spread is reciprocal: The smaller the wave-number spread, the larger the coherence distance.

To understand the physical meaning of the wave-number spectrum, we can expand the inner product of the wave-number vector $\vec{k}$ and position vector $\vec{r}$

$$<\vec{k},\vec{r}>> = \frac{2\pi f_c \|r\|}{c}(\cos\phi\sin\theta + \sin\phi\sin\theta + \cos\theta), \quad (3.64)$$

where ϕ and θ denote the azimuth AoA and elevation AoA, respectively. We can see that the wave-number spectrum models the power distribution of the different multiple propagation paths in the angular domain. Instead, we can directly look at the *power angle spectrum*, denoted as $S_h(\phi,\theta)$, that describes the power distribution of the received paths with respect to the AoA. Similarly, we can define the concept of an *angular spread* as the standard deviation of the power angle spectrum. The relationship between the coherence distance and angular spread is also reciprocal.

Assuming that the power angle spectra in the azimuth and elevation domains can be separated, we can generate power angle spectra for azimuth AoA and zenith AoA separately. Using the azimuth domain as an example, one popular model for the power angle spectrum is given by the *truncated Laplace distribution*:

$$S_h(\phi) = \frac{1}{(1 - e^{-\sqrt{2}\pi/\sigma})\sqrt{2}\sigma} e^{-|\frac{\sqrt{2}\phi}{\sigma}|}, \quad \phi \in [-\pi,\pi), \quad (3.65)$$

where σ is a scale parameter.

Spatial selectivity and angular spread have not been widely investigated for aerial channel modeling in prior art. In [126], the authors exploited the existing analytical and empirical results on the receive energy distribution in three-dimensional angular space to analyze angular spread, angular constriction, direction of maximum fading, among others, for AG channels. In general, with geometry-based stochastic channel models, it is possible to analyze or simulate spatial selectivity and angular spread for UAV multi-antenna systems. The basic idea behind the geometry-based stochastic channel models is to introduce scattering objects according to a given distribution in the propagation environment. One such model may be found in [127], which also derived the corresponding analytical density function of the AoA. Several other geometry-based stochastic channel models proposed for aerial channel modeling may be found in [128–132].

3.4.4 Envelope and Power Distributions

The statistical time varying envelope $|h|$ of a narrowband channel, or the envelope of a multipath component in a wideband channel, has been commonly modeled by the *Rayleigh* distribution with probability density function $f_{|h|}(x)$ given by

$$f_{|h|}(x) = \frac{x}{\sigma^2} e^{-\frac{x^2}{2\sigma^2}}, \quad x \geq 0, \tag{3.66}$$

where σ is a scale parameter. The physical meaning of σ is that the average channel gain is $2\sigma^2$. The corresponding power (i.e., squared magnitude $|h|^2$) distribution of the channel is exponentially distributed with probability density function $f_{|h|^2}(x)$ given by

$$f_{|h|^2}(x) = \frac{1}{\sigma^2} e^{-\frac{x}{\sigma^2}}, \quad x \geq 0. \tag{3.67}$$

The Rayleigh fading model is reasonable when there are a large number of statistically independent reflected and scattered paths contributing to a channel tap. It is a simple analytically tractable model. However, caution must be exercised when applying this model to aerial wireless channels, in which a specular LOS component often exists and there may not be sufficient number of statistically independent reflected and scattered paths.

Another frequently used model is the so-called *Rician* model, in which random independent reflected and scattered paths are superimposed on a stationary non fading component (such as the LOS component). The probability density function of the Rician distributed channel envelope is given by

$$f_{|h|}(x) = \frac{x}{\sigma^2} e^{-\frac{x^2+A^2}{2\sigma^2}} I_0\left(\frac{Ax}{\sigma^2}\right), \quad x \geq 0, \tag{3.68}$$

where $A \geq 0$ denotes the peak amplitude of the stationary non fading component and $I_0(\cdot)$ is the modified zero-order first kind Bessel function. The average channel gain in the Rician fading model equals $A^2 + 2\sigma^2$, which is the sum of the power of the stationary nonfading component and the power of the random multipaths. The ratio between them, i.e., $A^2/2\sigma^2$, is known as the K factor, which is often used to describe the Rician fading. As $K \to 0$, the Rician distribution degenerates to a Rayleigh distribution. As $K \to \infty$, the stationary nonfading component dominates and the multipath fading components vanish.

The Rician model has been widely used for aerial channel modeling due to the high likelihood of LOS [45, 50, 61, 62, 133]. For the NLOS case, the Rayleigh model might provide a better fit [45, 88, 133]. Table 3.4 lists Rician K factor values for different environments reported in the literature. In [50, 61, 62], the authors presented K factors for over fresh water and sea scenarios, hilly and mountainous settings, and also for near-urban and suburban environments, as summarized in Table 3.4. The Rician K factors were given in [45] for different phases of the flight (parking, taxiing, takeoff/landing, and en-route). The K factor of the en-route phase is the largest. This agrees with intuition and other measurement results. For example, as shown in [88], the K factors are smaller for small elevation angles compared to large elevation angles.

3.4 Small-Scale Propagation Effects

Table 3.4 Rician K factor values measured under different aerial propagation environments.

Reference[a]	Environment	Frequency (GHz)	K factor (dB)
[61]	Over water	5.030–5.091	27.3–31.3 (mean)
			11.1–12.4 (min)
			33.0–35.6 (max)
			1.8 (std dev.)
		0.960–0.977	12.5–12.8 (mean)
			8.7–9.4 (min)
			16.5–20.7 (max)
			1.2–1.5 (std dev.)
[62]	Hilly, mountainous	5.030–5.091	28.8–29.4 (mean)
			22.2–23.1 (min)
			35.3–40.5 (max)
			2.0–2.1 (std dev.)
		0.960–0.977	12.8–13.8 (mean)
			4.0–5.1 (min)
			16.6–16.9 (max)
			0.8–1.3 (std dev.)
[50]	Suburban, near-urban	5.030–5.091	27.5–29.8 (mean)
			7.9–12.7 (min)
			33.7–40.2 (max)
			1.8–2.4 (std dev.)
		0.960–0.977	12.4–14.9 (mean)
			−87.1–7.8 (min)
			14.7–27.5 (max)
			1.2–2.3 (std dev.)
[45]	Parking	5.8	1.5 (median)
			−2 (min)
			5 (max)
	Taxing	5.8	6 (median)
			5 (min)
			7 (max)
	Takeoff/landing	5.8	10 (median)
			5 (min)
			15 (max)
	En-route	5.8	≥ 17 (median)
			15 (min)
			≥ 20 (max)

[a] Refer to Table 3.1 for more information regarding the measurement setups in each reference.

A more general fading distribution, the Nakagami fading distribution, is given by:

$$f_{|h|}(x) = \frac{2m^m x^{2m-1}}{\Gamma(m) P_r^m} e^{-\frac{mx^2}{P_r}}, \quad x \geq 0, \tag{3.69}$$

where $\Gamma(\cdot)$ is the gamma function, P_r is the average received power, and $m \geq 0.5$ is a fading parameter. For $m = 1$, the Nakagami fading reduces to Rayleigh fading. For $m = (K+1)^2/(2K+1)$, the Nakagami fading approximates Rician fading with parameter

K. For $m \to \infty$, the channel becomes deterministic. Some works have also considered the Nakagami fading model for aerial channel modeling [87, 89, 134].

3.5 Waveform Design

A waveform is the shape and form of a wireless signal that carries information bits. Designing an appropriate waveform is essential for a UAV-enabled wireless communication system, as in any other wireless network. In this section, we start with reviewing the waveform basics, such as power spectral density (PSD), which is a key concept for spectrum management. Due to the diverse UAV applications outlined in Chapter 2, a single waveform choice unlikely fits all the UAV wireless communications systems. We chose a few example popular waveform choices, including OFDM, spread spectrum, and CPM, to discuss the main design considerations for UAV wireless communications and networking. All these waveform choices have been used for wireless communications in unmanned aircraft systems [135]. For more general and comprehensive treatments of waveform design in digital communications systems, we refer to the textbooks [136, 137].

3.5.1 Waveform Basics

We start with the canonical complex baseband equivalent representation for passband signals. A passband signal $s_p(t)$ can be written as

$$s_p(t) = \sqrt{2} s_I(t) \cos(2\pi f_c t) - \sqrt{2} s_Q(t) \sin(2\pi f_c t), \tag{3.70}$$

where $s_I(t)$ and $s_Q(t)$ are real-valued signals. The waveforms $s_I(t)$ and $s_Q(t)$ are referred to as the in-phase and quadrature components of the passband signal $s_p(t)$, respectively. The complex baseband equivalent representation $s_b(t)$ of the passband signal $s_p(t)$ is defined as

$$s_b(t) = s_I(t) + j s_Q(t). \tag{3.71}$$

It is easy to see that $s_p(t) = \Re\left(\sqrt{2} s_b(t) e^{j 2\pi f_c t}\right)$.

The information carried by a passband signal resides in its complex envelope, i.e., the amplitude and phase variations captured in the baseband equivalent representation. The predictable fast phase variation due to the fixed carrier frequency f_c does not carry information and is subtracted out in the baseband equivalent representation. The choice of f_c in a UAV communication system is mainly determined by the available spectrum allocated to the system [138].

Due to the scarcity of the radio spectrum and potential interference of waveforms in neighboring frequency bands, harmonious coexistence of different systems in the same band or adjacent bands is essential. For example, federal agencies and the military in the United States use the 1755–1850 MHz band for unmanned aerial systems, among others. The 1755–1780 MHz portion of the band was auctioned for commercial wireless

3.5 Waveform Design

services in 2015 [139]. Some federal systems will remain indefinitely in the 1755–1780 MHz portion, while others will compress operations into the 1780–1850 MHz portion of the band or relocate to anther band.

To facilitate coexistence, it is essential to determine the spectral occupancy of a waveform. To this end, we define the PSD for a finite-power signal $s(t)$ as follows:

$$S_s(f) = \lim_{T_w \to \infty} \frac{|S_{T_w}(f)|^2}{T_w}, \quad (3.72)$$

where $S_{T_w}(f)$ is the Fourier transform of $s_{T_w}(t) = s(t)I_{[-\frac{T_w}{2}, \frac{T_w}{2}]}(t)$. Here T_w is the length of an observation window and $I_A(x)$ is the indicator function of a set A: $I_A(x) = 1$ if $x \in A$ and $I_A(x) = 0$ otherwise.

Modulation deals with how to convert bits into a waveform that can be sent over a band-limited channel. Linear modulation is a fundamental technique whose baseband transmit waveform can be written as

$$u(t) = \sum_{n=-\infty}^{\infty} b[n] g_{tx}(t - nT), \quad (3.73)$$

where $\{b[n]\}$ are the data symbols taking values from a fixed constellation, $g_{tx}(t)$ is a fixed baseband waveform, and T is the symbol duration. Assuming that $\{b[n]\}$ are zero-mean and uncorrelated, the PSD of the linearly modulated signal $u(t)$ is given by

$$S_u(f) = \frac{|G_{tx}(f)|^2}{T} \mathbb{E}[|b[n]|^2], \quad (3.74)$$

where $G_{tx}(f)$ is the Fourier transform of $g_{tx}(t)$. We can see that for uncorrelated symbols, the PSD of a linearly modulated signal is determined by the spectrum of the modulating waveform. The shape of the PSD shall be designed such that the waveform complies with regulatory requirements on, for example, inband PSD and out-of-band spurious emissions.

Beyond going through the transmit filter $g_{tx}(t)$, the transmitted symbols must also go through a channel $g_{ch}(t)$ and a receive filter $g_{rx}(t)$. The noiseless waveform at the output of the receive filter is given by

$$r(t) = \sum_{n=-\infty}^{\infty} b[n] g(t - nT), \quad (3.75)$$

where $g(t) = (g_{tx} * g_{ch} * g_{rx})(t)$ is the composite filter characterizing the overall system response. If we sample the receive waveform $r(t)$ at rate $\frac{1}{T}$, we obtain $r(nT)$. A natural question arises: when is $r(nT) = b[n], \forall n$? The answer is the *Nyquist criterion*: The inter-symbol interference (ISI) can be avoided if $g(nT) = 1$ for $n = 0$ and $g(nT) = 0$ otherwise.

It is well known that the minimum bandwidth Nyquist waveform is $g(t) = \text{sinc}\left(\frac{t}{T}\right)$, where $\text{sinc}(x) = \frac{\sin(\pi x)}{\pi x}$. The sinc pulse decays with $\frac{1}{t}$, which is slow and may lead to large fluctuations in the signal $r(t)$. An example Nyquist waveform with a fast time decay of $\frac{1}{t^3}$ is the raised cosine pulse $g(t) = \text{sinc}\left(\frac{t}{T}\right) \frac{\cos(\pi a \frac{t}{T})}{1 - \left(\frac{2at}{T}\right)^2}$, where a is the fractional excess bandwidth due to the faster time decay. In practice, since the channel is not under control, the transmit and receive filters are designed such that the cascade $(g_{tx} * g_{rx})(t)$ satisfies Nyquist criterion. A typical choice is to set the transmit and receiver filters to be the square roots (in the frequency domain) of a Nyquist pulse.

3.5.2 Orthogonal Frequency Division Multiplexing

OFDM is a type of multicarrier modulation technique that divides the data stream into multiple substreams transmitted over different orthogonal subcarriers. OFDM is a popular scheme for wireless communications and has been adopted in the wireless systems such as 4G LTE, 5G New Radio (NR), and IEEE 802.11 specifications for WiFi. Many UAVs are already equipped with WiFi and LTE chips, making OFDM an integral part of UAV wireless communications systems. OFDM is used in the broadband aeronautical multi-carrier (B-AMC) system, which is an aeronautical data communications system intended to be operated in the L-band. The first option in the L-band digital aeronautical communication system (LDACS) also utilizes OFDM modulation techniques [135]. Many research works on UAV wireless communications and networking have assumed the use of OFDM as well [140, 141].

The transmit OFDM waveform in the time period $[0, T]$ can be written as follows:

$$u(t) = \sum_{k=0}^{N-1} B[k] e^{j2\pi \frac{k}{T} t} I_{[0,T]}(t), \tag{3.76}$$

where k is the subcarrier index, $B[k]$ is the symbol transmitted using the subcarrier k at frequency $\frac{k}{T}$. The Fourier transform of the waveform $g_{tx,k}(t) = e^{j2\pi \frac{k}{T} t} I_{[0,T]}(t)$ carrying $B[k]$ equals $G_{tx,k}(f) = T\text{sinc}(Tf - k) e^{-\frac{\pi}{T}}$. If T is large enough such that $\frac{1}{T}$ is small compared to the channel coherence bandwidth B_c, the channel gain seen by the k-th subcarrier is approximately constant $G_{ch}(f) \approx G_{ch}(\frac{k}{T})$. Then the waveform $g_{tx,k}(t)$ is an approximate eigenfunction of the channel, i.e.,

$$\left(g_{tx,k} * g_{ch}\right)(t) \approx G_{ch}(\frac{k}{T}) g_{tx,k}(t). \tag{3.77}$$

The approximation becomes exact when $T \to \infty$.

To demodulate $B[\ell]$, the receiver multiplies the receive waveform with $e^{-j2\pi \frac{\ell}{T} t}$ and integrates over the period $[0, T]$, yielding that

$$\frac{1}{T} \int_0^\infty ((u * g_{ch})(t)) e^{-j2\pi \frac{\ell}{T} t} dt \approx \frac{1}{T} \sum_{k=0}^{N-1} G_{ch}\left(\frac{k}{T}\right) B[k] \int_0^T e^{j2\pi \frac{k-\ell}{T} t} dt \tag{3.78}$$

$$= \sum_{k=0}^{N-1} G_{ch}\left(\frac{k}{T}\right) B[k] \frac{e^{j2\pi(k-\ell)} - 1}{j2\pi(k-\ell)} \tag{3.79}$$

$$= G_{ch}\left(\frac{\ell}{T}\right) B[\ell], \tag{3.80}$$

where the first line follows from (3.77). The last equality shows that the different subcarriers are orthogonal over an interval of length T if the frequencies are separated by an integer multiple of $\frac{1}{T}$.

The widespread use of OFDM is partly due to the cost-effective discrete implementation of OFDM using the fast Fourier transform (FFT) and inverse FFT (IFFT). The waveform $u(t)$ can be represented by the time samples $\{b[n]\}$ at the sampling rate $\frac{1}{T_s} = \frac{N}{T}$:

$$b[n] = u(nT_s) = \sum_{k=0}^{N-1} B[k] e^{j2\pi \frac{n}{N} k}, \qquad (3.81)$$

which is the inverse discrete Fourier transform (DFT) of the symbol sequence $\{B[k]\}$. The receiver can perform the reverse operation – DFT – on the received time samples $\{b[n]\}$ to recover the original data sequence $\{B[k]\}$.

Most ISI can be removed by choosing N large enough such that $\frac{1}{T} \ll B_c$. One popular approach to removing all ISI is to add a cyclic prefix of length N_{cp} (not shorter than the channel delay spread) after the IFFT at the transmitter. The cyclic prefix is removed at the receiver before the FFT. These operations effectively create a circular convolution of the discrete transmit signal and channel impulse response (which by nature is a linear convolution). Thus, the noiseless output of the FFT at the receiver is $Y[k] = H[k]B[k]$, where $H[k]$ is the FFT of the length-L channel impulse response $\{h[n]\}_{n=0}^{L-1}$. The data symbol $B[k]$ can be recovered by a one-tap frequency equalizer, i.e., $H^*[k]Y[k]$ removes the channel effect. This makes channel equalization in OFDM easy to implement for multipath wideband channels.

Adding the cyclic prefix leads to losses of a fraction $\frac{N_{cp}}{N_{cp}+N}$ of the time and a fraction $\frac{N_{cp}}{N_{cp}+N}$ of the average power not utilized for data communication. The length N_{cp} of the cyclic prefix is chosen to cover the typical channel delay spread encountered in a communication system. On one hand, the OFDM block length N should be chosen as large as possible to minimize the cyclic prefix overhead. On the other hand, the OFDM block length N should be small enough such that the channel is approximately constant over the block length N to avoid inter-carrier interference. For the design of UAV communications systems, Doppler shifts and Doppler spread must be taken into account. The Doppler effects mainly depend on the frequency, velocity of the aircraft, and geometry. A large Doppler shift caused by the motion of the aircraft can be estimated and compensated appropriately at the transceiver. The Doppler spread introduces uncertainty in the frequency of the received signal. A large enough subcarrier spacing can be used so that the Doppler spread effect is negligible [140, 142]. This is equivalent to constrain N such that the channel is approximately constant over an OFDM block length.

The OFDM waveform $u(t)$ can be seen as tight packing of N linearly modulated narrowband signals. By (3.74), the PSD of the OFDM waveform $u(t)$ will be:

$$S_u(f) = T \sum_{k=0}^{N-1} \mathbb{E}[|B[k]|^2] (\mathrm{sinc}(Tf - k))^2, \qquad (3.82)$$

where we assumed that the data symbols transmitted on different subcarriers are uncorrelated. The symbol rate is $\frac{N}{T}$, and the majority of the signal power is contained in the frequency range $\left[-\frac{N}{2T}, \frac{N}{2T}\right]$. The close to Nyquist signaling rate makes OFDM highly spectrally efficient.

The main drawback of OFDM is that the peak-to-average-power ratio (PAPR) of the transmit signal may be high. A high PAPR may require the power amplifiers to back off into a linear regime that leads to lower power efficiency. Many solutions have been proposed in the literature to mitigate the PAPR issue in OFDM, such as clipping and removing some transmitted sequences that result in high PAPR [143].

3.5.3 Direct Sequence Spread Spectrum

Spread spectrum is a modulation technique that increases the transmit signal bandwidth to mitigate ISI and narrowband interference. The inherent property of "hiding" the signal below the noise floor and resistance to narrowband jamming make the spread spectrum particularly desirable for military communication systems. DSSS and frequency hopping spread spectrum (FHSS) are two common spread spectrum techniques. In DSSS, the data signal is multiplied by a pseudo-random sequence known as *spreading code*. DSSS is the fundamental building block in the code division multiple access (CDMA) based 3G mobile systems. CDMA, together with OFDM, has also been used in the broad very high frequency (B-VHF) project, which is the first aeronautical communications system employing multicarrier technology [135]. Some research works on UAV wireless communications and networking have considered the use of CDMA with DSSS as well [144–146]. In FHSS, the center frequency of the waveform is hopped over different frequencies, determined by a pseudo-random sequence. The work [147] combined DSSS and FHSS to reduce interference in UAV communications and control.

Next, we will focus on DSSS as it is more commonly used. To send a symbol b in DSSS, a vector $b(c[0], ..., c[K-1])^T$ is sent, where $(c[0], ..., c[K-1])^T$ is a spreading code and K is the length of the spreading code. In other words, K "chips" are used to transmit a single symbol. The spreading waveform can be written as

$$c(t) = \sum_{k=0}^{K-1} c[k]\psi(t - kT_c), \qquad (3.83)$$

where T_c is the chip duration and $\psi(t)$ is the modulating chip waveform. We assume that the same spreading waveform is used for all the symbols $\{b[n]\}$. Then the transmit signal can be written as

$$u(t) = \sum_n b[n]c(t - nT) = \sum_n \sum_{k=0}^{K-1} c[k]b[n]\psi(t - kT_c - nT), \qquad (3.84)$$

where T is the symbol duration and $T = KT_c$. The data rate $\frac{1}{T}$ is typically much smaller than the chip rate $\frac{1}{T_c}$ that determines the transmission bandwidth. The ratio $K = \frac{T}{T_c}$ is often called the *processing gain* of the system.

Consider the multipath channel $h(t) = \sum_{\ell=0}^{L-1} a_\ell \delta(t - \tau_\ell)$, where a_ℓ and τ_ℓ are the complex gain and delay of the ℓ-th path, respectively. The noiseless receive signal will then be:

$$r(t) = (u * h)(t) = \sum_n b[n] \sum_{k=0}^{K-1} c[k] \sum_{\ell=0}^{L-1} a_\ell \psi(t - \tau_\ell - kT_c - nT). \quad (3.85)$$

Equalization at the receiver is not required for demodulation. This is due to the fact that ISI is negligible considering that the symbol duration T is typically large relative to the channel delay spread in a spread spectrum system. *Rake receiver* is a common DSSS demodulation receiver structure that ignores ISI. The Rake receiver involves correlating the receive signal $r(t)$ with L "fingers", each of which is a shifted version of the spreading waveform. We now consider a generic operation for the correlation of $r(t)$ and $c(t - \tau)$ where τ is the delay. The output of the correlation yields the estimation statistic $Z(\tau)$:

$$Z(\tau) = \int r(t) c^*(t - \tau) dt \quad (3.86)$$

$$= \sum_{k=0}^{K-1} c^*[k] \int r(t) \psi^*(t - \tau - kT_c) dt \quad (3.87)$$

$$= \sum_{k=0}^{K-1} c^*[k] (r * \psi_{mf})(kT_c + \tau), \quad (3.88)$$

where $\psi_{mf}(t) = \psi^*(-t)$ is the matched filter of the chip waveform $\psi(t)$. We can see that the de-spreading process involves a discrete correlation of the spreading code $\{c[k]\}$ and the output of the chip matched filter $\psi_{mf}(t)$ sampled at the time instances $\{kT_c + \tau\}$.

In the Rake receiver, for demodulating symbol $b[n]$, the received signal $r(t)$ is correlated with L "fingers": $a_\ell c(t - \tau_\ell), \ell = 0, ..., L-1$, and the L outputs are coherently combined, i.e.,

$$Z[n] = \sum_{\ell=0}^{L-1} a_\ell^* Z(nT + \tau_\ell), \quad (3.89)$$

where $Z[n]$ represents the estimation statistic for deciding on $b[n]$. The Rake receiver is essentially a form of time diversity combining of the signals from the L branches.

3.5.4 Continuous Phase Modulation

In CPM, the baseband representation is in the form of $e^{j\theta(t)}$, where $\theta(t)$ is a continuous function of time t that encodes the data. The transmit signal has constant envelope and is not sensitive to amplitude distortion. As a result, the power amplifier can work in a nonlinear regime to achieve high power efficiency. A prominent CPM example is the Gaussian minimum shift keying (GMSK) used in the 2G Global System for Mobile Communications (GSM). GMSK modulation is also used in the all-purpose multichannel aviation communications system (AMACS) proposed in 2007 and in the second option of LDACS [135]. Some recent research works also considered using GMSK for UAV wireless communications systems [148, 149].

The continuous phase function $\theta(t)$ can be related to the instantaneous frequency function $u(t)$ as follows:

$$\theta(t) = 2\pi \int_{-\infty}^{t} u(\tau)d\tau, \tag{3.90}$$

where the instantaneous frequency is linearly modulated

$$u(t) = \sum_{n} b[n]g(t - nT), \tag{3.91}$$

where $\{b[n]\}$ are the data symbols and $g(t)$ is the frequency pulse.

We can specify a CPM system by specifying the frequency pulse. One example is minimum shift keying (MSK) with symbols $\{b[n]\}$ taking binary values $\{+1, -1\}$. The frequency pulse is a rectangular one given by

$$g(t) = \frac{1}{4T}I_{[0,T]}(t). \tag{3.92}$$

For this MSK system, there are two possible frequency shifts of $\pm \frac{1}{4T}$. The frequency separation is $\frac{1}{2T}$, which is the minimum spacing needed to preserve orthogonality in coherent frequency shift keying (FSK).

To improve spectral efficiency, we can use smoother pulses. A common choice is the Gaussian pulse shape that has a transfer function of the form:

$$G(f) = e^{-\beta^2 f^2}, \tag{3.93}$$

where β is a parameter related to the 3 dB bandwidth B_{3dB} of $G(f)$ as $\beta = \frac{\left(\frac{1}{2}\log_e 2\right)^{\frac{1}{2}}}{B_{3dB}}$. The Gaussian pulse shape in the time domain is given by:

$$g(t) = \frac{\sqrt{\pi}}{\beta} e^{-\frac{\pi^2}{\beta^2} t^2}. \tag{3.94}$$

MSK with Gaussian pulse shape is known as GMSK.

CPM is a modulation technique with memory. A maximum-likelihood approach can be used for demodulation by exploiting the trellis structure of CPM and using the Viterbi algorithm. The maximum-likelihood approach can become quite cumbersome for a frequency-selective channel, since the number of the states of the extended trellis structure grows exponentially with the channel delay spread. To reduce the complexity, a popular approach is to use Laurent's representation of CPM signals that decomposes a CPM signal into a sum of parallel linearly modulated signals. The modulating pulse in the Laurent's representation is not, however, Nyquist pulse or square root of Nyquist pulse. Therefore, the ISI is built into the model, which is not surprising given the memory of the CPM. The dispersive channel induces further ISI. The advantage of Laurent's representation is that we could use the equalization techniques developed for linear modulation to demodulate the received CPM signal.

3.6 Chapter Summary

Aerial channel modeling and waveform design are undoubtedly two of the most fundamental aspects in UAV wireless communications and networking. In this chapter, we have provided in-depth knowledge and background on the key aspects of channel modeling and waveform design for UAV wireless communications systems, creating a foundation for the remainder of the book. Both channel modeling and waveform design are rich subjects in wireless communications. We refer interested readers to the classic textbooks [60, 64, 65, 136, 137] for general treatments of the subjects.

Compared to the terrestrial wireless channels, the characteristics of aerial wireless channels are quite different in many respects due to the different heights at which the UAV may be operated. This chapter has illuminated the salient characteristics of aerial wireless channels, provided fundamental treatment of the main areas of aerial radio propagation and channel modeling, and reviewed the key aerial wireless channel measurement results in the literature. We can see that, despite the distinct characteristics of aerial wireless channels, the basic channel modeling principles still largely follow the established theory and modeling methodologies of general wireless channels. The proliferation of UAV applications continues calling for new measurements and models that can aid the design of the corresponding UAV wireless communications systems.

This chapter has also reviewed the key basics of waveform design. We have discussed a few example popular waveform choices, including OFDM, spread spectrum, and CPM, which have been considered for UAV wireless communications and networking. The treatment of these example waveform choices has particularly focused on the corresponding design considerations for UAV wireless communications. Though a single waveform choice unlikely fits all the UAV wireless communications systems, OFDM has become the dominating waveform in the current major wireless standards and thus holds the greatest potential. As will be seen in the subsequent chapters, different types of channel models and waveforms may be used in different scenarios depending on the problem being solved, the UAV application being addressed, the role of the UAV (BS or UE), as well as the need for tractability.

4 Performance Analysis and Tradeoffs

In this chapter, we will focus on the performance limits and metrics for UAV BSs. In particular, in Section 4.1, we will first have a brief overview on performance analysis methodologies, such as stochastic geometry. Then, in Section 4.2, we will introduce several detailed case studies to analyze the performance limits of wireless communications with UAVs, while exposing the impact of various unique UAV features, such as altitude, mobility, line-of-sight communications, and elevation angle, on the various metrics. We particularly focus on UAV networks that also encompass an underlaid D2D network. We then conclude the chapter with a brief summary in Section 4.3.

4.1 UAV Network Modeling: Challenges and Tools

In order to characterize the effect of design parameters on UAV-based wireless communication systems, there is a need for a comprehensive analysis on the system performance [31, 150]. In this regard, the performance of communication systems that integrate UAVs should be evaluated considering various metrics, including latency, coverage probability, rate, and link reliability. In fact, the fundamental performance analysis can demonstrate key tradeoffs when designing wireless networks that encompass UAVs. While characterizing the performance of UAV-enabled wireless networks, the unique aspects of UAVs (e.g., those we discussed in Chapter 1), such as their mobility, altitude-depended channels (as exposed in Chapter 3), battery lifetime, and flight time constraints, need to be taken into account. In particular, the wireless communication performance of a UAV system is significantly affected by the UAV's flight time duration as well as its transmit power, as will be showcased throughout this book and, particularly, when we deal with deployment in Chapter 5, mobility in Chapter 6, and resource management in Chapter 7.

Meanwhile, performance analysis of UAVs in coexistence with any terrestrial network is essential for an efficient design of integrated aerial-terrestrial wireless networks. Characterizing the performance of a heterogeneous network of flying UAV BSs and terrestrial BSs is challenging due to the mutual interference between drones and the ground network. In fact, there is a need for powerful mathematical tools to thoroughly analyze the performance of drone-enabled wireless networks.

Stochastic geometry (SG) provides powerful mathematical tools, which are used in many diverse areas, such as ecology, geodesy, and cosmology. In recent years, SG

4.1 UAV Network Modeling: Challenges and Tools 69

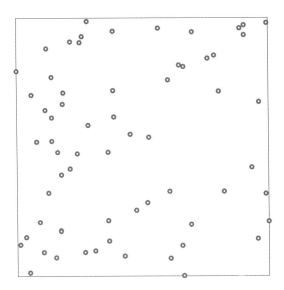

Figure 4.1 Illustration of 2D PPP for modeling the locations of users (gray points).

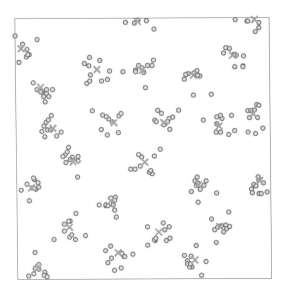

Figure 4.2 Illustration of 2D PCP for modeling the locations of users (gray points).

became a very popular tool for performance analysis of wireless cellular and ad hoc networks [151, 152]. A fundamental underlying technique for evaluation of key performance metrics, e.g., coverage, rate, and throughput, is to endow locations of wireless nodes as a random point process and then obtain statistical distribution or moments. Poisson point process (PPP), and Poisson cluster process (PCP) are well-known processes that have been significantly used for performance analysis cellular network, pictorial illustrations of these processes are shown in Figures 4.1 and 4.2. Given the effective use of SG in performance analysis of cellular network in 2D and existence of SG mathematical tools in n-dimensions, SG will provide a set of useful tools for performance analysis of 3D UAV networks [153]. To this end, suitable types of point process for modeling the locations of drones need to be adopted. For example, when deploying drones in hotspots for capacity enhancement, Poisson and Binomial cluster processes [152] can be used. Meanwhile, a process such as the Matern hard core process [151] can be adopted when a minimum distance between UAVs must be ensured for collision avoidance and interference management. In fact, by leveraging tools from stochastic geometry along with appropriate point process for UAVs and ground networks, the performance of UAV-enabled wireless networks can be analytically characterized. Such rigorous performance analysis is an essential step for design and deployment of UAVs in wireless networks, and it can reveal insights and inherent tradeoffs in UAV communications.

4.2 Downlink Performance Analysis for UAV BS

In this section, we investigate the performance of a UAV communication system that coexists with a ground D2D communication network. We evaluate the downlink performance of this network in terms of coverage and system sum-rate. In the considered model, a UAV BS serves its ground users spread over a geographical area. In addition to the ground users that will be served by the UAV BS, there are a number D2D users that operate in an underlay mode using the same band as the UAV BS. More precisely, two types of users are considered: (1) downlink users (DUs), which are served by the UAV BS over downlink transmission links, and (2) D2D users that directly communicate with one another while sharing resources with DUs. In this network, there is interference between D2D transmissions and UAV BS transmissions. For performance evaluation of this network, we consider two scenarios: (1) static UAV BS, and (2) mobile UAV BS. By exploiting tools from SG, we derive closed-form expressions that characterize the downlink coverage probabilities of DUs and D2D users. Moreover, we evaluate the overall system performance while capturing the effect of the UAV BS's height as well as the D2D users' density. For the static UAV BS scenario, we determine the optimal UAV BS's altitude for which the DU's coverage probability is maximized. Moreover, the impact of the UAV's altitude on the sum-rate of the network consisting of DUs and D2D users, is analyzed. In the mobile UAV scenario, we characterize the tradeoff between delay and coverage probability and derive the D2D's outage probability.

4.2.1 System Model

We study a circular geographical area that encompasses a number of DUs and D2D user. We use R_c to represent the radius of the considered geographical area. Here, a single UAV BS is used to provide wireless service for ground users in downlink. The densities of DUs and D2D users, which are uniformly distributed in the area based on homogeneous PPPs, are λ_{du} and λ_d. The average number of DUs and D2D users depends

4.2 Downlink Performance Analysis for UAV BS

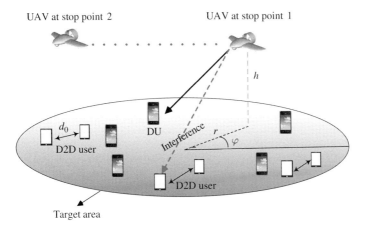

Figure 4.3 System model illustration.

on their density as well as the size of the geographical area. Moreover, a D2D transmitter and receiver pair are separated by a fixed distance [154]. A D2D receiver receives signals from its associated D2D transmitter, as well as interfering D2D transmitters and the UAV BS. A DU, which is served by the UAV BS, experiences interference from the D2D transmitters in the area.

For a given D2D receiver, we can write the signal-to-interference-plus-noise ratio (SINR) expression:

$$\gamma_d = \frac{P_{r,d}}{I_d^c + I_u + N}, \tag{4.1}$$

where $P_{r,d}$ represents the D2D user's received signal power and I_d^c is the aggregated interference from all interfering D2D users. The term I_u captures the interference caused by the UAV BS on the D2D transmitter while N captures the noise power. Moreover, we also have:

$$P_{r,d} = P_d d_0^{-\alpha_d} g_0, \tag{4.2}$$

$$I_d^c = \sum_{i \neq 0} P_d d_i^{-\alpha_d} g_i, \tag{4.3}$$

$$I_d = \sum_i P_d d_i^{-\alpha_d} g_i, \tag{4.4}$$

with g_0 being the channel gain between a D2D transmitter and its target receiver. Also, g_i is the channel gain between the D2D receiver and the ith interfering D2D transmitter. We also consider a Rayleigh fading channel for the small-scale fading at the terrestrial links [154, 155] and [156], and we define P_d as the transmit power of a D2D user. In addition, we use d_i to capture the distance between the D2D receiver and the interfering transmitter and d_0 to capture the fixed distance between each D2D transmitter/receiver pair. Finally, α_d is the path loss exponent for D2D communications.

The downlink SINR for a DU that is served by the UAV BS is:

$$\gamma_u = \frac{P_{r,u}}{I_d + N}, \qquad (4.5)$$

with $P_{r,u}$ and I_d being, respectively, the received signal power from the serving UAV BS and the sum interference from all ground D2D transmitters.

Considering the SINR metric, the downlink coverage probability for the DUs and D2D users can be expressed by:

$$P_{\text{cov},du}(\beta) = \mathbb{P}\left[\gamma_u \geq \beta\right], \qquad (4.6)$$
$$P_{\text{cov},d}(\beta) = \mathbb{P}\left[\gamma_d \geq \beta\right], \qquad (4.7)$$

where β is the SINR threshold needed for connectivity, γ_u is the SINR for a DU, and γ_d is the SINR for a D2D transmitter.

For UAV-to-ground channel model, we consider the commonly used probabilistic LOS/NLOS model in which LOS and NLOS links can accrue with specific probabilities [157]. Moreover, an NLOS link experiences higher attenuation compared to an LOS link. Based on the probabilistic AG channel model discussed in Chapter 3 and presented in 3.3.6, the received signal power of a ground user will be [158]:

$$P_{r,u} = \begin{cases} P_u |X_u|^{-\alpha_u} & \text{LOS link,} \\ \eta P_u |X_u|^{-\alpha_u} & \text{NLOS link,} \end{cases} \qquad (4.8)$$

where P_u represents the transmit power of the UAV BS, $|X_u|$ is the UAV BS-to-user distance, and α_u is the path loss exponent of the air-to-ground link. Also, η is an excessive attenuation factor associated with the NLOS link. Note that, as discussed in Chapter 3, the LOS probability can be a function of the propagation environment, obstacles, and elevation angle, as well as the positions of the UAV BS and the ground users. The probability of having an LOS link between the UAV BS and a ground user is [158]:

$$P_{\text{LoS}} = \frac{1}{1 + C\exp(-B[\theta - C])}, \qquad (4.9)$$

with C and B being environment-dependent constants. θ denotes the elevation angle between the UAV BS and a ground user, $\theta = \frac{180}{\pi} \times \sin^{-1}\left(\frac{h}{|X_u|}\right)$, $|X_u| = \sqrt{h^2 + r^2}$. The NLOS probability is given by $P_{\text{NLoS}} = 1 - P_{\text{LoS}}$.

Clearly, increasing the elevation angle leads to a higher probability for having an LOS link between the UAV BS and a ground user.

Based on the considered model, we investigate two scenarios for wireless networking with a UAV BS: (1) network with a static UAV BS (e.g., hovering or tethered), and (2) network with a mobile UAV BS. In particular, we will conduct a thorough performance analysis of these two UAV-based wireless communication system scenarios, in terms of coverage probability and average sum-rate. Moreover, the impact of a UAV BS's altitude as well as the density of D2D users on the overall system performance will be analyzed.

4.2.2 Network with a Static UAV

We will now analyze the performance of the network with a static UAV BS positioned at altitude h over the center of the considered circular area. We derive the coverage probability for DUs, which are served by the UAV BS, and for D2D users that operate in an underlay mode. Note that, when ground users are uniformly distributed on the geographical area, the downlink coverage performance can be maximized by deploying the UAV BS over the center of the area.

D2D User Coverage Probability

Let (r, φ) be the location of a typical D2D receiver in a polar coordinate system, with r and φ being, respectively, the radius and the angle of the location. We consider a fixed distance d_0 between each D2D transmitter/receiver pair. Next, based on our derivation in [31], we can state the following result pertaining to the coverage probability for a typical D2D transmitter.

THEOREM 4.1 *The coverage probability of a D2D receiver located at (r, ϕ) and connected to its associated transmitter at a fixed distance d_0 will be:*

$$P_{\mathrm{cov},d}(r,\varphi,\beta) = \exp\left(\frac{-2\pi^2 \lambda_d \beta^{2/\alpha_d} d_0^2}{\alpha_d \sin(2\pi/\alpha_d)} - \frac{\beta d_0^{\alpha_d} N}{P_d}\right)$$
$$\times \left[P_{\mathrm{LoS}} \exp\left(\frac{-\beta d_0^{\alpha_d} P_u |X_u|^{-\alpha_u}}{P_d}\right) \right.$$
$$\left. + P_{\mathrm{NLoS}} \exp\left(\frac{-\beta d_0^{\alpha_d} \eta P_u |X_u|^{-\alpha_u}}{P_d}\right) \right], \quad (4.10)$$

and $|X_u| = \sqrt{h^2 + r^2}$.

As we can see from this theorem, when the height of the UAV BS increases, the interference on the D2D users stemming from the UAV BS does not always decrease. In fact, when the altitude of the UAV BS increases, the coverage probability for the D2D user can decrease and increase. This is because as the UAV BS's height increases, both the UAV-to-ground distance and LOS probability increase, which have conflicting impact on the D2D coverage probability. Another observation is that when the transmit power of the D2D user increases, the coverage probability for D2D also increases. Meanwhile, increasing the transmit power of the UAV BS results in a lower D2D coverage probability due to a higher interference stemming from the UAV. To enhance the D2D coverage probability, three approaches can be considered: (1) increasing the D2D transmit power, (2) reducing the number of D2D users (i.e., D2D density), and (3) reducing the separation distance between D2D transmitter/receiver pairs. In a nutshell, this theorem clearly showcases how the presence of a flying UAV BS impacts ground network performance.

Now, using 4.1, we determine the average coverage probability for D2D users over the given geographical area. For a uniform distribution of the ground users, we have $f(r,\varphi) = \frac{r}{\pi R_c^2}$, $0 \leq r \leq R_c$, $0 \leq \varphi \leq 2\pi$. Subsequently, the average coverage probability for the D2D users is:

$$\bar{P}_{\text{cov},d}(\beta) = \mathbb{E}_{r,\varphi}\left[P_{\text{cov},d}(r,\varphi,\beta)\right]$$

$$= \exp\left(\frac{-2\pi^2 \lambda_d \beta^{2/\alpha_d} d_0^2}{\alpha_d \sin(2\pi/\alpha_d)} - \frac{\beta d_0^{\alpha_d} N}{P_d}\right)$$

$$\times \int_0^{R_c} \mathbb{E}_{I_u}\left[\exp\left(\frac{-\beta d_0^{\alpha_d} I_u}{P_d}\right)\right] f(r,\varphi) \mathrm{d}r \mathrm{d}\varphi$$

$$= \exp\left(\frac{-2\pi^2 \lambda_d \beta^{2/\alpha_d} d_0^2}{\alpha_d \sin(2\pi/\alpha_d)} - \frac{\beta d_0^{\alpha_d} N}{P_d}\right)$$

$$\times \int_0^{R_c} \mathbb{E}_{I_u}\left[\exp\left(\frac{-\beta d_0^{\alpha_d} I_u}{P_d}\right)\right] \frac{2r}{R_c^2} \mathrm{d}r. \quad (4.11)$$

As shown in (4.11), increasing the size of the area leads to a higher average coverage probability. While deploying the UAV BS over a larger geographical area, the UAV BS-D2D distance will increase, and, therefore, a D2D user experiences less interference from the UAV BS. We can now present a closed-form expression for D2D coverage probability in two spacial cases.

Remark 4.1 For $P_u = 0$ or $h \to \infty$, the average coverage probability for the D2D users is simplified to [159]:

$$\bar{P}_{\text{cov},d}(\beta) = \exp\left(\frac{-2\pi^2 \lambda_d \beta^{2/\alpha_d} d_0^2}{\alpha_d \sin(2\pi/\alpha_d)} - \frac{\beta d_0^{\alpha_d} N}{P_d}\right). \quad (4.12)$$

Clearly, (4.12) represents the D2D coverage probability in overlay mode where the UAV does not create interference on D2D users.

Downlink Users Coverage Probability

In this subsection, we find the upper and lower bounds for the coverage probability of DUs communicating with the UAV BS.

THEOREM 4.2 The bounds for the DU's coverage probability are as follows:

$$\bar{P}^L_{\text{cov,du}}(\beta,h) = \int_0^{R_c} P_{\text{LoS}}(r,h) L_I\left(\frac{P_u |X_u|^{-\alpha_u}}{\beta} - N\right) \frac{2r}{R_c^2} \mathrm{d}r$$

$$+ \int_0^{R_c} P_{\text{NLoS}}(r,h) L_I\left(\frac{\eta P_u |X_u|^{-\alpha_u}}{\beta} - N\right) \frac{2r}{R_c^2} \mathrm{d}r, \quad (4.13)$$

$$\bar{P}^U_{\text{cov,du}}(\beta, h) = \int_0^{R_c} P_{\text{LoS}}(r,h) U_I\left(\frac{P_u|X_u|^{-\alpha_u}}{\beta} - N\right) \frac{2r}{R_c^2} dr$$

$$+ \int_0^{R_c} P_{\text{NLoS}}(r,h) U_I\left(\frac{\eta P_u|X_u|^{-\alpha_u}}{\beta} - N\right) \frac{2r}{R_c^2} dr, \quad (4.14)$$

where $\beta N < P_u \|X_u\|^{-\alpha_u}$. Also, for $T > 0$ we have:

$$L_I(T) = \left[1 - \frac{2\pi \lambda_d \Gamma(1+2/\alpha_d)}{\alpha_d - 2}\left(\frac{T}{P_d}\right)^{-2/\alpha_d}\right]$$

$$\times \exp\left(-\pi \lambda_d \left(\frac{T}{P_d}\right)^{-2/\alpha_d} \Gamma(1+2/\alpha_d)\right), \quad (4.15)$$

$$U_I(T) = \exp\left(-\pi \lambda_d \left(\frac{T}{P_d}\right)^{-2/\alpha_d} \Gamma(1+2/\alpha_d)\right). \quad (4.16)$$

with $\Gamma(t) = \int_0^\infty x^{t-1} e^{-x} dx$ representing the gamma function [160].

Proof This proof stems from our work in [31] and is presented here to showcase the key steps in establishing such a fundamental result. First, we can find that the DU's coverage probability, whenever the user is located at (r, φ), can be given by:

$$P_{\text{cov,du}}(r, \varphi, \beta) = \mathbb{P}\left[\gamma_u \geq \beta\right] = P_{\text{LoS}}(r) \mathbb{P}\left[\frac{P_u r^{-\alpha_u}}{I_d + N} \geq \beta\right]$$

$$+ P_{\text{NLoS}}(r) \mathbb{P}\left[\frac{\eta P_u r^{-\alpha_u}}{I_d + N} \geq \beta\right]$$

$$= P_{\text{LoS}}(r) \mathbb{P}\left[I_d \leq \frac{P_u r^{-\alpha_u} - \beta N}{\beta}\right]$$

$$+ P_{\text{NLoS}}(r) \mathbb{P}\left[I_d \leq \frac{\eta P_u r^{-\alpha_u} - \beta N}{\beta}\right]. \quad (4.17)$$

Since the CDF of the total interference generated by D2D users does not have closed-form expression [161] and [162], we present the upper bound and lower bound on the CDF. Let us consider two categories of D2D transmitters [163]:

$$\begin{cases} \Phi_1 = \{\Phi_B | P_d d_i^{-\alpha_d} g_i \geq T\}, \\ \Phi_2 = \{\Phi_B | P_d d_i^{-\alpha_d} g_i \leq T\}, \end{cases} \quad (4.18)$$

in which T represents a threshold used for finding the CDF of the D2D users' interference.

Let I_{d,Φ_1} and I_{d,Φ_2} be, respectively, the power of interference from D2D users that belong to sets Φ_1 and Φ_2. Subsequently:

$$\mathbb{P}\left[I_d \leq T\right] = \mathbb{P}\left[I_{d,\Phi_1} + I_{d,\Phi_2} \leq T\right] \leq \mathbb{P}\left[I_{d,\Phi_1} \leq T\right]$$

$$= \mathbb{P}\left[\Phi_1 = 0\right] = \mathbb{E}\left[\prod_{\Phi_B} \mathbb{P}(P_d d_i^{-\alpha_d} g_i < T)\right]$$

$$= \mathbb{E}\left[\prod_{\Phi_B} \mathbb{P}(g_i < \frac{T d_i^{\alpha_d}}{P_d})\right] \stackrel{(a)}{=} \mathbb{P}\left[\prod_{\Phi_B} 1 - \exp(-\frac{T d_i^{\alpha_d}}{P_d})\right]$$

$$\stackrel{(b)}{=} \exp\left(-\lambda_d \int_0^\infty \exp(-\frac{T r^{\alpha_d}}{P_d}) r dr\right)$$

$$= \exp\left(-\pi \lambda_d \left(\frac{T}{P_d}\right)^{-2/\alpha_d} \Gamma(1 + 2/\alpha_d)\right), \quad (4.19)$$

in (a), we use the properties of Rayleigh fading, and (b) follows from the PGFL of the PPP.

The upper bound on the CDF of the D2D users' interference can now be derived as follows:

$$\mathbb{P}\left[I_d \leq T\right] = 1 - \mathbb{P}\left[I_d \geq T\right]$$

$$= 1 - \left(\mathbb{P}\left[I_d \geq T | I_{d,\Phi_1} \geq T\right] \mathbb{P}\left[I_{d,\Phi_1} \geq T\right]\right.$$

$$\left. + \mathbb{P}\left[I_d \geq T | I_{d,\Phi_1} \leq T\right] \mathbb{P}\left[I_{d,\Phi_1} \leq T\right]\right)$$

$$= 1 - \left(\mathbb{P}\left[I_{d,\Phi_1} \geq T\right] + \mathbb{P}\left[I_d \geq T | I_{d,\Phi_1} \leq T\right]\right.$$

$$\left. \times \mathbb{P}\left[I_{d,\Phi_1} \leq T\right]\right)$$

$$= 1 - \left(1 - \mathbb{P}\left[\Phi_1 = 0\right] + \mathbb{P}\left[I_d \geq T | I_{d,\Phi_1} \leq T\right]\right.$$

$$\left. \times \mathbb{P}\left[\Phi_1 = 0\right]\right)$$

$$= \mathbb{P}\left[\Phi_1 = 0\right]\left(1 - \mathbb{P}\left[I_d \geq T | \Phi_1 = 0\right]\right). \quad (4.20)$$

We also have:

$$\mathbb{P}\left[I_d \geq T | \Phi_1 = 0\right] \stackrel{(a)}{\leq} \frac{\mathbb{E}\left[I_d \geq T | \Phi_1 = 0\right]}{T}$$

$$= \frac{1}{T} \mathbb{E}\left[\sum_{\Phi} P_d d_i^{-\alpha_d} g_i \mathbb{1}(P_d d_i^{-\alpha_d} g_i \leq T)\right]$$

$$= \frac{1}{T} \mathbb{E}_{d_i}\left[\sum_{\Phi} P_d d_i^{-\alpha_d} \mathbb{E}_{g_i}\left[g_i \mathbb{1}(g_i \leq \frac{T d_i^{\alpha_d}}{P_d})\right]\right]$$

$$= \frac{1}{T} \mathbb{E}_{d_i}\left[\sum_{\Phi} P_d d_i^{-\alpha_d} \int_0^{\frac{T d_i^{\alpha_d}}{P_d}} g e^{-g} dg\right]$$

$$= \frac{2\pi P_d \lambda_d}{T} \int_0^\infty r^{-\alpha_d} \left(\int_0^{\frac{Tr^{\alpha_d}}{P_d}} g e^{-g} dg \right) r dr$$

$$= \frac{2\pi \lambda_d \Gamma(1 + 2/\alpha_d)}{\alpha_d - 2} \left(\frac{T}{P_d} \right)^{-2/\alpha_d}, \qquad (4.21)$$

in (a), we use Markov's inequality, in which $P(X \geq L) \leq \frac{\mathbb{E}[X]}{L}$ with $X \geq 0$ being a random variable and $L > 0$. $\mathbb{1}(.)$ represents the indicator function with value of 0 or 1.

Therefore, we can present the lower bound (L_I) and the upper (U_I) of the CDF of D2D user's interference as follows:

$$L_I(T) = \left[1 - \frac{2\pi \lambda_d \Gamma(1 + 2/\alpha_d)}{\alpha_d - 2} \left(\frac{T}{P_d} \right)^{-2/\alpha_d} \right]$$
$$\times \exp \left(-\pi \lambda_d \left(\frac{T}{P_d} \right)^{-2/\alpha_d} \Gamma(1 + 2/\alpha_d) \right), \qquad (4.22)$$

$$U_I(T) = \exp \left(-\pi \lambda_d \left(\frac{T}{P_d} \right)^{-2/\alpha_d} \Gamma(1 + 2/\alpha_d) \right), \qquad (4.23)$$

which leads to $L_I(T) \leq \mathbb{P}\{I_d \leq T\} \leq U_I(T)$.

Given (4.17), (4.22), and (4.23), the bounds on the DU's average coverage probability are given by:

$$\bar{P}^L_{\text{cov,du}}(\beta) = \int_0^{R_c} P_{\text{LoS}}(r) L_I \left(\frac{P_u |X_u|^{-\alpha_u}}{\beta} - N \right) \frac{2r}{R_c^2} dr$$
$$+ \int_0^{R_c} P_{\text{NLoS}}(r) L_I \left(\frac{\eta P_u |X_u|^{-\alpha_u}}{\beta} - N \right) \frac{2r}{R_c^2} dr, \qquad (4.24)$$

$$\bar{P}^U_{\text{cov,du}}(\beta) = \int_0^{R_c} P_{\text{LoS}}(r) U_I \left(\frac{P_u |X_u|^{-\alpha_u}}{\beta} - N \right) \frac{2r}{R_c^2} dr$$
$$+ \int_0^{R_c} P_{\text{NLoS}}(r) U_I \left(\frac{\eta P_u |X_u|^{-\alpha_u}}{\beta} - N \right) \frac{2r}{R_c^2} dr. \qquad (4.25)$$

This proves Theorem 4.2.

□

According to Theorem 4.2, when $T >> P_d$, we can observe that $U_I(T) = L_I(T) \approx 1 - \pi \lambda_d \left(\frac{T}{P_d} \right)^{-2/\alpha_d} \Gamma(1 + 2/\alpha_d)$. That is, as the D2D transmit power decreases, the upper and lower bounds will be tighter. In addition, for $\lambda_d \to \infty$, we have $U_I = L_I = 0$. In this case, $\bar{P}_{\text{cov,du}} = 0$ due to a significant amount of interference from D2D transmitters on the AG links of the DUs. Another observation is that deploying the UAV BS at a higher altitude can result in a higher coverage probability due to a higher LOS chance. However, since $|X_u|$ increases by increasing h, the DU's coverage probability can decrease considering the fact that both L_I and U_I decrease. In fact, the DU's coverage probability can be maximized by properly adjusting the height of the UAV BS.

From Theorem 4.2, we can also see that as R_c increases, the DU's coverage probability decreases. In contrast, for larger values of R_c, we can see that a higher coverage probability will be achieved for the D2D users. Meanwhile, the DU's average coverage probability can improve by reducing the number of D2D transmitters. Next, based on our result in [31], we can state the following proposition that determines the coverage probability for the DU when there are no interfering D2D users in the network.

PROPOSITION 1 The DU's coverage probability when $P_d \to 0$ or $\lambda_d \to 0$ can be written by:

$$\bar{P}_{\text{cov,du}}(\beta) = \int_0^{\min\left\{\left[\left(\frac{P_u}{\beta N}\right)^{2/\alpha_u} - h^2\right]^{0.5}, R_c\right\}} P_{\text{LoS}}(r) \frac{2r}{R_c^2} dr$$
$$+ \int_0^{\min\left\{\left[\left(\frac{\eta P_u}{\beta N}\right)^{2/\alpha_u} - h^2\right]^{0.5}, R_c\right\}} P_{\text{NLoS}}(r) \frac{2r}{R_c^2} dr. \qquad (4.26)$$

Note that (4.26) represents the maximum DU's coverage probability, which can be obtained in the absence of D2D interference.

System Sum-Rate

Here, we present the average achievable transmission rates for the DUs (i.e., the air-to-ground transmission rates) as well as the D2D users [164]:

$$\bar{C}_{du} = W\log_2(1+\beta)\bar{P}_{\text{cov},du}(\beta), \qquad (4.27)$$
$$\bar{C}_d = W\log_2(1+\beta)\bar{P}_{\text{cov},d}(\beta), \qquad (4.28)$$

with W being the transmission bandwidth available for DUs and D2D users.

The average sum-rate in the network with all DUs and D2D users can be given by:

$$\bar{C}_{\text{sum}} = R_c^2 \pi \lambda_{du} \bar{C}_{du} + R_c^2 \pi \lambda_d \bar{C}_d. \qquad (4.29)$$

Considering $\mu = \frac{\lambda_{du}}{\lambda_d}$ leads to:

$$\bar{C}_{\text{sum}} = \lambda_d R_c^2 \pi \left[\mu \bar{P}_{\text{cov},du}(\beta) + \bar{P}_{\text{cov},d}(\beta)\right] W\log_2(1+\beta), \qquad (4.30)$$

where the number of DUs and D2D users in the considered circular geographical area are, respectively, given by $R_c^2 \pi \lambda_d$ and $R_c^2 \pi \lambda_{du}$.

We can now see that the average sum-rate $\bar{C}_{\text{sum}}$ is an increasing function of λ_d as well as the coverage probabilities of DUs and D2D links. However, the coverage probabilities decrease when the number of D2D users increases. Hence, a higher λ_d will not necessarily result in a higher sum-rate. In fact, the system sum-rate can be maximized by optimally adjusting the density of the D2D users. For example, in a practical cellular system, the network operator can properly schedule the D2D transmissions in order to control this density.

Meanwhile, $\bar{C}_{\text{sum}}$ depends on the coverage probability and the threshold (β). While the coverage probability decreases by increasing β, the logarithmic function in $\bar{C}_{\text{sum}}$ increases when β increases. Given such a tradeoff, the sum-rate can be maximized for an optimum value of the threshold.

4.2.3 Mobile UAV BS Scenario

Now we consider a network in which the UAV BS can move. In this scenario, the UAV BS's mobility is exploited to provide full coverage for users in the given geographical area. Specifically, the UAV moves and stops at some pre-defined locations known as stop points in order to provide wireless connectivity for DUs. In this wireless network with a mobile UAV BS, we want to find the minimum number of UAV BS stop points (M) as well as the locations of stop points such that the coverage requirement for all DUs within the area is met. The design problem is essentially to entirely cover the considered circular area with a minimum movement for the UAV BS. To tackle this problem, we use the *disk covering problem* [165] framework from mathematics. The disk covering problem deals with the problem of covering a big disk with the minimum number of small disks. In particular, this problem aims at finding how to completely cover a given disk with a minimum number of smaller equally-sized disks that have a specified radius.

Figure 4.4 shows an illustrative example for the disk covering problem that can be used to analyze the mobile UAV BS scenario. Here, each small disk's center corresponds to a UAV BS stop point, and its radius represents the coverage radius of the UAV BS. Table 4.1 shows the minimum coverage radius that the mobile UAV BS needs in order to fully cover the circular coverage area, considering a varying number of UAV BS

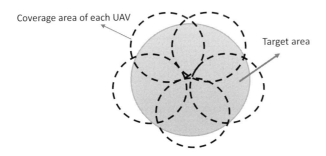

Figure 4.4 Five disks covering problem.

Table 4.1 Number of disks and radius of each disk in covering problem.

Number of stop points	Minimum required coverage radius ($R_{\min}$)
$M = 1, 2$	R_c
$M = 3$	$\frac{\sqrt{3}}{2} R_c$
$M = 4$	$\frac{\sqrt{2}}{2} R_c$
$M = 5$	$0.61 R_c$
$M = 6$	$0.556 R_c$
$M = 7$	$0.5 R_c$
$M = 8$	$0.437 R_c$
$M = 9$	$0.422 R_c$
$M = 10$	$0.398 R_c$
$M = 11$	$0.38 R_c$
$M = 12$	$0.361 R_c$

stop points over the area [165, 166]. Subsequently, given the size of the geographical area as well as the coverage range of the UAV BS, the minimum number of stop points (and their locations) needed for providing full coverage for DUs within the area can be determined. In the sequel, we discuss the disk covering-based performance analysis for the considered mobile UAV BS scenario.

We need to find the maximum UAV BS's coverage radius for which the DU's coverage requirement is satisfied. The UAV BS's coverage radius corresponds to the maximum range within which the DU's coverage probability exceeds the target, ε. Therefore, all the DUs located within the coverage range of the UAV BS will be considered as covered by the UAV BS and will have at least an ϵ coverage probability. The maximum UAV BS's coverage range can now be defined as follows:

$$R_m = \max\{R | P_{\text{cov},du}(\beta, R) \geq \varepsilon, P_u, h\} = P_{\text{cov},du}^{-1}(\beta, \varepsilon), \tag{4.31}$$

where h and P_u are the altitude and transmit power of the UAV. Also, ε represents the DU's coverage probability requirement.

Now, we determine the minimum number of UAV BS stop points needed for full coverage of the area:

$$\begin{cases} L = \min\{M\}, \\ P_{\text{cov},du}(r, \varphi, \beta) \geq \varepsilon, \end{cases} \tag{4.32}$$

where M is the number of stop points and L is the minimum value of M. We also have:

$$R_{\min,L} \leq R_m \leq R_{\min,L-1} \rightarrow \min\{M\} = L. \tag{4.33}$$

Given Table 4.1, $R_{\min,L-1}$ and $R_{\min,L}$ represent the minimum radius needed for completely covering the geographical circular area using $L - 1$ and L disks.

In this case, in order to ensure that UAV BS uses a minimum transmit power while covering the area, we should have:

$$P_{u,\min} = \underset{P_u}{\operatorname{argmin}}\{P_{\text{cov,du}}^{-1}(\beta, \varepsilon) = R_{\min,L}|h\}, \tag{4.34}$$

in (4.34), $P_{u,\min}$ represents the minimum UAV's transmit power. Therefore, the area is entirely covered by the mobile UAV BS with the corresponding, minimum UAV BS's transmit power as well as the stop points.

In the sequel, we analyze how the number of stop points can affect the coverage time of the area for serving DUs and the D2D user's outage probability.

Let us now consider the mobile UAV BS case within a total of M time instances. At each time instance, the UAV BS and D2D links have simultaneous transmissions. During these M time instances, the flying UAV BS stops at M stop points and provides full downlink coverage for the DUs located on the ground. In this case, by increasing M the total coverage time of the UAV BS will increase since the UAV BS needs to move more for a higher number of stop points. Here, the total time that the UAV BS flies while serving DUs is referred to as *delay* or latency. This delay is a function of the distance between stop points, the speed of the UAV BS, and the UAV transmission time at each stop point. This delay can be given by:

$$\tau = T_{tr} + MT_s, \tag{4.35}$$

where T_{tr} represents the total flight time of the UAV BS and M is the number of stop points. Also, T_s is the UAV BS transmission time at each stop point. As discussed in earlier chapters, the flight time is a function of various factors such as the UAV BS's speed, the size of the area, and the locations of the stop points. For instance, we can show that the flight times for $M = 3$ and $M = 4$ are, respectively, $\frac{\sqrt{3}R_c}{v}$ and $\frac{3R_c}{v}$, with v being the UAV BS's velocity and R_c being the radius of the considered circular area. The transmission time (T_s) is affected by the multiple access scheme. In case of a time division multiple access (TDMA), the transmission time can be approximated by:

$$T_s \approx T_{s,1} \frac{R_{\min}^2(M)}{R_c^2} U, \tag{4.36}$$

where $T_{s,1}$ captures the UAV BS transmission time needed to service each DU and U is the total number of DUs. The coverage range of the UAV BS is given by $R_{\min}$, which is a function of M as well as the size of the geographical area.

In the frequency division multiple access (FDMA) case, all ground users can be served at the same time; hence, we have $T_s = T_{s,1}$. Figure 4.5 shows how the total delay changes by varying the number of stop points. Here, the UAV BS's speed is 10 m/s, and two transmission times are considered. As we can infer from Figure 4.5, having a higher number of stop points yields a larger delay. Furthermore, the delay increases when the transmission time at each stop point increases. For instance, Figure 4.5 shows that, for $T_{s,1} = 20$ s, by increasing the number of stop point from 3 to 10, the delay can increase by a factor of 2.

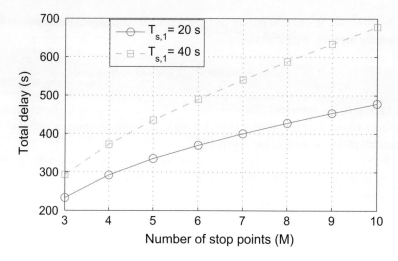

Figure 4.5 Coverage delay.

We now seek to find the D2D user's outage probability during M transmissions of the UAV BS and D2D users. The outage probability concept is defined as the probability that, among M D2D transmissions, at least one transmission becomes unsuccessful. Let (r_i, h_i) be the location of stop point i relative to a typical D2D user, with r_i being the horizontal distance between the UAV BS and the D2D user. We assume that the Rayleigh fading is independent at different time instances. Nevertheless, there is a spatial/temporal correlation in D2D interference at different time instances. Next, we find the overall D2D user's outage probability (the proof is found in [31]).

THEOREM 4.3 For a wireless network with underlaid D2D transmissions and a mobile UAV BS, the overall outage probability of the D2D users is:

$$P_{out,d} = 1 - \exp\left(-\lambda_d \int_{R^2} \left[1 - \left(\frac{1}{1 + \frac{\beta |x|^{-\alpha_d}}{d_0^{-\alpha_d}}}\right)^M\right] dx\right)$$

$$\times \prod_{i=1}^{M} \mathbb{E}_{I_{u,i}}\left[\exp\left(\frac{-d_0^{\alpha_d} \beta I_{u,i}}{P_d}\right)\right] \exp\left(\frac{-d_0^{\alpha_d} \beta M N}{P_d}\right), \quad (4.37)$$

where $I_{u,i}$ represents the UAV interference on a D2D user. Also, $\mathbb{E}_{I_{u,i}}(.)$ is given by:

$$\mathbb{E}_{I_{u,i}}\left[\exp\left(\frac{-d_0^{\alpha_d} \beta I_{u,i}}{P_d}\right)\right] = P_{\text{LoS},i} \exp\left(\frac{-\beta d_0^{\alpha_d} P_u |X_{u,i}|^{-\alpha_d}}{P_d}\right)$$

$$+ P_{\text{NLoS},i} \exp\left(\frac{-\beta d_0^{\alpha_d} \eta P_u |X_{u,i}|^{-\alpha_d}}{P_d}\right). \quad (4.38)$$

Table 4.2 Simulation parameters.

Description	Parameter	Value
UAV BS transmit power	P_u	5 W
D2D transmit power	P_d	100 mW
Path loss coefficient	K	-30 dB
Path loss exponent for UAV-user link	α_d	2
Path loss exponent for D2D link	α_u	3
Noise power	N	-120 dBm
Bandwidth	W	1 MHz
D2D pair fixed distance	d_0	20 m
Excessive attenuation factor for NLOS	η	20 dB
Parameters for a dense urban environment	B, C	0.136, 11.95

According to Theorem 4.3, the outage probability increases when M increases. This is due to the fact that, for a higher number of stop points and UAV BS transmissions, the D2D users will experience a more severe interference from the UAV BS. As a result, we can see that $P_{out,d}$ approaches one, for large values of M. Meanwhile, increasing the number of UAV BS stop points leads to a higher DU's coverage probability. Therefore, there is an inherent tradeoff between the coverage probability of DUs and the overall outage probability for D2D users, while changing the number of stop points.

4.2.4 Representative Simulation Results

In Table 4.2, we provide the simulation parameters based on [158] and [164]. We evaluate the performance of the considered UAV BS-D2D network while capturing the effect of different parameters, such as the UAV BS's height, the number of D2D users, and the SINR target value.

Figure 4.6 plots the coverage probability for D2D users versus the SINR threshold. Figure 4.7 also shows the lower bound and upper bound for the coverage probability of DUs as a function of the SINR threshold. In Figures 4.6 and 4.7, we can compare the analytical results with the simulation results. As we can see, the D2D coverage probability and the DU's coverage probability decrease by increasing the SINR target value.

In Figure 4.8, we show how the system sum-rate is affected by the density of D2D users. We can first observe that the D2D interference decreases when the number of D2D transmitters decreases. Nevertheless, a lower D2D user density results in a lower system sum-rate. Although decreasing the number of D2D users improves the coverage probability for the ground D2D users, this decrease will negatively impact the system sum-rate, which is directly proportional to the number of D2D users. From Figure 4.8, we can see that there is an optimal value (that leads to a maximum sum-rate) for the density of D2D users. For example, for $\lambda_{du} = 10^{-4}$, the sum-rate is maximized when λ_d is equal to 0.9×10^{-4}.

In Figure 4.9, we analyze how the sum-rate of the considered UAV-enabled wireless D2D system varies by changing the SINR threshold. Here, the bandwidth is 1 MHz,

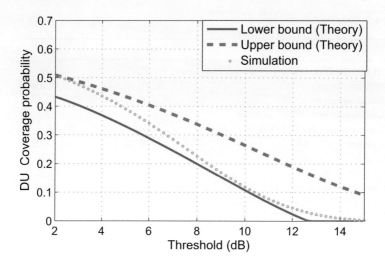

Figure 4.6 DU coverage probability versus SINR target.

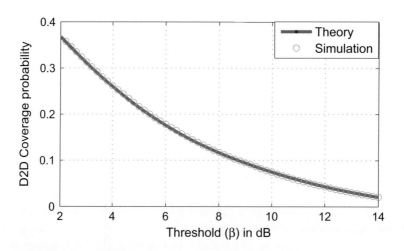

Figure 4.7 D2D coverage probability versus SINR target.

the UAV BS's altitude is 500 m, and $\lambda_{du} = 10^{-4}$. As β increases, satisfying the SINR requirement becomes less likely; hence, the coverage probability decreases. However, given (4.27) and (4.28), increasing β leads to an increase in $\log_2(1 + \beta)$. Considering the overall impact of the SINR threshold on the system sum-rate performance, we can observe that the sum-rate approaches to zero when $\beta \to \infty$.

In order to capture the impact of D2D separation distance, d_0, on the performance, in Figure 4.10 we examine the system sum-rate versus d_0 and the density of the D2D users. By decreasing d_0, the system sum-rate increases since the D2D coverage probability also increases. Another observation is that, for higher values of d_0, the optimal density

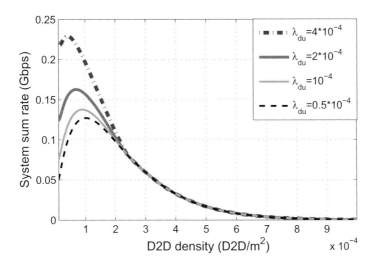

Figure 4.8 Average sum-rate versus the density of D2D users.

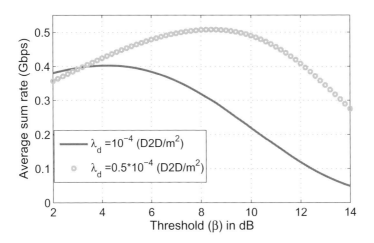

Figure 4.9 Average sum-rate versus SINR target.

of the D2D users, for which the sum-rate is maximized, increases. For example, when d_0 increases from 5 m to 8 m, the optimal D2D density decreases by about 60%.

Figure 4.11 illustrates the impact of the UAV BS's height on the coverage probabilities of the D2D users and DUs. For DUs, it is desirable to deploy the UAV BS at an optimal height for which their coverage probability is maximized. The optimal UAV BS altitude can be determined based on the tradeoff between LOS probability and distance between the UAV BS and the DUs. From Figure 4.11, we can see that the maximum coverage probability for DUs can be attained when the UAV BS is placed at a 500 m height. For D2D users, however, the UAV is a source of interference. Hence, a wide

86 Performance Analysis and Tradeoffs

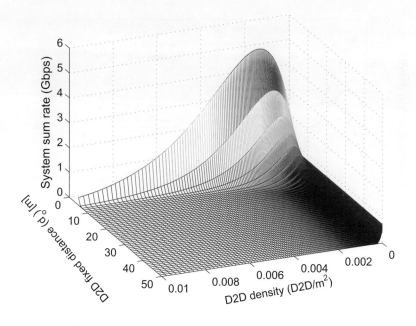

Figure 4.10 Average sum-rate versus d_0 and λ_d.

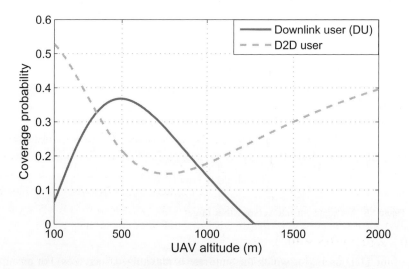

Figure 4.11 Coverage probability versus the altitude of UAV.

UAV BS coverage range is not desirable for the D2D users. As intuitively expected, the coverage probability for D2D users is maximized when the UAV BS is placed at a very high altitude where the interference stemming from the UAV BS on the D2D links becomes negligible. Meanwhile, for some UAV BS altitudes (e.g., 800 m), the D2D users experience poor coverage due to a strong interference generated by the UAV BS's transmissions.

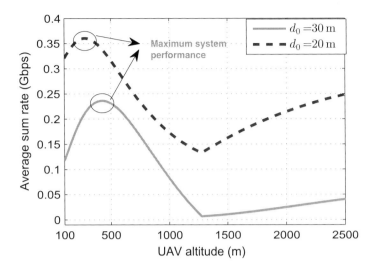

Figure 4.12 Average sum-rate versus the altitude of the UAV BS.

Figure 4.12 investigates the impact of the UAV BS's altitude on the wireless network's sum-rate. As shown in this figure, the system sum-rate is maximized at altitudes 300 m, 350 m, and 400 m for D2D separation distances of 20 m, 25 m, and 30 m. From Figure 4.12, we can clearly observe that the sum-rate increases when the UAV BS's altitude exceeds 1300 m. Above this altitude, the DUs are not covered by the UAV while D2D users still have coverage. By increasing the UAV's altitude, the coverage performance of the D2D users improves, which increases the system sum-rate.

We will now evaluate the performance of the UAV-D2D network in the mobile UAV BS scenario. Figure 4.13 analyzes the effect of the number of D2D users on the minimum number of UAV BS stop points needed to ensure the DUs' coverage requirement. We can directly see that, by increasing the density of D2D users, the DUs receive stronger interference from the D2D transmitters. Therefore, the number of UAV BS stop points should increase to meet the SINR requirement of the DUs. For example, the number of stop points, M, increases from 3 to 8 while increasing the D2D density from 0.2×10^{-4} to 0.8×10^{-4}.

In Figure 4.14, we demonstrate the fundamental tradeoff between the DUs' coverage probability and the total UAV BS coverage time. Here, the time needed to fully cover the geographical area is referred to as delay, which depends on the number of stop points. By increasing the number of stop points, the UAV BS can provide better coverage (i.e., higher SINR) for its DUs. For instance, for 10^{-4} D2D density, to improve the coverage probability of DUs from 0.4 to 0.7, the number of UAV BS stop points needs to be increased by a factor of 4.6. This, in turn, results in a higher total UAV BS coverage time.

Figure 4.15 analyzes the impact of the number of transmission instances, M on the D2D users' outage probability. By increasing M, the outage probability of D2D users

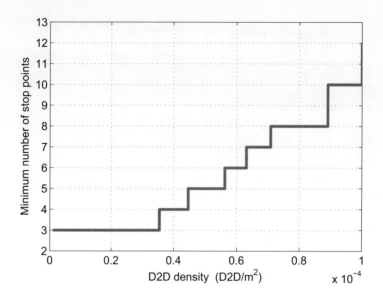

Figure 4.13 Number of stop points versus density of D2D users.

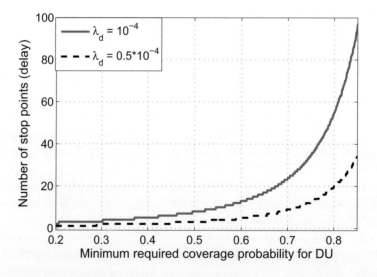

Figure 4.14 Minimum number of stop points versus the coverage probability of DUs.

also increases for two reasons. First, with a higher number of transmissions, the probability of having one D2D transmission failure increases. Second, as M increases, the interference on D2D receivers generated by the UAV BS increases, thus degrading SINR of D2D users. As an example, Figure 4.15 shows that by increasing the M from 3 to 7, the D2D outage probability increases from 20% to 38%.

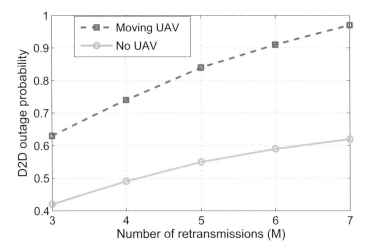

Figure 4.15 Overall D2D outage probability versus M.

4.3 Chapter Summary

In this chapter, we have studied the performance limits, metrics, and tradeoffs for a wireless network that is complemented by a single UAV BS that can be static or mobile. We have provided a brief overview on stochastic geometry as a powerful tool for performance analysis. Then, we have introduced several detailed case studies to analyze the performance limits of wireless communications with a UAV BS, while exposing the impact of various unique UAV BS features, such as altitude, mobility, and line-of-sight communications. In particular, we have analyzed the performance of a network that composed of a UAV BS, DUs, which are served by the UAV, and D2D users that coexist with the UAV and DUs. We have evaluated the performance of such UAV-D2D network in both static and mobile UAV BS in terms of key metrics, such as coverage probability, sum-rate, outage probability, and coverage delay. In particular, in the static UAV scenario, we have derived the coverage probability and system sum-rate for DUs and D2D users. Meanwhile, in the mobile UAV scenario, using the disk covering problem framework, we have identified the locations (stop points) of the mobile UAV for fully covering the given circular geographical area. Furthermore, we have determined the outage probability of D2D users and captured the impact of the number of UAV BS stop points on the performance of D2D users and DUs. The performance analysis done in this chapter provides a fundamental basis for analyzing more complex and varied scenarios for wireless communications and networking with UAVs. Such scenarios can incorporate the various UAV roles discussed in Chapter 1 within a variety of UAV use cases, such as those exposed in Chapter 2. Meanwhile, the fundamental performance of UAV-based wireless communication systems (in UAV BS and UAV UE scenarios) can be further studied while capturing UAV's mobility aspects, temporal channel variations, and antenna patterns.

5 Deployment of UAVs for Wireless Communications

Having provided insights on the performance of wireless networks with UAVs in the previous chapter, our next step is to investigate how UAV BSs can be deployed for enhancing wireless capacity and coverage. Indeed, the 3D placement of UAVs (in all of their roles) is an important design challenge in UAV-enabled wireless communication systems. The adaptive altitude of UAVs as well as their mobility feature allows using effective deployment schemes. Hence, the problem of optimal placement of UAVs has attracted remarkable attention in the literature [109, 158, 167–175]. Clearly, the performance of wireless networks with UAVs, in general, and wireless networks with UAV BSs, in particular, is significantly affected by how the UAVs are deployed in a given region. In general, optimizing UAV deployment is challenging due to the fact that it is a function of various parameters, such as UAV channel gain, interference between UAVs, and deployment environment. Furthermore, one must consider the onboard battery limitation of UAVs that naturally affect the system performance. In fact, there is a need for comprehensive studies on the UAV deployment while designing UAV communication systems.

Hence, in this chapter, we study the problem of UAV deployment for wireless communication purposes. In particular, we focus on the deployment of UAV BSs whose locations will strongly impact the performance that they can deliver. To this end, in Section 5.1, we start by providing a broad overview on the analytical tools that can be used to develop optimized deployment strategies for wireless networks with UAVs. Then, in Section 5.2, we provide a comprehensive study on how UAV BSs can be deployed for optimizing the wireless coverage for a ground network of wireless devices that seek to communicate with UAV BSs in the downlink. We shed important light on how to deploy the UAV BSs, by determining their number and locations, in a way to maximize network performance under various constraints, such as power. Next, in Section 5.3, inspired from the IoT application of UAVs in Chapter 2, we study the problem of optimally deploying UAV BSs for collecting data, in the uplink, from ground IoT devices. In this regard, we show how UAV BSs can be deployed and operate in an energy-efficient manner to service IoT data collection tasks. Then, in Section 5.4, we turn our attention to the optimized deployment of UAV BSs that have the ability to cache popular content and to track the mobility of ground users. In particular, we leverage tools from machine learning to proactively deploy such UAV BSs and to enable them to predictively cache content. We then see how one can use learning techniques to address the joint problems of deployment, cache management, and resource allocation in a network with UAV BSs. We then conclude the chapter in Section 5.5.

5.1 Analytical Tools for UAV Deployment

Here, we describe key analytical tools that are needed to address UAV deployment problems in wireless networks.

5.1.1 Centralized Optimization Theory

Due to the variety of UAV applications in wireless networks, there will be a need for solving many complex optimization problems. Depending on each specific application of UAVs, one must optimize the 3D UAVs' positions so as to achieve the maximum network performance considering various metrics, such as rate, coverage, and energy consumption. While optimizing the locations of UAVs, traditional convex and non-convex optimization methods can be utilized. Next, we discuss two tools from optimization theory that can be used to study UAV deployment optimization problems.

Facility Location Theory
The facility location problem deals with the problem of optimal placement of facilities such that the transportation costs between the facilities and customers are minimized. This problem is called the location-allocation problem when the demand of all customers should be satisfied by multiple facilities. The main components of the facility location problem include customers (that use certain facilities), facilities that should be optimally placed, the location space of facilities and customers, and the objective function, which can depend on the transportation time and distance as well as other factors [176]. In a general form of the facility location problem, the objective is to determine the optimal number of facilities as well as their optimal locations, which lead to the minimum total costs. The costs include the transportation costs, which is a function of distance between the facilities and customers, and the costs of building the facilities. The facility location problem can be modeled based on single or multiple facility cases, capacitated or uncapacitated facilities, and continuous or discrete location space. Moreover, based on the objective function, the facility location problems can be classified into minisum, minimax, and covering problems. These terms are briefly defined as follows [177]:

- *Continuous vs. discrete:* In the discrete location problems, the location of facilities must be chosen from a set of discrete candidate points. However, in the continuous case, the facilities are located over a continuous space.
- *Capacitated vs. uncapacitated:* In the capacitated facility location, the facilities have a limited capacity to service the users. In the uncapacitated case, the capacity of facilities is unlimited.

- *Minisum problem:* The objective is to minimize the sum of distances/costs to all the customers.
- *Minimax problem:* The goal is to minimize the distance/cost to the farthest customer.
- *Covering problem:* The objective is to place the facilities in a way to maintain the distance between each customer and the corresponding facility below a desired, specified threshold.

Next, we describe some of the most common facility location problems.

Categories of facility location problems

Facility location problems can be defined in discrete or continuous space. However, due to the complexity of continuous domain analysis, most of the literature is focused on the discrete facility location problems.

- *Median problem:* In the median problem, the objective is to minimize the average distance of the facilities to the customers.
- *Center problem:* In the center problem, the maximum distance between a customer and its corresponding facility is minimized. This type of problem is suitable for deploying the emergency stations, and their distance to the farthest demand point should not exceed a specified value.
- *Covering problem:* In the covering problems, the goal is to provide maximum coverage for the users. In fact, unlike the median and center problems, the covering problems do not deal with minimizing the distance between facilities and users. Instead, the covering problems try to ensure that the distance between each user and one of the facilities is less than a predefined threshold. Maximizing the coverage for a given number of facilities is called the maximum covering location problem (MCLP).

    If the capacity of the facilities is limited, then there is no guarantee that all the points are covered simultaneously. In this case, the probability of availability is considered for each facility and the expected value of coverage is maximized.

    Another type of the covering problems is the location set covering problem (LSCP) in which the number of facilities required to completely cover the desired area is minimized.
- *Stochastic facility location problem:* The facility location problems can be categorized into deterministic and stochastic problems. In the deterministic case, all the parameters are deterministic and known. However, in the stochastic case, different parameters such as customers' demands are random. One approach to capture the uncertainties in the system is to use the scenario planning models.
- *Multi-objective facility location problem:* In the multi-objective facility location problem, the facilities are placed while considering different criteria and objective functions simultaneously. In this case, optimal location of facilities should be determined in the presence of tradeoffs between different objectives. Typically, in multi-objective optimization, it is not possible to find an optimal solution for each single objective. Consequently, efficient solutions, in which none of the objectives can be improved without degrading others, are usually presented [177].

- *Mobile facility location problem:* In the mobile facility location problem, the goal is to move the facilities to new locations such that the total costs, including the transportation costs between the facilities and customers as well as the costs of the facilities' movements, are minimized [178]. In fact, by moving the facilities, the average distance between the customers and the facilities can be reduced, however, with the cost of the facilities' movements.

Circle Packing and Covering Problems

Packing theory is a class of optimization problems that aims to pack objects together into containers. The main objective in a packing problem is to pack a single container as densely as possible [179]. A packing problem has a dual covering problem in which the number of non-overlapping objects required to completely cover every region of the container is minimized. More specifically, in geometry, circle packing is the study of the arrangement of equal or unequal circles on a given 2D shape such that: (1) no overlapping occurs between circles, and (2) the maximum packing density is achieved.

The packing density corresponds to the proportion of the given surface covered by circles. In two-dimensional Euclidean space, the maximum circle packing density is $\frac{\pi\sqrt{3}}{6}$, which obtains a hexagonal packing arrangement. Figures 5.1 and 5.2 show an illustrative example for circle packing and circle covering on a square area. As will be seen in subsequent sections, circle packing and covering problems can be used to tackle important UAV placement problems. The generalization of a circle packing problem is referred to as a sphere packing problem, which usually considers identical spheres.

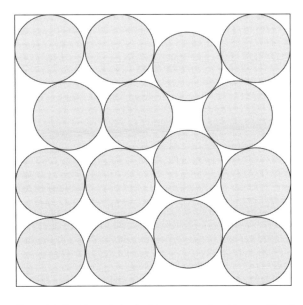

Figure 5.1 Packing circles inside a square (packing problem).

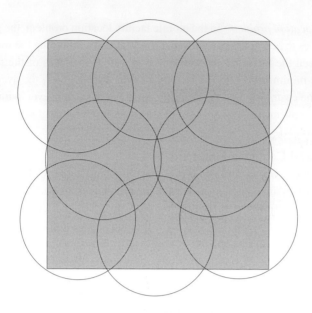

Figure 5.2 Covering a square with circles (covering problem).

5.2 Deployment of UAV BSs for Optimized Coverage

In this section, we study the optimal 3D placement of UAV BSs to maximize the wireless coverage in downlink, inspired from our work in [168]. Moreover, while maximizing the coverage, we aim to minimize the transmit power of UAV BSs so as to reduce their energy consumption as well as any potential interference on ground networks. More precisely, a framework for optimizing the positions of UAV BSs is developed while factoring in the size of the geographical area, the number of available UAV BSs (equipped with directional antennas), and the coverage requirements of ground users. First, as a function of each UAV's altitude and the antenna beamwidth, the UAV coverage probability for serving ground users is derived. Then, a method for optimal placement of multiple UAV BSs is designed based on the notion of circle packing theory [180].

5.2.1 Deployment Model

As we can see in Figure 5.3, we study a wireless network problem in which M UAV BSs must be placed over a given circular geographical area in order to provide wireless connectivity for ground users. The UAV BSs are stationary LAPs (e.g., rotary-wing UAV BSs) and have the same altitude and transmit power. The UAV BS's antenna gain, which uses a directional antenna with a half beamwidth θ_B, is given by [181]:

$$G = \begin{cases} G_{3\text{dB}}, & \frac{-\theta_B}{2} \leq \varphi \leq \frac{\theta_B}{2}, \\ g(\varphi), & \text{otherwise,} \end{cases} \quad (5.1)$$

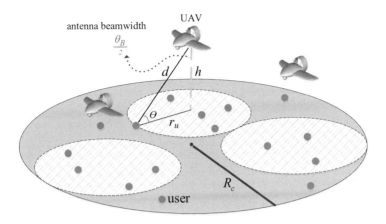

Figure 5.3 System model.

where $G_{3dB} \approx \frac{29000}{\theta_B^2}$ with θ_B in degrees is the main lobe gain, and φ is the sector angle [182]. In (5.1), $g(\varphi)$ represents the antenna gain outside of the main lobe of the directional antenna. For the AG channel between the UAV BS and a ground user, we consider the probabilistic LOS/NLOS model described in Section 3.3.6. In this case, the received signal power for a user served by UAV BS j can be expressed by [157]:

$$P_{r,j}(dB) = \begin{cases} P_t + G_{3dB} - L_{dB} - \psi_{\text{LOS}}, & \text{LOS link}, \\ P_t + G_{3dB} - L_{dB} - \psi_{\text{NLOS}}, & \text{NLOS link}, \end{cases} \quad (5.2)$$

where P_t is the transmit power of each UAV BS, $P_{r,j}$ is the received signal power, and G_{3dB} is the UAV BS antenna gain measured in dB. L_{dB} represents the large-scale path loss for UAV BS-to-user communications:

$$L_{dB} = 10n\log\left(\frac{4\pi f_c d_j}{c}\right), \quad (5.3)$$

where c is the speed of light, f_c is the carrier center frequency, d_j is the distance between a ground user and UAV BS j, and $n \geq 2$ represents the path loss exponent in AG communications. Moreover, $\psi_{\text{LOS}} \sim N(\mu_{\text{LOS}}, \sigma_{\text{LOS}}^2)$ and $\psi_{\text{NLOS}} \sim N(\mu_{\text{NLOS}}, \sigma_{\text{NLOS}}^2)$ represent shadow fading with normal distribution for LOS and NLOS links, separately. These shadow fading normal distributions have the following mean and variance values: $(\mu_{\text{LOS}}, \sigma_{\text{LOS}}^2)$, and $(\mu_{\text{NLOS}}, \sigma_{\text{NLOS}}^2)$. Naturally, the variance is a function of the elevation angle and type of the environment, which is given by [157]:

$$\sigma_{\text{LOS}}(\theta_j) = k_1 \exp(-k_2 \theta_j), \quad (5.4)$$
$$\sigma_{\text{NLOS}}(\theta_j) = g_1 \exp(-g_2 \theta_j), \quad (5.5)$$

where $\theta_j = \sin^{-1}(h/d_j)$ is the elevation angle between UAV j and a ground user, and k_1, k_2, g_1, and g_2 are constants that depend on the type of environment. In this model, the LOS probability is: [157]:

$$P_{\text{LOS},j} = \alpha\left(\frac{180}{\pi}\theta_j - 15\right)^\gamma, \quad (5.6)$$

where α and γ are constants values that capture the environment effects. Also, the probability of NLOS link is $P_{\text{NLOS},j} = 1 - P_{\text{LOS},j}$.

5.2.2 Deployment Analysis

First, we determine the coverage radius of each UAV BS based on the coverage requirement of ground users. To find the coverage radius, based on our result in [168], in Theorem 5.1, we derive the downlink coverage probability when serving a ground user. Then, an efficient approach for placing multiple flying UAV BSs is proposed with the goal of maximizing the total downlink coverage performance.

THEOREM 5.1 The downlink coverage probability for a ground user served by UAV BS j is:

$$P_{\text{cov}} = P_{\text{LOS},j} Q\left(\frac{P_{\min} + L_{dB} - P_t - G_{\text{3dB}} + \mu_{\text{LOS}}}{\sigma_{\text{LOS}}}\right)$$
$$+ P_{\text{NLOS},j} Q\left(\frac{P_{\min} + L_{dB} - P_t - G_{\text{3dB}} + \mu_{\text{NLOS}}}{\sigma_{\text{NLOS}}}\right), \quad (5.7)$$

where $r \leq h.\tan(\theta_B/2)$ is the horizontal distance of the ground user from the projection of the UAV on a geographical area, $P_{\min} = 10\log(\beta N + \beta \bar{I})$ is the minimum required received power for a successful detection at the ground receiver, and N is the noise power. β represents the SINR threshold, and $\bar{I}$ is the mean interference power received from the nearest interfering UAV, which can be written by:

$$\bar{I} \approx P_t g(\varphi_k) \left[10^{\frac{-\mu_{\text{LOS},k}}{10}} P_{\text{LOS},k} + 10^{\frac{-\mu_{\text{NLOS},k}}{10}} P_{\text{NLOS},k}\right] \left(\frac{4\pi f_c d_k}{c}\right)^{-n}. \quad (5.8)$$

Note that $Q(.)$ is the Q function.

Proof We present this proof from [168] in order to provide the readers with a step-by-step discussion on how to show such fundamental results on UAV BS deployment. First, we note that the downlink coverage probability for a ground user while considering the mean interference between UAV BSs can be derived as follows:

$$P_{\text{cov}} = \mathbb{P}\left[\frac{P_{r,j}}{N + \bar{I}} \geq \beta\right] = \mathbb{P}\left[P_{r,j}(dB) \geq P_{\min}\right]$$
$$= P_{\text{LOS},j} \mathbb{P}\left[P_{r,j}(\text{LOS}) \geq P_{\min}\right] + P_{\text{NLOS},j} \mathbb{P}\left[P_{r,j}(\text{NLOS}) \geq P_{\min}\right]$$
$$\stackrel{(a)}{=} P_{\text{LOS},j} \mathbb{P}\left[\psi_{\text{LOS}} \leq P_t + G_{\text{3dB}} - P_{\min} - L_{dB}\right]$$
$$+ P_{\text{NLOS},j} \mathbb{P}\left[\psi_{\text{NLOS}} \leq P_t + G_{\text{3dB}} - P_{\min} - L_{dB}\right]$$
$$\stackrel{(b)}{=} P_{\text{LOS},j} Q\left(\frac{P_{\min} + L_{dB} - P_t - G_{\text{3dB}} + \mu_{\text{LOS}}}{\sigma_{\text{LOS}}}\right)$$
$$+ P_{\text{NLOS},j} Q\left(\frac{P_{\min} + L_{dB} - P_t - G_{\text{3dB}} + \mu_{\text{NLOS}}}{\sigma_{\text{NLOS}}}\right), \quad (5.9)$$

where $\mathbb{P}[.]$ is the probability notation, and $P_{\min} = 10\log(\beta N + \beta \bar{I})$. Clearly, due to the use of directional antennas, interference received from the nearest UAV k is dominant. Hence, $\bar{I}$ can be written as:

$$\bar{I} \approx P_{\text{LOS},k}\mathbb{E}\left[P_{r,k}(\text{LOS})\right] + P_{\text{NLOS},k}\mathbb{E}\left[P_{r,k}(\text{NLOS})\right]$$

$$= P_t g(\varphi_k)\left[10^{\frac{-\mu_{\text{LOS}}}{10}} P_{\text{LOS},k} + 10^{\frac{-\mu_{\text{NLOS}}}{10}} P_{\text{NLOS},k}\right]\left(\frac{4\pi f_c d_k}{c}\right)^{-n}.$$

where $\mathbb{E}[.]$ is the expectation function taken over the received interference power. (a) follows from (7.1), and (b) is the result of the complementary cumulative distribution function (CCDF) of a Gaussian random variable. Finally, we use $r \leq h.\tan(\theta_B/2)$, which shows a user is covered by a UAV BS when it is located within its coverage beam. This proves the theorem. □

From Theorem 5.1, we can see that increasing a UAV BS's height results in a higher path loss and LOS probability, thus a longer coverage radius. Moreover, by increasing the number of UAV BSs, the interference stemming from the nearest UAV BS increases since the UAV BSs are placed closer to each other. The coverage radius of each UAV is defined as r_u, which essentially represents the maximum range within which the coverage probability for a ground user exceeds a given threshold (ϵ). The coverage radius of each UAV BS is a function of the antenna beamwidth, the transmit power, the coverage threshold, the number of UAV BSs, as well as the locations of UAV BSs. Mathematically, a UAV BS's coverage radius is expressed by:

$$r_u = \max\{r|P_{\text{cov}}(r, P_t, \theta_B) \geq \varepsilon\}. \tag{5.10}$$

Using the result in (5.10), we now wish to see how to deploy UAV BSs such that the total coverage is maximized, while avoiding any overlap between the UAV BSs' coverage areas. Moreover, UAV BSs use a minimum transmit power so as to maximize their coverage lifetime. Then, we can formally pose our UAV placement problem:

$$(\vec{r}_j^*, h^*, r_u^*) = \underset{i\in\{1,...,M\}}{\arg\max} M.r_u^2, \tag{5.11}$$

$$\text{st. } ||\vec{r}_j - \vec{r}_k|| \geq 2r_u, \ j \neq k \in \{1, ..., M\}, \tag{5.12}$$

$$||\vec{r}_j + r_u|| \leq R_c, \tag{5.13}$$

$$r_u \leq h.\tan(\theta_B/2), \tag{5.14}$$

where R_c is the radius of the considered circular area, and M is the number of UAV BSs. $\vec{r}_j$ represents the 2D position of UAV BS j projected on the geographical area, and r_u shows the maximum UAV's coverage radius. (5.12) guarantees that overlap between the coverage areas of the UAV BSs is avoided, and (5.13) ensures that UAV BSs only cover the inside of the desired geographical area.

Solving (5.11) is a complex and challenging task given the high number of unknowns as well as the highly nonlinear nature of the optimization problem. We tackle (5.11) using the notion of a *circle packing problem* [180] discussed in the previous section. In this problem, multiple circles need to be placed inside a given plane in a way that packing density with these non-overlapping circles is maximized. In Figure 5.4, we illustrate the optimal packing of three identical circles within a big circle. In Table 5.1, we list the radius of each small circle for which the maximum packing density is achieved when placing multiple small circles inside a given big circular area [180]. Clearly, the radius

Table 5.1 Covering a circular area with radius R_c using identical UAVs – the *circle packing in a circle* approach.

Number of UAV BSs	Coverage radius of each UAV BS	Maximum total coverage
1	R_c	1
2	$0.5R_c$	0.5
3	$0.464R_c$	0.646
4	$0.413R_c$	0.686
5	$0.370R_c$	0.685
6	$0.333R_c$	0.666
7	$0.333R_c$	0.778
8	$0.302R_c$	0.733
9	$0.275R_c$	0.689
10	$0.261R_c$	0.687

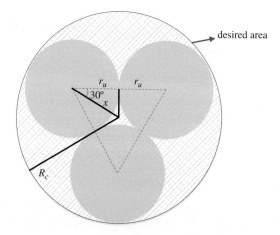

Figure 5.4 Packing problem in a circle with three circles.

of each circle decreases as the number of circles increases. Note that the total coverage corresponds to the packing density, which is the maximum portion of the circular area covered by small circles. As an example, we derive the optimal packing method for $M = 3$. Consider a circular area with radius R_c. To maximize the packing density, every two small circles must be tangent to each other and all circles be bounded by the big circle in Figure 5.4. In this case, the small circles' centers are on the vertices of an equilateral triangle. Considering Figure 5.4, we have $x = \frac{r_u}{\cos(30^\circ)}$ and

$$R_c = r_u + x = r_u \left(1 + \frac{2}{\sqrt{3}}\right) \rightarrow r_u = \frac{\sqrt{3}R_c}{2 + \sqrt{3}} \approx 00.464R_c.$$

Here, each circle (or disk) represents the coverage area of each UAV BS. Then the total coverage can be maximized by maximizing the packing density. This corresponds to the problem of maximizing the coverage area with non-overlapping smaller circles. Subsequently, based on the size of the given geographical area and the number of UAV BSs, one can find the 3D UAV BSs' positions along with their coverage radius. Meanwhile, the height of the UAV BSs is related to the beamwidth and the coverage radius by $h = \frac{r_u}{\tan(\theta_B/2)}$.

5.2.3 Representative Simulation Results

The following simulation results are based on $f_c = 2\,\text{GHz}$, $\alpha = 0.6$, $\gamma = 0.11$, $k_1 = 10.39$, $k_2 = 0.05$, $g_1 = 29.06$, $g_2 = 0.03$, $\mu_{\text{LOS}} = 1\,\text{dB}$, $\mu_{\text{NLOS}} = 20\,\text{dB}$, and $n = 2.5$ [157]. Also, we consider $\epsilon = 0.80$, $\beta = 5$, and $N = -120\,\text{dBm}$.

Figure 5.5 examines the optimal altitude of the UAV BSs as a function of their number. We can observe that the UAV BSs' height decreases by increasing their number. To deploy more UAV BSs, they should be deployed at a lower altitude in order to avoid potential overlap (and hence interference) between their coverage areas. As we can see from Figure 5.5, while increasing the number of UAV BSs from 3 to 6, the UAV BSs' altitude decreases from 2000 m to 1300 m. Moreover, this figure shows that the UAV BSs are placed at lower altitudes when using directional antennas with higher beamwidths.

Figure 5.6 illustrates the minimum number of UAV BSs needed to meet a given coverage requirement for ground users within a geographical area. The coverage threshold represents the portion of the entire area that must be covered by multiple UAV BSs. Here, we consider $P_t = 35\,\text{dBm}$ and $\theta_B = 80°$. From Figure 5.6, we can see that, in order to meet a 0.7 coverage requirement, we can deploy one or more than 6 UAV BSs. Note that the minimum number of required UAV BSs depends on the size of the target geographical area. For instance, for $R_c < 5400\,\text{m}$, deploying one UAV BS meets a 0.6 coverage requirement, however, more UAV BSs will be required to cover a larger area.

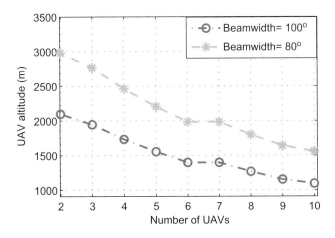

Figure 5.5 Drone altitude versus number of UAV BSs.

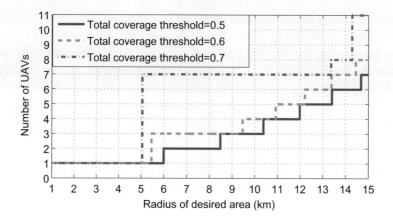

Figure 5.6 Number of required UAV BSs versus radius of the ground area that must be covered.

Hence, the minimum number of UAV BSs needed to provide a desired coverage is a function of the coverage threshold as well as the size of the geographical area.

5.2.4 Summary

In summary, we have introduced an efficient framework that can be employed for optimal placement of UAV BSs in order to deliver wireless connectivity to a given geographical area. We first have derived the downlink coverage probability for ground users. Next, by exploiting circle packing theory, we have presented a framework for 3D placement of identical UAVs to provide a maximum wireless coverage with minimum transmit power. In particular, we have described key design aspects of UAV BS deployment considering drone altitude, transmit power, antenna bandwidth, and the number of UAV BSs.

5.3 Deployment of UAV BSs for Energy-Efficient Uplink Data Collection

As discussed in Chapter 2, UAV BSs can play an important role in the IoT, particularly when considering IoT applications in which the IoT devices are small, battery-constrained devices, including radio frequency identification devices (RFIDs) and sensors [28, 29]. These low-power devices may not be able to have long-range communication for sending their data [29]. In this case, flying UAV BSs can be intelligently used to effectively collect IoT data from ground devices.

Although the operations of a UAV BS for such an IoT data collection application were discussed in detail in Chapter 2, in this section, we develop a framework for optimizing the 3D placement and mobility of UAV BSs to enable energy-efficient uplink communications for terrestrial IoT devices. Here, UAV BSs are used to successfully collect data from IoT devices in an energy-efficient manner. In fact, by optimizing the deployment and locations of UAV BSs along with the device-UAV BS association rule and transmit power of each IoT device, the wireless IoT system can guarantee reliable uplink communications for its devices, while also minimizing the total power (hence operating in an energy-efficient manner). Building on our work in [109], for this IoT data collection use case, we study two scenarios for the use of UAV BSs: (1) static case, in which the set of active IoT devices does not change, and (2) dynamic case, in which a time-varying activation process is considered for the ground IoT devices.

5.3.1 System Model and Problem Formulation

For our model, we investigate an IoT system that encompasses a set $\mathcal{L}$ of L ground (low-power) IoT devices. Moreover, a set $\mathcal{K}$ of K flying UAV BSs are utilized in order to collect data from IoT devices using uplink communication links. In the considered system, an IoT device is served by a UAV BSs if its uplink SINR exceeds a predefined threshold. We consider an FDMA multiple access scheme with R orthogonal channels. We use $E_{\max}$ to designate each UAV BS's energy used for mobility. The positions of device $i \in \mathcal{L}$ and UAV BS $j \in \mathcal{K}$ are (x_i, y_i) and $v_j = (x_j^{\text{uav}}, y_j^{\text{uav}}, h_j)$, as illustrated in Figure 5.7. Note that we consider a cloud server for managing the UAV BSs' positions, the device-UAV BS cell association, as well as each IoT device's transmit power.

Here, the IoT network is analyzed within a time interval $[0, T]$ during which the devices can be activated. In this interval, the locations of the UAV BSs and device-UAV BS associations are updated according to the positions of active devices. We use the term *update time* to represent the time instances for updating UAV BSs' locations and associations. The update times are denoted by t_n, $1 \leq n \leq N$ shows update time n, where N is the number of updates. Each IoT device that becomes active within $[t_{n-1}, t_n)$ is serviced by the UAV BSs during $[t_n, t_{n+1})$. These update times are design parameters that depend on the devices' activation. Given this UAV BS-IoT network, our goal is to

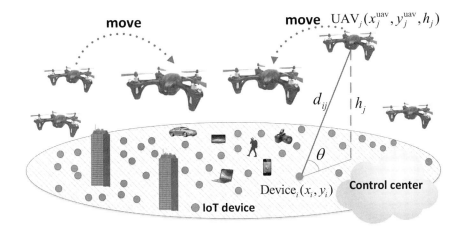

Figure 5.7 Model for an IoT application in which UAV BSs are engaged in uplink data collection of IoT device data.

find the optimal 3D drones' positions and device associations at each update time t_n such that the total devices' transmit power is minimized while satisfying SINR requirements of all devices. In addition, we present a framework for finding the update times and the drones' mobility in the considered time-varying IoT network.

5.3.2 Ground-to-Air Channel Model

For ground-to-air communications, we consider the probabilistic path loss model, described in Section 3.3.6 of Chapter 3, with the following LOS probability: [31, 158, 170]:

$$P^{ij}_{\text{LOS}} = \frac{1}{1 + \psi \exp(-\beta [\theta_{ij} - \psi])}, \quad (5.15)$$

where ψ and β are a function of the carrier frequency and environment. θ_{ij} represents the elevation angle, $\theta = \frac{180}{\pi} \times \sin^{-1}\left(\frac{h_j}{d_{ij}}\right)$, where $d_{ij} = \sqrt{(x_i - x_j^{\text{uav}})^2 + (y_i - y_j^{\text{uav}})^2 + h_j^2}$ is the distance between UAV BS j and device i.

Now, the path loss between device i and UAV BS j is [158]:

$$L_{ij} = \begin{cases} \eta_1 \left(\frac{4\pi f_c d_{ij}}{c}\right)^\alpha, & \text{LOS link,} \\ \eta_2 \left(\frac{4\pi f_c d_{ij}}{c}\right)^\alpha, & \text{NLOS link,} \end{cases} \quad (5.16)$$

where α is the path loss exponent, f_c is the carrier frequency, η_1 and η_2 are the excessive path loss coefficients for LOS and NLOS links, and c is the light's speed. Also, $P^{ij}_{\text{NLOS}} = 1 - P^{ij}_{\text{LOS}}$.

Subsequently, the average path loss between device i and UAV BS j will be:

$$\bar{L}_{ij} = P^{ij}_{\text{LOS}} \eta_1 \left(\frac{4\pi f_c d_{ij}}{c}\right)^\alpha + P^{ij}_{\text{NLOS}} \eta_2 \left(\frac{4\pi f_c d_{ij}}{c}\right)^\alpha = \left[P^{ij}_{\text{LOS}} \eta_1 + P^{ij}_{\text{NLOS}} \eta_2\right] (K_o d_{ij})^\alpha, \quad (5.17)$$

where $K_o = \frac{4\pi f_c}{c}$.

5.3.3 Activation Model of IoT devices

IoT devices can be active based on the services they provide. For example, IoT devices may periodically report their data in weather monitoring and smart grid applications. In contrast, in health monitoring applications, IoT devices can be randomly activated. In such time-varying IoT systems, UAV BSs should be dynamically deployed for data collection according to the activation process of IoT devices. Naturally, the optimal locations of the UAV BSs and their update times depend on the activation process of the IoT devices. Here, we focus on the case of random activation of IoT devices. In the random activation scenario, the simultaneous transmissions of massive IoT devices within a short period of time can result in bursty traffic [183]. In order to capture such traffic characteristics, 3GPP suggests the use of a *beta* distribution for the activation of

IoT devices [184]. In this model, each IoT device becomes active at time $t \in [0, T]$ based on the beta distribution [184–186]:

$$f(t) = \frac{t^{\kappa-1}(T-t)^{\omega-1}}{T^{\kappa+\omega-1}B(\kappa,\omega)}, \quad (5.18)$$

where κ and ω are the beta distribution's parameters, $[0, T]$ is the activation time interval of IoT devices, and $B(\kappa, \omega) = \int_0^1 t^{\kappa-1}(1-t)^{\omega-1} dt$ represents the beta function [187].

In the following, for each update time t_n, we present a joint optimization problem that can be used to determine the 3D positions of UAV BSs, the cell association rule between IoT devices and UAV BSs, and the transmit power of all active IoT devices:

(OP):

$$\min_{\boldsymbol{v}_j, \boldsymbol{c}, \boldsymbol{P}} \sum_{i=1}^{L_n} P_i, \quad \forall i \in \mathcal{L}_n, \forall j \in \mathcal{K}, \quad (5.19)$$

$$\text{s.t.} \quad \frac{P_i \bar{g}_{ic_i}(\boldsymbol{v}_{c_i})}{\sum_{k \in \mathcal{Z}_i} P_k \bar{g}_{kc_i}(\boldsymbol{v}_{c_i}) + \sigma^2} \geq \gamma, \quad (5.20)$$

$$0 < P_i \leq P_{\max}, \quad (5.21)$$

where $\mathcal{L}_n$ is the set of IoT devices' indices at t_n, and L_n is the number of active devices. $\boldsymbol{P}$ represents a transmit power vector whose each element P_i is the transmit power of device i. The 3D location of UAV BS j is denoted by $\boldsymbol{v}_j$. $\boldsymbol{c}$ is a vector of device-drone associations whose each element c_i is the index of the UAV BS, which is associated with device i. The maximum transmit power of each IoT device is limited to $P_{\max}$, the noise power is denoted by σ^2, and $\bar{g}_{ic_i}(\boldsymbol{v}_{c_i})$ represents the average channel gain between device i and its associated UAV BS. Moreover, $\bar{g}_{kc_i}(\boldsymbol{v}_{c_i})$ shows the average channel gain between interfering device k and UAV BS c_i. $\mathcal{Z}_i$ represents the set of interferer devices that transmit over the same channel as device i. γ is the SINR threshold for IoT devices, and (5.21) captures the maximum transmit power constraint for IoT devices. Here, **(OP)** is referred to as the *original problem*. It should be noted that solving (5.19) is a challenging task. First, the optimization variables are mutually dependent. Second, the optimization problem is highly nonlinear and non-convex. In the sequel, a practical framework for solving **(OP)** is presented.

In Figure 5.8, we illustrate a block diagram of the key steps needed for solving the original optimization problem at each update time. In the first step, for fixed locations of the UAV BSs, the device-UAV BS associations and transmit power of each IoT device are optimized. In the second step, considering fixed device-UAV BS associations, the UAV BSs' positions and the transmit power of devices are determined. These steps are performed iteratively until the solution converges.

5.3.4 UAV BS Placement and Device Association with Power Control

We seek to minimize the total IoT devices' transmit power by optimizing the UAV BSs' locations, device-UAV BS associations, and uplink transmit power of each IoT devices. To this end, **(OP)** is decomposed into two subproblems that need to be solved separately.

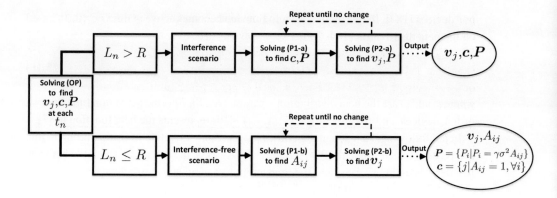

Figure 5.8 Block diagram for the proposed solution.

Device Association and Power Control

Given the positions of the UAV BSs, **(OP)** can be represented by:

(P1-a):

$$\min_{c,P} \sum_{i=1}^{L_n} P_i, \quad \forall i \in \mathcal{L}_n, \forall j \in \mathcal{K}, \tag{5.22}$$

$$\text{s.t.} \quad \frac{P_i \bar{g}_{ic_i}}{\sum_{k \in \mathcal{Z}_i} P_k \bar{g}_{kc_i} + \sigma^2} \geq \gamma, \tag{5.23}$$

$$0 < P_i \leq P_{\max}. \tag{5.24}$$

By solving **(P1-a)**, the device-UAV BS associations as well as the transmit power of each active IoT device are determined. Given the SINR constraints, the feasibility of this optimization problem is affected by the UAV BSs' locations. Next, we provide upper and lower bounds on the height of UAV BS j that serves device i (the proof is based on [109]).

PROPOSITION 2   The lower bound and upper bound for the height of drone j for successfully serving device i, are:

$$d_{ij} \sin\left(\frac{1}{\beta} \ln\left(\frac{\psi Q}{1-Q}\right) + \psi\right) \leq h_j \leq \left(\frac{P_{\max}}{\gamma K_o^\alpha \sigma^2 \eta_1}\right)^{1/\alpha}, \tag{5.25}$$

where d_{ij} is the distance between UAV j and device i, and $Q = \frac{P_{\max}}{\gamma d_{ij}^\alpha K_o^\alpha \sigma^2(\eta_1-\eta_2)} - \frac{\eta_2}{\eta_1-\eta_2}$.

Considering fixed locations for the UAV BSs, problem **(P1-a)** can be transformed to the classical joint user association and uplink power control problem in cellular networks. Hence, we can use the algorithm proposed in [188] and [189] to find the joint optimal user-UAV BS association and uplink power control given the SINR requirement and the device's transmit power constraint. In **(P1-a)**, the IoT devices can be considered terrestrial users, and the UAVs are obviously acting as BSs. Algorithm 1 provides the detailed steps for solving **(P1-a)**. In step 3, an initial value for active devices' transmit power is considered. In step 4, $\rho_{ij}^{(t)}$ is computed at iteration t. Then, based on step 5, we compute the minimum transmit power of device i when connecting to its serving UAV BSs. Then, the index of the best UAV BSs, which is assigned to device i, is given in step 6. In step 7, the transmit power of device i is updated so as to meet SINR threshold γ. Finally, for all IoT devices, steps 4–7 are repeated until the optimal solution is achieved.

Algorithm 1 Iterative algorithm for joint power control and device-UAV BSs association

---

1: **Inputs:** Locations of UAV BSs and IoT devices
2: **Outputs:** Device association vector (c), and transmit power of all IoT devices (P).
3: Set $t = 0$, and initialize $\boldsymbol{P}^{(0)} = \left(P_1^{(0)}, ..., P_K^{(0)}\right)$.
4: Define $\rho_{ij}^{(t)} = \dfrac{\sigma^2 + \sum\limits_{k \in \mathcal{Z}_i} P_k^{(t)} \bar{g}_{kj}}{\bar{g}_{ij}}$.
5: Find $S_i(\boldsymbol{P}^{(t)}) = \min\limits_{j \in \mathcal{K}} \rho_{ij}^{(t)}$.
6: Compute $c_i(\boldsymbol{P}^{(t)}) = \arg\min\limits_{j \in \mathcal{K}} \rho_{ij}^{(t)}$.
7: Update $P_i^{(t+1)} = \min\left\{\gamma S_i(\boldsymbol{P}^{(t)}), P_{\max}\right\}, \forall i \in \mathcal{L}_n$.
8: Repeat steps 4 to 7 for all IoT devices until $\boldsymbol{P}^{(t)}$ converges.
9: $\boldsymbol{P} = \boldsymbol{P}^{(t)}, \boldsymbol{c} = \left[c_i(\boldsymbol{P}^{(t)})\right], \forall i \in \mathcal{L}_n$.

For any given locations of UAV BSs, the solution of **(P1-a)** provides the optimal devices' transmit powers as well as the device-UAV BS associations. These results are inputs to the second subproblem that optimizes the 3D positions of UAV BSs.

Optimizing the Deployment Locations of UAV BSs

Now, we aim to optimize the 3D locations of UAV BSs in order to minimize the total transmit power of IoT devices in uplink.

By fixing the device-UAV BS associations, the UAV BSs' locations and transmit powers of IoT devices can be found by solving the following optimization problem:

(P2-a):

$$\min_{\boldsymbol{v}_j, \boldsymbol{P}} \sum_{i=1}^{L_n} P_i, \quad \forall i \in \mathcal{L}_n, \forall j \in \mathcal{K}, \tag{5.26}$$

$$\text{s.t.} \quad \frac{P_i \bar{g}_{ij}(\boldsymbol{v}_j)}{\sum\limits_{k \in \mathcal{Z}_i} P_k \bar{g}_{kj}(\boldsymbol{v}_j) + \sigma^2} \geq \gamma, \tag{5.27}$$

$$0 < P_i \leq P_{\max}, \tag{5.28}$$

where $\boldsymbol{v}_j = (x_j^{\text{uav}}, y_j^{\text{uav}}, h_j)$ is the position of UAV BS j. This problem is difficult to address because of its non-convexity.

To solve (**P2-a**), we separately optimize the position of each UAV BS. First, we optimize the location of each UAV BS based on the locations of its associated devices. Next, the transmit power (i.e., P_i^*) of each associated device is updated based on the new position of its serving UAV BS. Therefore, we update the UAV BS's location as well as the transmit power of its devices. After computing P_i^* at each iteration, $P_{\max} = P_i^*$ is considered for the next iteration. This guarantees that the devices' transmit powers do not increase in multiple iterations. Note that this procedure must be done for all UAV BSs one after another, until convergence is achieved.

Considering a single UAV BS j and the set of its associated IoT devices $\mathcal{C}_j$, we need to solve the following optimization problem:

$$\min_{\mathbf{v}_j} \sum_{i \in \mathcal{C}_j} F_i(\mathbf{v}_j), \tag{5.29}$$

$$\text{s.t.} \quad F_i(\mathbf{v}_j) = \gamma \left(\eta_1 P_{\text{LOS}}^{ij} + \eta_2 P_{\text{NLOS}}^{ij} \right) (K_o d_{ij})^\alpha$$

$$\left[\sum_{k \in Z_i} \frac{P_k}{\left(\eta_1 P_{\text{LOS}}^{kj} + \eta_2 P_{\text{NLOS}}^{kj} \right) (K_o d_{kj})^\alpha} + \sigma^2 \right], \tag{5.30}$$

$$F_i(\mathbf{v}_j) \leq P_i^*, \quad \forall i \in \mathcal{C}_j, \tag{5.31}$$

In order to efficiently solve (5.29), we transform the problem into a quadratic programming form. More formally, we use the sequential quadratic programming (SQP) technique, which is suitable for tackling differentiable and large-scale non-linear optimization problems [190]. In this case, we linearize the constraints of the optimization problem and approximate the objective function by a quadratic function. Then, several quadratic subproblems are solved in order to solve the original optimization problem.

Up until now, we analyzed the IoT network at one update time during $[0, T]$. Next, we consider a time-varying IoT network in $[0, T]$ time period in which the activation pattern of IoT devices changes and drones dynamically update their positions in different update times.

5.3.5 Update Time Analysis

Here, we analyze the impact of update times on the IoT data collection performance. Naturally, the mobility of UAV BSs and update times are affected by the activation pattern of IoT devices. Considering the fact that the set of active IoT devices varies over time, UAV BSs must dynamically update their positions while collecting IoT data. Note that the locations of UAV BSs are updated at specific update times. At different update times, the number of active IoT devices can be different. The required transmit power IoT devices and the energy consumption of drones for their movements depend on the number of update times.

Increasing the number of updates will yield a shorter time duration between each two consecutive updates. Therefore, fewer active IoT need to be served at each update time.

With fewer number of active devices, the uplink interference stemming from IoT devices will be lower. Hence, IoT devices will require lower transmit power to send their data to UAV BSs while satisfying their SINR constraint. Nevertheless, increasing the number of updates leads to more movements and higher energy consumption for the UAV BSs. In a probabilistic activation model of IoT devices (e.g., in health monitoring applications), each IoT device can be active at time $t \in [0, T]$ according to the beta distribution. In this case, we can state the following theorem from [109], which derives the relation between update times and the average number of active IoT devices at each update time.

THEOREM 5.2   The average number of active IoT devices, a_n, at update time t_n is given by:

$$t_n = T \times I^{-1}\left(\frac{a_n}{L} + I_{\frac{t_{n-1}}{T}}(\kappa, \omega), \kappa, \omega\right), \quad n > 1, \tag{5.32}$$

$$t_1 = T \times I^{-1}\left(\frac{a_1}{L}, \kappa, \omega\right), \tag{5.33}$$

where $I_x(.)$ and $I^{-1}(.)$ are, respectively, the regularized incomplete beta function and inverse of it. L represents the number of IoT devices in the network. Also, $[0, T]$ is an interval that represents the entire time period within which the IoT devices can be activated.

As we can see from 5.2, we need to find the update times according to the activation of the IoT devices. In this case, t_n is a function of the total number of IoT devices as well as their activation pattern. Another observation is that the update times that the UAV BSs adopt must be adjusted based on the number of active IoT devices at each update time. In fact, the number of update times is a design parameter that impacts the transmit power of IoT devices, interference between devices, and energy consumption of the UAV BSs.

5.3.6 Representative Simulation Results

We simulate an IoT system with 500 devices on a 1 km × 1 km square geographical area. We assume that these IoT devices are uniformly distributed over the studied area. Other simulation parameters are $\psi = 11.95$ and $\beta = 0.14$, and $f_c = 2$ GHz [158]. In Table 5.2, we show the various parameters that we used in our simulation scenarios. For update time analysis, we consider $\kappa = 3$, and $\omega = 4$ for the beta distribution [184]. For benchmark comparison, we consider stationary UAV BSs whose locations are predetermined, and they are not dynamically optimized based on the locations of active IoT devices.

In Figure 5.9, we show a snapshot of the 3D locations of the deployed UAV BSs as well as their associated IoT devices. We consider five UAV BSs that collect data from 100 IoT devices. Here, the UAV BSs deployment strategy and the device-UAV BS associations are determined such that the total transmit that the IoT devices select for sending their uplink data to their serving UAV BS is minimized.

Given the SINR and the maximum transmit power constraints, it may not be possible to serve all IoT devices in the network. Therefore, we evaluate the system's reliability,

Table 5.2 Simulation parameters for our IoT system with UAV BSs.

Parameter	Description	Value
P_{max}	Maximum transmit power of each device	200 mW
α	Path loss exponent for LOS links	2
σ^2	Noise power	-130 dBm
γ	SINR threshold	5 dB
L	Total number of IoT devices	500
η_1	Additional path loss to free space for LOS	3 dB
η_2	Additional path loss to free space for NLOS	23 dB

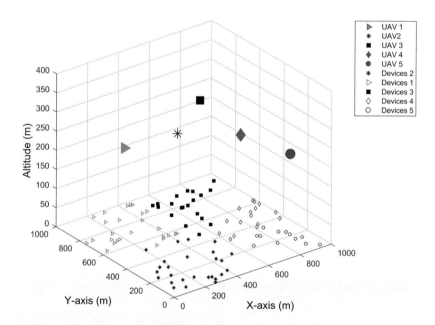

Figure 5.9 Positions of UAV BSs and their associated ground IoT devices for one illustrative snapshot of our considered simulation setting.

which represents the probability of serving all IoT devices. Figure 5.10 shows the reliability versus P_{max} in the considered IoT network with UAV BSs. As we can observe from Figure 5.10, the wireless network reliability increases by increasing P_{max}. With a higher P_{max}, each device will have a higher possibility to send its data to a UAV BS. Figure 5.10 shows that the proposed approach results in a higher reliability compared to the stationary case in which UAV BSs are pre-deployed. In fact, we can achieve a higher reliability by optimizing the deployment positions of the UAV BS according to the IoT devices' positions. For example, Figure 5.10 demonstrates that our introduced solution can improve the reliability by 28% compared to the benchmark stationary case.

5.3 Deployment of UAV BSs for Energy-Efficient Uplink Data Collection

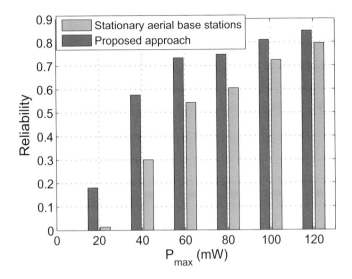

Figure 5.10 Reliability comparison between our optimized UAV BS deployment approach and pre-deployed stationary UAV BSs using 5 UAVs.

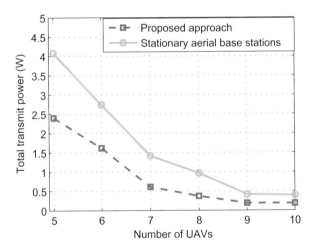

Figure 5.11 Total transmit power of devices vs. number of UAV BSs in the presence of interference.

Figure 5.11 illustrates how the total transmit power of IoT devices will vary as the number of UAV BSs changes. As expected, by using more UAV BSs for data collection, the power consumption of the IoT devices can be decreased. In this example, for 100 IoT devices and 20 channels, using our solution approach, the devices' total transmit power decreases by 91% while deploying 10 UAV BSs compared with 5 UAV BSs. Also, compared to the stationary deployment scenario, our presented approach leads to a 45% lower IoT devices' transmit power.

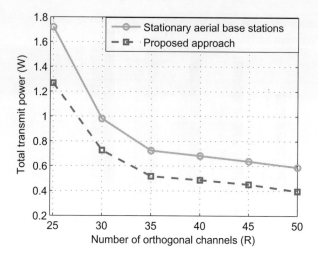

Figure 5.12 Total transmit power of devices vs. number of orthogonal channels.

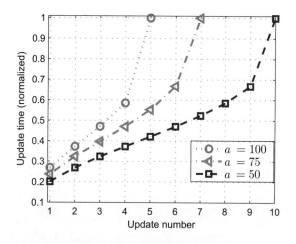

Figure 5.13 Update times for different average number of active devices.

In Figure 5.12, we show the total transmit power of IoT devices as a function of the number of channels. By increasing the number of channels, the devices can reduce their transmit powers while sending their data to drones. In fact, more orthogonal channels leads to uplink lower interference between the IoT devices. Hence, to meet the SINR requirement, each device can use a lower transmit power. For instance, as shown in Figure 5.12, the IoT devices' total transmit power decreases by 68% when the number of orthogonal channels increases from 25 to 50.

In Figure 5.13, we examine how the average number of active devices, a, is related to the update times. By increasing the number of update times, the time interval between two consecutive updates decreases, thus fewer devices will be active. For instance, as shown in Figure 5.13, in order to have $a = 100$, the fifth update should happen at

5.3 Deployment of UAV BSs for Energy-Efficient Uplink Data Collection

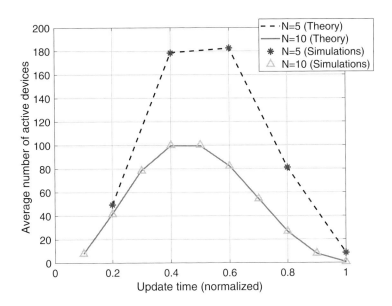

Figure 5.14 Average number of active devices at update times for the probabilistic activation.

$t_n = 0.41$. In this case, to decrease the average number of active devices from 100 to 50, the number of update times must be increased by a factor of two.

Figure 5.14 showcases the average number of active IoT devices that needs to be serviced by UAV BSs at each update time. Here, we normalized the update times by the entire activation duration, T. From this figure, we observe that, for $N = 10$, a_n decreases when the normalized update time is higher than 0.5. By increasing the number of update times, the UAV BSs will have to serve fewer IoT devices. For example, for $t_n = 0.6$, by increasing the number of updates from 5 to 10, the average number of active devices can decrease by 55%. This, in turn, reduces the interference between devices at the cost of more locations' updates and movements for the UAV BSs.

5.3.7 Summary

In summary, we have described an optimization-based method for an efficient deployment of UAV BSs for IoT data collection tasks whereby information must be transmitted from terrestrial IoT devices to the flying UAV BSs. We have identified the optimal locations at which the UAV BSs must be deployed along with the optimal device-UAV BSs associations and the associated uplink transmit power of each IoT device. The goal was to maximize performance while minimizing the total transmit power consumption of the IoT devices. Moreover, we have analyzed the dynamic placement of UAV BSs for serving IoT devices in a time-varying IoT system. In this scenario, we have characterized the relationship between update times of UAV BSs and the number of active IoT devices based on the activation pattern of the devices.

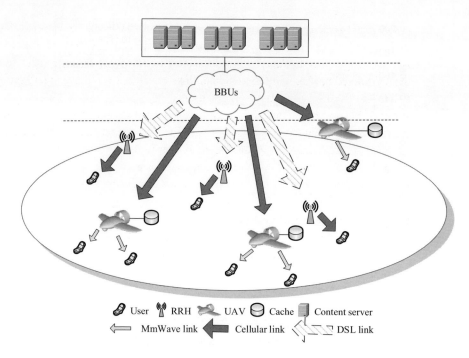

Figure 5.15 A CRAN with cache-enabled UAVs.

5.4 Proactive Deployment with Caching

In this section, we introduce a potential application of cache-enabled UAV BSs for capacity enhancement as well as traffic offloading in cellular networks. We particularly focus on a cellular architecture that relies on a cloud radio access network (CRAN) [191] in which a central cloud that learns a variety of users' information is considered. We show how the concept of networking with human-in-the-loop can be exploited for designing CRANs that are assisted by UAV BSs. In particular, we describe a framework for deploying cache-enabled UAV BSs while maximizing the CRAN users' quality of experience (QoE). This framework, based on our work in [192], exploits user-centric information including users' mobility patterns and their content request distribution. To characterize QoE, various factors, such as transmission delay and the type of devices, are taken into account. To effectively deploy cache-enabled UAV BSs, the content request distribution along with the mobility patterns of ground users are predicted by a cloud center using the machine learning tools of recurrent neural networks. These predictions will be used for placing the UAV BSs that will assist the CRAN.

5.4.1 Model

We consider a CRAN that provides wireless services to a set $\mathcal{U}$ of U ground users using a set $\mathcal{R}$ of R remote radio heads (RRHs). Along with the terrestrial RRHs, a set $\mathcal{K}$ of K cache-enabled UAV BSs are used to service ground users. These UAV BSs use a different frequency from the terrestrial network. RRHs uses the cellular band, and they connect to the cloud's pool of the baseband units (BBUs) through fronthaul links. Also, through fiber backhaul links, the cloud is connected to the content servers. In the considered model, wireless fronthaul links for drones can create interference on the RRHs while transmitting to the ground users.

Let $\mathcal{N}$ be a set of N content that needs to be stored for the ground users with L being the size of each content. Let $\mathcal{C}_k$ be the set of C cached contents at the storage of UAV BS k. We assume that $C \leq N$ and $k \in \mathcal{K}$. A user's content is requested within time slot τ, and Δ_τ is the time slot duration. In this case, the UAV BSs' contents can be updated in every T time slots. The main notations of this section are listed in Table 5.3.

Ground Users' Mobility Model

For users' mobility, we consider a periodic model in which a user can regularly go to specific locations, such as to work premises or to a coffee shop. In the considered CRAN, every H time duration, BBUs gather information about the users' locations. Furthermore, mobile users move with a constant speed. The users' mobility pattern will be used to effectively deploy our caching-equipped UAV BSs for serving the ground users.

The UAV BS-user and RRH-user associations are determined based on the users' QoE requirements.

Transmission Model

Here, we present the transmission models for UAV BS-User, BBU-UAV BS, and RRH-user communications.

For a UAV BS-user link, we consider a probabilistic LOS/NLOS model described in Chapter 3, Section 3.3.6. In this case, the path loss for LOS and NLOS links between UAV BS k and ground user i at time t can be given by [193]:

$$l_{t,ki}^{\text{LOS}}\left(\mathbf{w}_{\tau,t,k}, \mathbf{w}_{\tau,t,i}\right) = \\ L_{FS}(d_0) + 10\mu_{\text{LOS}} \log\left(d_{t,ki}\left(\mathbf{w}_{\tau,t,k}, \mathbf{w}_{\tau,t,i}\right)\right) + \chi_{\sigma_{\text{LOS}}}, \quad (5.34)$$

$$l_{t,ki}^{\text{NLOS}}\left(\mathbf{w}_{\tau,t,k}, \mathbf{w}_{\tau,t,i}\right) = \\ L_{FS}(d_0) + 10\mu_{\text{NLOS}} \log\left(d_{t,ki}\left(\mathbf{w}_{\tau,t,k}, \mathbf{w}_{\tau,t,i}\right)\right) + \chi_{\sigma_{\text{NLOS}}}, \quad (5.35)$$

where $(x_{\tau,k}, y_{\tau,k}, h_{\tau,k})$ is location of drone k, and $\mathbf{w}_{\tau,t,k} = [x_{t,i}, y_{t,i}]$ represents the location of user i at time t. $L_{FS}(d_0) = 20 \log(d_0 f_c 4\pi/c)$ is the free-space propagation. d_0 and f_c are, respectively the reference path loss distance and the carrier frequency. Also, c is the speed of light, and $d_{t,ki}(\mathbf{w}_{\tau,t,k}, \mathbf{w}_{\tau,t,i}) = \sqrt{(x_{t,i} - x_{\tau,k})^2 + (y_{t,i} - y_{\tau,k})^2 + h_{\tau,k}^2}$ represents the drone-user distance. For LOS and NLOS communications, the path loss exponents are defined as μ_{LOS} and μ_{NLOS}. Also, $\chi_{\sigma_{\text{LOS}}}$ and $\chi_{\sigma_{\text{NLOS}}}$ are zero-mean Gaussian random variables. We can define the average path loss between UAV BS k and user i by [194]:

$$\bar{l}_{t,ki}\left(\mathbf{w}_{\tau,t,k}, \mathbf{w}_{\tau,t,i}\right) = \Pr\left(l_{t,ki}^{\text{LOS}}\right) \times l_{t,ki}^{\text{LOS}} + \Pr\left(l_{t,ki}^{\text{NLOS}}\right) \times l_{t,ki}^{\text{NLOS}}, \quad (5.36)$$

Table 5.3 List of notations.

Notation	Description	Notation	Description
U	Number of users	C	Number of contents stored at the cache storage of a UAV BS
K	Number of UAV BSs	F	Number of intervals in each time slot
R	Number of RRHs	H	Number of time slots to collect user mobility
P_R	Transmit power of RRHs	$P_{t,ki}$	Transmitted power of UAV BS or RRH
N	Number of contents	τ, Δ_τ	Time slot index, Time slot duration
$l_{t,ki}$	Path loss of UAVs-users	$d_{t,ki}$	Distance between RRHs or UAV BSs and users
$x_{\tau,k}, y_{\tau,k}, h_{\tau,k}$	Coordinates of UAV BSs	$\delta_{S_i,n}$	Rate requirement of device type
L_{FS}	Free-space path loss	d_0	Free-space reference distance
f_c	Carrier frequency	$l^F_{t,ki}$	Path loss of fronthaul links
μ_{LOS}, μ_{NLOS}	Path loss exponents	$\chi_{\sigma_{LOS}}, \chi_{\sigma_{NLOS}}$	Shadowing random variable
$\gamma^V_{t,ki}, \gamma^H_{t,ki}$	SINR of user i	$L^{LOS}_{t,k}, L^{NLOS}_{t,k}$	LOS/NLOS path loss from the BBUs to UAV k
t, Δ_t	Small interval, interval duration	$l^{LOS}_{t,k}, l^{NLOS}_{t,k}$	LOS/NLOS path loss from UAV k to users
c	Speed of light	$h_{t,ki}$	Channel gains between RRHs k and user i
$\bar{D}_{\tau,i,n}$	Delay	$C^F_{\tau,ki}$	Fronthaul rate of UAV or RRH k
$C^V_{\tau,ki}$	Rate of UAV BSs-user link	$C^H_{\tau,qi}$	Rate of RRH-user link
$Q_{\tau,i,n}$	QoE of each user i	T	Number of time slots for caching update
$x_{t,i}, y_{t,i}$	Coordinates of users	P_B	Transmit power of the BBUs

5.4 Proactive Deployment with Caching

with $\Pr\left(l_{t,ki}^{\text{LOS}}\right)$ $\Pr\left(l_{t,ki}^{\text{NLOS}}\right)$ being the LOS and NLOS probabilities (as defined in Chapter 3) between UAV BS k and user i.

Subsequently, the SNR for the UAV BS-user link will be:

$$\gamma_{t,ki}^{\text{V}} = \frac{P_{t,ki}}{10^{\bar{l}_{t,ki}(\mathbf{w}_{\tau,t,k},\mathbf{w}_{\tau,t,i})/10}\sigma^2}, \tag{5.37}$$

where σ^2 is the noise power, and $P_{t,ki}$ is the transmit power used by a UAV BS k while serving user i. We assume that the total bandwidth available for each UAV BS is B_V, which is equally divided among the associated users. The transmission rate between UAV BS k and user i is written by:

$$C_{\tau,ki}^{\text{V}} = \frac{1}{F_{\tau,i}} \sum_{t=1}^{F_{\tau,i}} \frac{B_V}{U_k} \log_2\left(1 + \gamma_{t,ki}^{\text{V}}\right), \tag{5.38}$$

where B_V is the total UAV BS bandwidth and U_k is the number of users served by this UAV BS. Also, the number of time intervals allocated to a CRAN user is defined as $F_{\tau,i}$.

For BBU-UAV BS communications, we adopt the following probabilistic LOS/NLOS channel model:

$$L_{t,k}^{\text{LOS}} = d_{t,ki}\left(\mathbf{w}_{\tau,t,k}, \mathbf{w}_{\tau,t,B}\right)^{-\beta}, \tag{5.39}$$

$$L_{t,k}^{\text{NLOS}} = \eta d_{t,ki}\left(\mathbf{w}_{\tau,t,k}, \mathbf{w}_{\tau,t,B}\right)^{-\beta}, \tag{5.40}$$

where $\mathbf{w}_{\tau,t,B} = [x_B, y_B]$ denotes the BBU's location. Also, β shows the path loss exponent.

Here, we consider E clusters of RRHs. In this case, the signal received by a user that connects to cluster q of RRHs can be expressed by:

$$\mathbf{b}_{t,q} = \sqrt{P_R} \mathbf{H}_{t,q} \mathbf{F}_{t,q} \mathbf{a}_{t,q} + \mathbf{n}, \tag{5.41}$$

where $\mathbf{H}_{t,q} \in \mathbb{R}^{U_q \times R_q}$ represents a path loss matrix. U_q is the number of users connected to the RRHs, R_q denotes the number of antennas for RRHs, and P_R the RRH's transmit power. $\mathbf{a}_{t,q} \in \mathbb{R}^{U_q \times 1}$ is the continent received by a user interval t, and $\mathbf{n}_{t,q} \in \mathbb{R}^{U_q \times 1}$ is the white noise component. In addition, $\mathbf{F}_{t,q} = \mathbf{H}_{t,q}^{\text{H}} \left(\mathbf{H}_{t,q} \mathbf{H}_{t,q}^{\text{H}}\right)^{-1} \in \mathbb{R}^{R_q \times U_q}$ represents the matrix used for beamforming [195].

Now, the SINR for user i is given by:

$$\gamma_{t,qi}^{\text{H}} = \frac{P_R \|\mathbf{h}_{t,qi} \mathbf{f}_{t,qi}\|^2}{\underbrace{\sum_{j=1, j\neq q}^{E} \sum_{u \in \mathcal{U}_j} P_R \|\mathbf{h}_{t,ji} \mathbf{f}_{t,ju}\|^2}_{\text{other cluster RRHs interference}} + \underbrace{P_B g_{t,Bi} d_{t,Bi}^{-\beta}}_{\text{wireless fronthaul interference}} + \sigma^2},$$

where $\mathcal{M}_j$ and $\mathcal{U}_j$ are, respectively, the sets of the RRHs and their associated users. Also, $\mathbf{h}_{t,qi} \in \mathbb{R}^{1 \times R_q}$ represents the channel gain of the RRHs-user link, $h_{t,ki} =$

$g_{t,ki}d_{t,ki}(x_i,y_i)^{-\beta}$, $g_{t,ki}$, and $d_{t,ki}(x_i,y_i) = \sqrt{(x_{t,k}-x_{t,i})^2 + (y_{t,k}-y_{t,i})^2}$ shows user-RRHs distance. Moreover, the beamforming vector is defined as $f_{t,qi} \in \mathbb{R}^{R_q \times 1}$. Subsequently, the transmission rate of RRHs when serving user i is:

$$C_{\tau,qi}^{\text{H}} = \frac{1}{F_{\tau,i}} \sum_{t=1}^{F_{\tau,i}} B\log_2\left(1 + \gamma_{t,qi}^{\text{H}}\right). \tag{5.42}$$

Problem Formulation

Consider the transmission between a UAV BS k positioned at $w_{\tau,t,k}$ and a user i positioned at coordinates $w_{\tau,t,i}$.

We can now present the problem formulation of efficient placement of cache-enabled UAV BSs to meet the user's QoE requirements with a minimum UAV BS transmit power. In particular, we proactively determine the optimal UAV BS-user associations, contents that must be cached at UAV BSs, as well as the 3D deployment positions of the UAV BSs.

$$\min_{\mathcal{C}_k, \mathcal{U}_{\tau,k}, w_{\tau,t,k}} \sum_{\tau=1}^{T} \sum_{k \in \mathcal{K}} \sum_{i \in \mathcal{U}_{\tau,k}} \sum_{t=1}^{F_{\tau,i}} P_{\tau,t,ki}^{\min}\left(w_{\tau,t,k}, \delta_{i,n}^R, n_{\tau,i}\right), \tag{5.43}$$

$$\text{s. t. } h_{\min} \leq h_{\tau,k}, k \in \mathcal{K}, \tag{5.43a}$$

$$m \neq j, m, j \in \mathcal{C}_k, \mathcal{C}_k \subseteq \mathcal{N}, k \in \mathcal{K}, \tag{5.43b}$$

$$0 < P_{\tau,t,ki}^{\min} \leq P_{\max}, i \in \mathcal{U}, k \in \mathcal{K}, \tag{5.43c}$$

where $P_{t,ki}^{\min}\left(w_{\tau,t,k}, \delta_{i,n}^R, n\right) = \left(2^{\delta_{i,n}^R U_k / B_V} - 1\right) \sigma^2 10^{\bar{l}_{t,ki}(w_{\tau,t,k}, w_{\tau,t,i})/10}$.

Also, $\delta_{i,n}^R$ represents the QoE requirement of each user, $n_{\tau,i}$ shows the user's content, and $h_{\min}$ is the minimum height for each drone.

We use $\mathcal{U}_{\tau,k}$ to denote the set of ground users assigned to UAV BS k.

5.4.2 Optimal Deployment and Content Caching for UAV BSs

The content request distribution and movement patterns of ground users can be predicted using the machine learning tools of echo state networks, which are essentially recurrent neural networks. More specifically, we will use a conceptor-based echo state network (ESN) approach [192]. A conceptor-based ESN separates the users' behaviors into several patterns and learns them independently. This, in turn, will significantly enhance the prediction's accuracy compared to classical ESN algorithms (the fundamentals of ESNs will be revisited and discussed in Chapter 6). In our setting, users that are not served by RRHs will be connected to the drones. The remaining users are clustered into K clusters, and each UAV BS provides service for one cluster. Here, we optimize the UAV BSs' deployment positions and determine the caching contents at each UAV BS.

Optimal Content Caching for UAV BSs

To find the optimal content caching, we first determine the user-UAV BS association using a K-mean clustering algorithm [196]. In this case, the ground users are grouped into K clusters and each cluster is served by one of the UAV BSs in the CRAN. Given the UAV BS-user association, the optimal set of contents to cache at the UAV BSs can be determined. The optimal contents to store at the UAV BS storage lead to a maximum reduction in the UAV BS's transmit power. The reduction of UAV BS transmit power is caused by the decrease of the delay requirement. We define vector $\boldsymbol{p}_{j,i} = [p_{j,i1}, p_{j,i2}, \ldots, p_{j,iN}]$ as the content request distribution of user i during period j that consists of H time slots. The optimal contents that will be stored at each UAV BS cache can be determined based on the following theorem that we proved in [192].

THEOREM 5.3 The optimal set of contents $\mathcal{C}_k$ to cache at each UAV BS k during period T is:

$$\mathcal{C}_k = \arg\max_{\mathcal{C}_k} \sum_{j=1}^{T/H} \sum_{\tau=1}^{H} \sum_{i \in \mathcal{U}_{\tau,k}} \sum_{n \in \mathcal{C}_k} \left(p_{j,in} \Delta P_{j,\tau,ki,n} \right), \tag{5.44}$$

where $\Delta P_{j,\tau,ki,n} =$

$$\begin{cases} P_{\tau,ki}^{\min}\left(C_{\tau,ki}^R\right)_{n \notin \mathcal{C}_k} - P_{\tau,ki}^{\min}\left(C_{\tau,ki}^R\right)_{n \in \mathcal{C}_k}, & C_{\tau,ki,n \notin \mathcal{C}_k}^R \geq \delta_{S_i,n}, \\ P_{\tau,ki}^{\min}\left(\delta_{S_i,n}\right)_{n \notin \mathcal{C}_k} - P_{\tau,ki}^{\min}\left(C_{\tau,ki}^R\right)_{n \in \mathcal{C}_k}, & \delta_{S_i,n} > C_{\tau,ki,n \notin \mathcal{C}_k}^R, \end{cases}$$

with $C_{\tau,ki,n}^R$ being the transmission delay of drone k at time slot τ while sending content n. Also, $P_{\tau,ki}^{\min}\left(w_{\tau,t,k}, C_{\tau,ki}^R, n\right)$ is represented by $P_{\tau,ki}^{\min}\left(C_{\tau,ki}^R\right)$.

Theorem 5.3 implies that for an equal fronthaul rate for the users, the optimal content to cache at the UAV BSs will be $\mathcal{C}_k = \arg\max_{\mathcal{C}_k} \sum_{j=1}^{T/H} \sum_{\tau=1}^{H} \sum_{i \in \mathcal{U}_{\tau,k}} \sum_{n \in \mathcal{C}_k} p_{j,in}$. Furthermore, Theorem 5.3 shows that the content caching is a function of the users' content request distribution and their cell association.

Optimal Deployment Locations for the UAV BSs

We aim to optimize the deployment locations of UAV BSs such that they can service their users while using a minimum transmit power.

In the following, from [192], we provide a closed-form solution for the optimal position of UAV BS k with respect to the users' locations.

THEOREM 5.4 The optimal position of UAV BS k at relatively low or high altitudes (compared to its horizontal distance to users) will be:

$$x_{\tau,k} = \frac{\sum_{i \in \mathcal{U}_{\tau,k}} \sum_{t=1}^{F_{\tau,i}} x_{t,i} \psi_{t,ki}}{\sum_{i \in \mathcal{U}_{\tau,k}} \sum_{t=1}^{F_{\tau,i}} \psi_{t,ki}}, y_{\tau,k} = \frac{\sum_{i \in \mathcal{U}_{\tau,k}} \sum_{t=1}^{F_{\tau,i}} y_{t,i} \psi_{t,ki}}{\sum_{i \in \mathcal{U}_{\tau,k}} \sum_{t=1}^{F_{\tau,i}} \psi_{t,ki}}, \tag{5.45}$$

Table 5.4 System parameters used for simulating our CRAN with cache-enabled UAV BSs.

Parameter	Value	Parameter	Value	Parameter	Value
F	1000	Y	0.13	P_B	30 dBm
X	11.9	N	25	P_R	20 dBm
$X_{\sigma\text{LOS}}$	5.3	H	10	$P_{\max}$	20 W
N_{tr}	1000	d_0	5 m	σ^2	-95 dBm
N_s	12	λ	0.01	$h_{\min}$	100 m
N_x	4	β	2	B	1 MHz
μ_{LOS}	2	μ_{NLOS}	2.4	$\delta_{S_i,n}$	5 Mbit/s
χ	15	ζ_1	0.5	f_c	38 GHz
$X_{\sigma\text{NLOS}}$	5.27	η	100	B_y	1 GHz
K	5	C	1	L	1 Mbit
T	120	ζ_2	0.5	N_w	1000

where $\psi_{t,ki} = \left(2^{\delta_{i,n}^R/B} - 1\right)\sigma^2 10^{(L_{FS}(d_0)+\chi_\sigma)/10}$ and $\sigma = \begin{cases} \sigma_{\text{NLOS}}, & \text{for case } a), \\ \sigma_{\text{LOS}}, & \text{for case } b). \end{cases}$

Theorem 5.4 allows finding the optimal location of UAV BS k when it is deployed at high or low altitudes. In order to solve the UAV BS's location optimization problem in a general case, an efficient learning algorithm can be used, [197] and [198], which provides a suboptimal solution.

5.4.3 Representative Simulation Results

For simulating the considered CRAN, we consider a circular geographical area of radius 500 m, 70 ground users, and 20 RRHs. We assume that users and RRHs are uniformly distributed on the considered area. Table 5.4 shows the main simulation parameters.

Here, we consider the following benchmark cases: (1) optimal algorithm, which uses precise information about the mobility of users and their content request distribution, (2) ESN algorithm proposed by [199], and (3) random caching with ESN algorithm proposed by [199] to predict content request distribution.

In Figure 5.16, we examine how the rate required for meeting the QoE requirement of ground users changes with the wireless fronthaul rate. Here, the black plot corresponds to the user-UAV BS link and the blue plot is for the user-BBUs link. Clearly, by increasing the fronthaul rate, the rate needed for satisfying the QoE of users decreases. Nonetheless, for the user-UAV BS link, the rate does not considerably change while increasing the fronthaul rate. In this case, caching at the drones allows for the reduction of the required rate for guaranteeing the QoE requirements of users for low values of the wireless fronthaul rate.

Figure 5.17 presents the total transmit power of UAV BSs resulting from various approaches as function of the number of CRAN users. As shown in this figure, compared to a classical baseline ESN algorithm (which can only predict one

5.4 Proactive Deployment with Caching

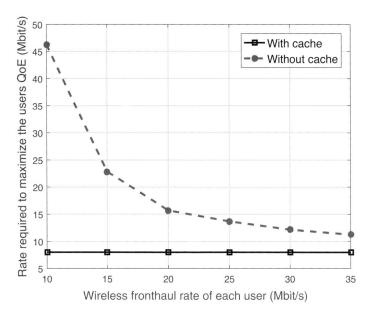

Figure 5.16 Rate required to maximize the users' QoE as the fronthaul rate of each user change.

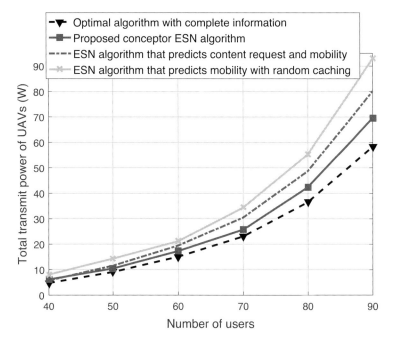

Figure 5.17 Total transmit power as the number of users varies ($K = 5$ and $C = 1$).

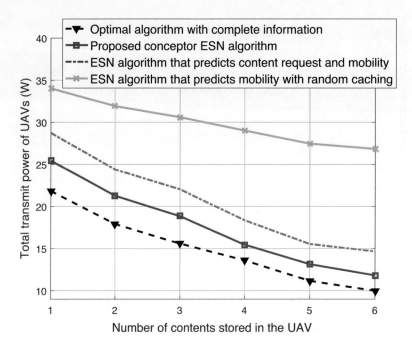

Figure 5.18 Total transmit power as the number of the contents stored in a UAV cache varies ($U = 70$ and $K = 5$).

nonlinear system and thus has lower accuracy compared to concocter-based ESN), the UAV BSs' transmit power can be decreased by 17% while using the conceptor-based ESN approach. This is due to the fact that the conceptor-based ESN uses more precise information about users' mobility than the ESN algorithm. This figure also shows that with the optimal deployment of cache-enabled UAV BSs, the transmit power of the drones can be reduced by 32% compared to the random caching case.

In Figure 5.18, we examine the total UAV BSs' transmit power versus the number of contents that are stored at the UAV BS cache. From Figure 5.18, we observe that increasing the number of storage units allows the UAV BSs to transmit with a lower power and, thus, reduce energy consumption. This is because the chance of storing the users' request contents at UAV BSs increases if more storage space is available. Therefore, the UAV BSs can effectively service their associated users while using a minimum transmit power.

Figure 5.20 shows the average minimum drones' transmit power as a function of the number of UAV BSs. By deploying more UAV BSs, the transmit power of each UAV BS can be reduced. For instance, increasing the number of UAV BSs from 3 to 7 results in an 86% lower transmit power for each drone. This is because when using more cache-enabled drones, each UAV BS will need to serve fewer CRAN users. Thus, the average transmit power that each UAV BS needs in order to communicate with the ground users can be reduced.

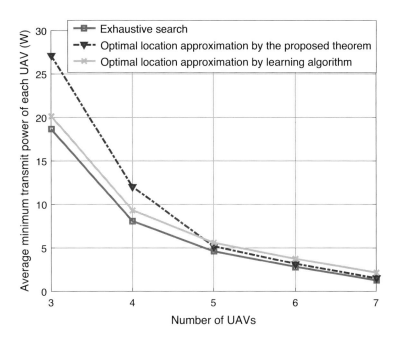

Figure 5.19 Average minimum transmit power and average altitude vs. the number of UAV BSs.

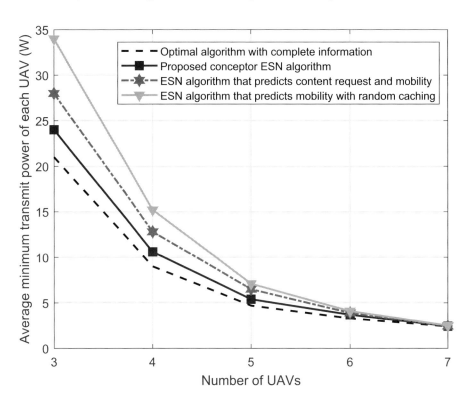

Figure 5.20 Average minimum transmit power as the number of UAV BSs changes ($U = 70$ and $C = 1$).

5.4.4 Summary

In this section, we have presented a framework for deploying cache-enabled UAV BSs that provide wireless service to ground CRAN users. In particular, we have optimized the deployment locations of the UAV BSs using a proactive approach that leverages machine learning tools from ESNs. The developed solution ensures that the CRAN users' QoE is met while using a minimum UAV BSs' transmit power. To efficiently deploy cache-enabled UAV BSs, various user-centric information, such as the mobility pattern of users and their content request distribution, were used. We have also provided simulation results to show the effectiveness and performance gain of employing cache-enabled UAV BS-assisted wireless CRANs.

5.5 Chapter Summary

In this chapter, we have focused on the deployment challenges of UAVs, particularly, UAV BSs. This chapter has provided a detailed development of various deployment case studies. In Section 5.1, we have presented the key analytical tools needed for performing UAV deployment in various scenarios. In Section 5.2, we have studied an efficient placement of UAV BSs for providing wireless connectivity to a given geographical area. We first have derived the downlink coverage probability for ground users. Next, by exploiting circle packing theory, we have presented a framework for 3D placement of identical UAV BSs to provide a maximum wireless coverage with minimum transmit power. In Section 5.3, we have described a method for an efficient deployment of drones for data collection from terrestrial IoT devices. We have identified the optimal locations of drones, device-drone associations, as well as the uplink transmit power of each IoT device that ensures a minimum total transmit power consumption of the devices. Moreover, we have analyzed the dynamic placement of drones for serving IoT devices in a time-varying IoT network. Finally, in Section 5.4, we have provided a framework for deploying cache-enabled UAV BSs that can provide cache-assisted connectivity to CRAN ground users. We have optimized the locations of multiple cache-enabled UAV BSs such that the users' QoE is satisfied while using minimum UAV transmit power.

6 Wireless-Aware Path Planning for UAV Networks

While the previous chapter focused on the deployment of UAVs, particularly UAV BSs, the next step is to analyze the mobility of UAVs. In particular, this chapter delves into the problem of wireless-aware path planning for UAVs with a focus on UAV UEs and their ability to connect with ground cellular networks. To this end, we present a very focused study on interference-aware path planning for cellular-connected UAV UEs in which each UAV aims to achieve a tradeoff between various QoS and mission goals, such as minimizing wireless latency and interference caused on the ground network. To this end, we start this chapter with Section 6.1, which motivates the need for wireless-aware path planning for UAV UEs. Then, in Section 6.2, we provide a comprehensive system model for a wireless network with UAV UEs, and we then formally pose the wireless-aware path planning problem for UAV UEs. Subsequently, in Sections 6.3 and 6.4, we show how tools from game theory and reinforcement learning can be merged to design autonomous, self-organizing path planning mechanisms for UAV UEs that can balance the various wireless and mission objectives of the drones. We also show how some of the unique features of UAV UEs, such as their altitude and their ability to establish LOS will have significant impact on the way in which their trajectory is designed. Using both theoretical and simulation results (detailed in Section 6.5), we study the impact of various parameters on the performance of both UAV UEs and ground users. We conclude the chapter in Section 6.6 with some general insights.

6.1 Need for Wireless-Aware Path Planning

UAV trajectory optimization and path planning are instrumental in mitigating interference toward ground users while adapting their movement based on their performance needs as well as the needs of ground UEs. Naturally, the trajectory of UAVs will impact their own communication performance as well as the performance of ground UEs. This interplay between UAV path planning and wireless network performance arises for both the UAV BS and UAV UE use cases. Although path planning and trajectory optimization are fundamental problems for UAV systems, the synergies between the trajectories of the UAVs and the wireless communication system performance bring forth novel challenges that are not typically addressed in classical UAV navigation and trajectory optimization works. Indeed, the prior art in UAV path planning focuses mainly on non-UAV UE applications [210–213]. For instance, in [210], a distributed path planning algorithm is proposed for multiple UAVs to deliver delay-sensitive information, whereas a UAV's trajectory is optimized in an energy-efficient manner in [211]. In [213], a fog-networking-based system architecture coordinates a network of UAVs for video services in sporting events. Nevertheless, while interesting, these works do not account for UAV UEs and their associated wireless challenges, which make them inadequate for cellular-connected UAV UEs.

Indeed, when dealing with path planning for cellular-connected UAV UEs, a number of new challenges must be addressed. First and foremost, the flying nature of UAV UEs allows them to establish LOS connectivity with ground BSs. This, in turn, is both a blessing and a curse. On the one hand, the ability to establish LOS connectivity enables UAV UEs to achieve high QoS. However, this comes at the expense of generating higher interference on the links of the ground UEs, due to the LOS nature of UAV UE interference. Hence, when dealing with trajectory optimization for UAV UEs, one must balance the tradeoff between improving the QoS of UAV UEs and minimizing their interference to ground UEs. Second, when dealing with UAV UE path planning, it is imperative for any trajectory design to explicitly take into account not only the wireless performance but also the mission time of the UAVs. Third, the coexistence of ground UEs and UAV UEs will require new ways to optimize and manage the network, while explicitly taking into account the highly dynamic nature of the UAV UEs.

To this end, in the remainder of this chapter, we will present an in-depth study on path planning for UAV UEs. We particularly focus on wireless-aware path planning whereby the trajectory of the UAV UEs is designed while taking into account both mission time and wireless performance. We then show how the use of game theory and learning can help in overcoming the challenges of the dynamic UAV UE environment. We finally shed light on how various network parameters impact the performance of wireless-aware path planning for UAV UEs.

6.2 Wireless-Aware Path Planning for UAV UEs: Model and Problem Formulation

We focus on the uplink of a cellular system having a total system bandwidth B and composed of S ground BSs in set $\mathcal{S}$, Q ground UEs in set $\mathcal{Q}$, and J cellular-connected UAV UEs in set $\mathcal{J}$. Each ground BS s will communicate with K_s UEs and N_s UAV UEs. We divided the bandwidth B into C resource blocks (RBs) and each UAV UE $j \in \mathcal{N}_s$ is allocated a set $\mathcal{C}_{j,s} \subseteq \mathcal{C}$ of $C_{j,s}$ RBs. Moreover, BS s allocates a set $\mathcal{C}_{q,s} \subseteq \mathcal{C}$ of $C_{q,s}$ RBs for each one of its ground UEs $q \in \mathcal{K}_s$. Note that, at the level of every BS s, a given RB c is associated with *at most* one UAV UE j or ground UE q. We then define (x_j, y_j, h_j) as the 3D coordinate of a UAV UE j and $(x_q, y_q, 0)$ as the 3D coordinate of a ground UE q. In our model, the UAV UEs will fly at a fixed altitude h_j. However, the horizontal coordinates (x_j, y_j) of each UAV UE j will be dynamically changing over time. Every UAV UE j must move from an initial location o_j to a final destination d_j while transmitting data (e.g., surveillance videos, images, sensor readings, etc.) in an online manner.

For simplicity, we consider a virtual grid for the UAV UEs' mobility whereby we discretize the space into A equally sized unit areas. UAV UEs move along the center of the areas $c_a = (x_a, y_a, z_a)$ and, hence, for each UAV UE j, we have a finite set of possible paths p_j. Here, the path p_j of each UAV UE j is defined as a sequence of area units $p_j = (a_1, a_2, \cdots, a_l)$ such that $a_1 = o_j$ and $a_l = d_j$. We choose a sufficiently small area size for the discretized area units $(a_1, a_2, \cdots, a_A) \in \mathcal{A}$ so that, within each area, the UAV UEs' positions can be seen as approximately constant at the maximum UAV UE's velocity. We consider a constant velocity $0 < V_j \leq \widehat{V}_j$ for each UAV UE where $\widehat{V}_j$ is defined as the maximum velocity of UAV UE j. For the AG channel model, aligned with Chapter 3, we use the following model for the path loss $\xi_{j,s,a}$ between a UAV UE j at location a, and BS s is given by [214]:

$$\xi_{j,s,a}(\text{dB}) = 20 \log_{10}(d_{j,s,a}) + 20 \log_{10}(\hat{f}) - 147.55. \quad (6.1)$$

Here, $\hat{f}$ represents the center frequency of our system and $d_{j,s,a}$ represents the Euclidean distance between UAV UE j at location a and BS s. A Rician channel model is assumed for the small-scale fading between UAV j and ground BS s reflecting the LOS and multipath scatterers, which can be a reasonable assumption for tractability, as discussed in Chapter 3. For the links between ground UEs and their BSs, the channel is assumed to follow Rayleigh fading. Hence, given a carrier frequency $\hat{f} = 2$ GHz, we can define the path loss between a ground UE q and its serving BS s as follows [215]:

$$\zeta_{q,s}(\text{dB}) = 15.3 + 37.6 \log_{10}(d_{q,s}), \quad (6.2)$$

with $d_{q,s}$ being the distance between BS s and ground UE q.

The average SINR, $\Gamma_{j,s,c,a}$, achieved by a UAV UE j (at location a) at its serving ground BS s over RB c is:

$$\Gamma_{j,s,c,a} = \frac{P_{j,s,c,a} h_{j,s,c,a}}{I_{j,s,c} + B_c N_0}, \quad (6.3)$$

with $P_{j,s,c,a} = \widehat{P}_{j,s,a}/C_{j,s}$ being the per-RB transmit power of UAV UE j over RB c and $\widehat{P}_{j,s,a}$ being the total transmit power of UAV UE j. For tractability, we assume that a UAV UE j will divided its total transmit power equally among all of its associated RBs. In (6.3), $h_{j,s,c,a} = g_{j,s,c,a} 10^{-\xi_{j,s,a}/10}$ is defined as the channel gain between UAV UE j and BS s on RB c at location a with $g_{j,s,c,a}$ being the Rician fading parameter. B_c is the bandwidth of RB c, and N_0 is the power spectral density of the noise. $I_{j,s,c} = \sum_{r=1, r\neq s}^{S} (\sum_{k=1}^{K_r} P_{k,r,c} h_{k,s,c} + \sum_{n=1}^{N_r} P_{n,r,c,a'} h_{n,s,c,a'})$ is the total interference power on UAV UE j at BS s when using RB c with $\sum_{r=1, r\neq s}^{S} \sum_{k=1}^{K_r} P_{k,r,c} h_{k,s,c}$ and $\sum_{r=1, r\neq s}^{S} \sum_{n=1}^{N_r} P_{n,r,c,a'} h_{n,s,c,a'}$ being, respectively, the interference from the K_r UEs and the N_r UAV UEs (at their transmission locations a') serviced by neighboring BSs r and employing the same RB c as UAV UE j. We define $h_{k,s,c} = m_{k,s,c} 10^{-\zeta_{k,s}/10}$ as the channel gain between ground UE k and its associated BS s when using RB c. Here, $m_{k,s,c}$ is the Rayleigh fading parameter. As a result, the data rate that UAV UE j (located at a and served by BS s) achieves will be $R_{j,s,a} = \sum_{c=1}^{C_{j,s}} B_c \log_2(1 + \Gamma_{j,s,c,a})$.

We are particularly interested in the latency achieved by the UAV UEs since it is an important performance metric for many of the UAV UE applications discussed in previous chapters. To model this latency, we consider an M/D/1 queue at each UAV UE and, then, we can find the latency of the link between UAV UE j and its serving BS as follows [216]:

$$\tau_{j,s,a} = \frac{\lambda_{j,s}}{2\mu_{j,s,a}(\mu_{j,s,a} - \lambda_{j,s})} + \frac{1}{\mu_{j,s,a}}, \quad (6.4)$$

with $\mu_{j,s,a} = R_{j,s,a}/\nu$ being the service rate over link (j,s) at location a where ν is the packet size. Here, $\lambda_{j,s}$ is defined as the average packet arrival rate stemming from UAV UE j over link (j,s). For the ground UEs, we are interested in the data rate as a key performance metric. In this context, for a ground UE q that is connected to a BS s, we can define the data rate as:

$$R_{q,s} = \sum_{c=1}^{C_{q,s}} B_c \log_2\left(1 + \frac{P_{q,s,c} h_{q,s,c}}{I_{q,s,c} + B_c N_0}\right), \quad (6.5)$$

where $h_{q,s,c} = m_{q,s,c} 10^{-\zeta_{q,s}/10}$ is the channel gain between ground UE q and its BS s over RB c and $m_{q,s,c}$ is the Rayleigh fading parameter. Using an equal power allocation assumption over the RBs, we define $\widehat{P}_{q,s,c} = \widehat{P}_{q,s}/C_{q,s}$ as the transmit power of UE q to its serving BS s on RB c with $\widehat{P}_{q,s}$ being the total transmit power of ground UE q. $I_{q,s,c} = \sum_{r=1, r\neq s}^{S}(\sum_{k=1}^{K_r} P_{k,r,c} h_{k,s,c} + \sum_{n=1}^{N_r} P_{n,r,c,a'} h_{n,s,c,a'})$ is the total interference power experienced by ground UE q at BS s over RB c where $\sum_{r=1, r\neq s}^{S} \sum_{k=1}^{K_r} P_{k,r,c} h_{k,s,c}$ and $\sum_{r=1, r\neq s}^{S} \sum_{n=1}^{N_r} P_{n,r,c,a'} h_{n,s,c,a'}$ correspond, respectively, to the interference from the K_r UEs and the N_r UAV UEs (at their respective locations a') associated with the neighboring BSs r and that employ the same RB c as ground UE q.

6.2.1 Problem Formulation

The objective is to find the optimal wireless-aware path for each UAV UE j based on its mission goals and its interference toward the ground cellular system. In other words, the interference level that each UAV UE causes on the ground UEs and other UAV UEs, the transmission delay, and the time needed to reach the destination should be minimized. To do this, at each location $a \in \mathcal{A}$, the paths of the UAV UEs' will be optimized jointly with the cell association vector and the power control vector. A directed graph $G_j = (\mathcal{V}, \mathcal{E}_j)$ is defined for each UAV UE j where $\mathcal{V}$ is the set of vertices corresponding to the centers of the unit areas $a \in \mathcal{A}$ and $\mathcal{E}_j$ is the set of edges formed along the path of UAV UE j. We define $\widehat{\boldsymbol{P}}$ as the transmission power vector with each element $\widehat{P}_{j,s,a} \in [0, \overline{P}_j]$ representing the transmission power level of UAV UE j to its associated BS s at location a where $\overline{P}_j$ is the maximum transmission power of UAV UE j. We define a path formation vector $\boldsymbol{\alpha}$ whose element $\alpha_{j,a,b} \in \{0,1\}$ represents, for each UAV UE j, whether or not a directed link is formed from area a toward area b. We also define a UAV UE to BS association vector $\boldsymbol{\beta}$ with each element $\beta_{j,s,a} \in \{0,1\}$ indicating whether or not, at location a, UAV UE j is associated with ground BS s. Formally, we can formulate an optimization problem to determine the path, cell association vector, and power control strategies for each UAV UE located at area a (along its path $\boldsymbol{p}_j$):

6.2 Wireless-Aware Path Planning for UAV UEs: Model and Problem Formulation

$$\min_{\widehat{P},\alpha,\beta} \vartheta \sum_{j=1}^{J}\sum_{s=1}^{S}\sum_{c=1}^{C_{j,s}}\sum_{a=1}^{A}\sum_{r=1,r\neq s}^{S} \frac{\widehat{P}_{j,s,a}h_{j,r,c,a}}{C_{j,s}} + \varpi \sum_{j=1}^{J}\sum_{a=1}^{A}\sum_{b=1,b\neq a}^{A} \alpha_{j,a,b}$$

$$+ \phi \sum_{j=1}^{J}\sum_{s=1}^{S}\sum_{a=1}^{A} \beta_{j,s,a}\tau_{j,s,a}, \tag{6.6}$$

$$\sum_{b=1,b\neq a}^{A} \alpha_{j,b,a} \leq 1 \quad \forall j \in \mathcal{J}, a \in \mathcal{A}, \tag{6.7}$$

$$\sum_{a=1,a\neq o_j}^{A} \alpha_{j,o_j,a}=1 \quad \forall j \in \mathcal{J}, \quad \sum_{a=1,a\neq d_j}^{A} \alpha_{j,a,d_j}=1 \quad \forall j \in \mathcal{J}, \tag{6.8}$$

$$\sum_{a=1,a\neq b}^{A} \alpha_{j,a,b} - \sum_{f=1,f\neq b}^{A} \alpha_{j,b,f} = 0 \quad \forall j \in \mathcal{J}, b \in \mathcal{A} \ (b \neq o_j, b \neq d_j), \tag{6.9}$$

$$\widehat{P}_{j,s,a} \geq \sum_{b=1,b\neq a}^{A} \alpha_{j,b,a} \quad \forall j \in \mathcal{J}, s \in \mathcal{S}, a \in \mathcal{A}, \tag{6.10}$$

$$\widehat{P}_{j,s,a} \geq \beta_{j,s,a} \quad \forall j \in \mathcal{J}, s \in \mathcal{S}, a \in \mathcal{A}, \tag{6.11}$$

$$\sum_{s=1}^{S} \beta_{j,s,a} - \sum_{b=1,b\neq a}^{A} \alpha_{j,b,a} = 0 \quad \forall j \in \mathcal{J}, a \in \mathcal{A}, \tag{6.12}$$

$$\sum_{c=1}^{C_{j,s}} \Gamma_{j,s,c,a} \geq \beta_{j,s,a}\overline{\Gamma}_j \quad \forall j \in \mathcal{J}, s \in \mathcal{S}, a \in \mathcal{A}, \tag{6.13}$$

$$0 \leq \widehat{P}_{j,s,a} \leq \overline{P}_j \quad \forall j \in \mathcal{J}, s \in \mathcal{S}, a \in \mathcal{A}, \tag{6.14}$$

$$\alpha_{j,a,b} \in \{0,1\}, \beta_{j,s,a} \in \{0,1\} \quad \forall j \in \mathcal{J}, s \in \mathcal{S}, a,b \in \mathcal{A}. \tag{6.15}$$

Our objective function here captures the total interference level that the UAV UEs cause on neighboring BSs along their paths, the length of the paths of the UAV UEs, and their wireless transmission delay. ϑ, ϖ, and ϕ are multi-objective weights used to control the tradeoff between the three performance metrics that we considered. These weights can be adjusted to meet the requirements of each UAV UE's mission. For instance, the time to reach the destination is critical in search and rescue applications while the latency is important for online video streaming applications (as inferred from Chapter 2, the importance of these metrics vary across different UAV UE applications). Constraint (6.7) guarantees that each area a is visited by UAV UE j at most once along its path p_j. Constraints (6.8) ensure that the trajectory chosen by each UAV UE j begins at its initial location o_j and ends at its final destination d_j (i.e., the mission destination and origin will not change). (6.9) ensures that, whenever a UAV UE j visits area b, this UAV UE should also leave from area b ($b \neq o_j, b \neq d_j$). Using constraints (6.10) and (6.11), we ensure that UAV UE j will transmit to BS s at area a with power $\widehat{P}_{j,s,a} > 0$ only if UAV UE j actually visits area a, i.e., $a \in \boldsymbol{p}_j$ and where UAV UE j is associated with BS s at location a. (6.12) guarantees that each UAV UE j can be served by one BS s at every location a along its path $\boldsymbol{p}_j$. Constraint (6.13) represents an upper limit, $\overline{\Gamma}_j$, on the SINR value $\Gamma_{j,s,c,a}$ of the transmission link between UAV UE j (located at a) and ground BS s, over RB c. This constraint will thereby guarantee a successful decoding of the packets transmitted by the UAV UE to its serving BS.

Solving the centralized optimization problem is challenging due to the various parameters and objectives involved. In addition, developing centralized wireless-aware path planning solutions for UAV UEs is undesirable given the inherently distributed nature of the system and the fact that UAV UEs will not belong to the wireless network operator. This, in turn, calls for a distributed solution in which each UAV UE j autonomously learns its path $\boldsymbol{p}_j$, transmission power level, and association vector at each location a. To develop such a solution, in the next section, we resort to tools from game theory and machine learning.

6.3 Self-Organizing Wireless-Aware Path Planning for UAV UEs

6.3.1 Path Planning as a Game

Our overarching goal is to develop a distributed path planning solution that enables each UAV UE to take actions in a self-organizing and online manner. This multi-agent path planning problem can be properly modeled as a finite dynamic noncooperative game model $\mathcal{G}$ with perfect information [217]. In particular, the UAV UE path planning game is defined by a tuple $\mathcal{G} = (\mathcal{J}, \mathcal{T}, \mathcal{Z}_j, \mathcal{V}_j, \Pi_j, u_j)$ with the set $\mathcal{J}$ of UAV UEs being the set of agents/players. $\mathcal{T}$ is defined as a finite set of game stages that represent the steps needed by all UAV UEs to arrive at their missions' destinations. For each UAV UE j, $\mathcal{Z}_j$ is the set of actions that this UAV UE can select at each given time $t \in \mathcal{T}$. Moreover, we define $\mathcal{V}_j$ as the set of all network states that are observed by UAV UE j up to stage T of the game, and we define Π_j as a set of probability distributions defined over all $z_j \in \mathcal{Z}_j$. In our game, u_j represents the individual utility function of UAV UE j. At each stage t of the game, the UAV UEs will take actions simultaneously. To this end, each UAV UE j will seek to determine its most preferred path $\boldsymbol{p}_j$ (from origin to destination) while also determining the corresponding optimal transmit power and cell association vector for each location $a \in \mathcal{A}$ along its path $\boldsymbol{p}_j$. Hence, at each time step t, UAV UE j selects an action tuple $z_j(t) = (\boldsymbol{a}_j(t), \widehat{P}_{j,s,a}(t), \boldsymbol{\beta}_{j,s,a}(t))$, where $\boldsymbol{a}_j(t)$={left, right, forward, backward, no movement} represents a fixed step size, $\tilde{a}_j$, in a given direction. For every UAV j, $\widehat{P}_{j,s,a}(t) = [\widehat{P}_1, \widehat{P}_2, \cdots, \widehat{P}_O]$ represents O different maximum transmit power levels, and $\boldsymbol{\beta}_{j,s,a}(t)$ represents the association vector between UAV UEs and their BSs.

For every UAV UE j, we define a set $\mathcal{L}_j$ that includes the L_j BSs that are nearest to this UAV UE. We can now formally define the network state $\boldsymbol{v}_j(t)$ that UAV UE j observes at stage t:

$$\boldsymbol{v}_j(t) = \left[\{\delta_{j,l,a}(t), \theta_{j,l,a}(t)\}_{l=1}^{L_j}, \theta_{j,d_j,a}(t), \{x_j(t), y_j(t)\}_{j \in \mathcal{J}} \right], \quad (6.16)$$

where $\delta_{j,l,a}(t)$ represents the Euclidean distance between UAV UE j, located at a, to BS l during stage t, $\theta_{j,l,a}$ represents the orientation angle (in the two-dimensional, horizontal, xy-plane) from UAV UE j at location a to BS l expressed by $\tan^{-1}(\Delta y_{j,l}/\Delta x_{j,l})$ [218] with $\Delta y_{j,l}$ and $\Delta x_{j,l}$ being the difference in the x and y coordinates of UAV UE j and BS l. Moreover, $\theta_{j,d_j,a}$ is the orientation angle in the xy-plane from UAV UE j at location a to its destination d_j defined as $\tan^{-1}(\Delta y_{j,d_j}/\Delta x_{j,d_j})$, and $\{x_j(t),y_j(t)\}_{j\in\mathcal{J}}$ are the horizontal coordinates of all UAV UEs at stage t. Moreover, different range intervals are considered for mapping each of the orientation angle and distance values, respectively, into different states.

Based on our optimization formulation in (6.6)–(6.15) and incorporating the Lagrangian penalty method into the utility function definition for the SINR constraint (6.13), the utility function for UAV UE j at stage t, $u_j(v_j(t), z_j(t), z_{-j}(t))$ can be expressed by:

$$u_j(v_j(t), z_j(t), z_{-j}(t)) = \begin{cases} \Phi(v_j(t), z_j(t), z_{-j}(t))+C, & \text{if } \delta_{j,d_j,a}(t) < \delta_{j,d_j,a'}(t-1), \\ \Phi(v_j(t), z_j(t), z_{-j}(t)), & \text{if } \delta_{j,d_j,a}(t) = \delta_{j,d_j,a'}(t-1), \\ \Phi(v_j(t), z_j(t), z_{-j}(t))-C, & \text{if } \delta_{j,d_j,a}(t) > \delta_{j,d_j,a'}(t-1), \end{cases} \quad (6.17)$$

where $\Phi(v_j(t), z_j(t), z_{-j}(t))$ is defined as:

$$\Phi(v_j(t),z_j(t),z_{-j}(t)) = -\vartheta' \sum_{c=1}^{C_{j,s}(t)} \sum_{r=1,r\neq s}^{S} \frac{\widehat{P}_{j,s,a}(v_j(t))h_{j,r,c,a}(t)}{C_{j,s}(t)}$$
$$- \phi' \tau_{j,s,a}(v_j(t),z_j(t),z_{-j}(t))$$
$$- \varsigma (\min(0, \sum_{c=1}^{C_{j,s}(t)} \Gamma_{j,s,c,a}(v_j(t),z_j(t),z_{-j}(t)) - \overline{\Gamma}_j))^2, \quad (6.18)$$

subject to (6.7)–(6.12), (6.14) and (6.15). ς is the penalty coefficient for (6.13), and C is a constant parameter. a' and a are the locations of UAV j at $(t-1)$ and t where $\delta_{j,d_j,a}$ is the distance between UAV UE j and its destination d_j. The action space of each UAV UE j and, thus, the complexity of our game $\mathcal{G}$ increases exponentially when updating the 3D coordinates of the UAV UEs. Nevertheless, each UAV UE's altitude must be bounded to guarantee an SINR threshold (for the UAV UE) and a minimum achievable data rate (for the terrestrial UEs). Next, we present an upper and lower bound for the optimal altitude of any given UAV UE j that is based on the proof done in [219]:

THEOREM 6.1 For all values of ϑ', ϕ', and ς, a given network state $v_j(t)$, and an action $z_j(t)$, the upper and lower bounds on the altitude of any UAV UE j are, respectively:

$$h_j^{\max}(v_j(t), z_j(t), z_{-j}(t)) = \max(\chi, \hat{h}_j^{\max}(v_j(t), z_j(t), z_{-j}(t))), \quad (6.19)$$

$$h_j^{\min}(v_j(t), z_j(t), z_{-j}(t)) = \max(\chi, \hat{h}_j^{\min}(v_j(t), z_j(t), z_{-j}(t))), \quad (6.20)$$

where χ is the minimum altitude at which a UAV UE can be allowed to fly. $\hat{h}_j^{\max}(v_j(t), z_j(t), z_{-j}(t))$ and $\hat{h}_j^{\min}(v_j(t), z_j(t), z_{-j}(t))$ are expressed as:

$$\hat{h}_j^{\max}(v_j(t), z_j(t), z_{-j}(t)) =$$

$$\sqrt{\frac{\widehat{P}_{j,s,a}(v_j(t))}{C_{j,s}(t) \cdot \overline{\Gamma}_j \cdot \left(\frac{4\pi\hat{f}}{\hat{c}}\right)^2} \cdot \sum_{c=1}^{C_{j,s}(t)} \frac{g_{j,s,c,a}(t)}{I_{j,s,c}(t) + B_c N_0}} - (x_j - x_s)^2 - (y_j - y_s)^2, \quad (6.21)$$

and

$$\hat{h}_j^{\min}(v_j(t), z_j(t), z_{-j}(t)) = \max_r \hat{h}_{j,r}^{\min}(v_j(t), z_j(t), z_{-j}(t)), \quad (6.22)$$

where $\hat{h}_{j,r}^{\min}(v_j(t), z_j(t), z_{-j}(t))$ represents the minimum altitude that UAV UE j should operate at with respect to a particular neighboring BS r and is given by:

$$\hat{h}_{j,r}^{\min}(v_j(t), z_j(t), z_{-j}(t))$$

$$= \sqrt{\frac{\widehat{P}_{j,s,a}(v_j(t)) \cdot \sum_{c=1}^{C_{j,s}(t)} g_{j,r,c,a}(t)}{C_{j,s}(t) \cdot \left(\frac{4\pi\hat{f}}{\hat{c}}\right)^2 \cdot \sum_{c=1}^{C_{j,s}(t)} \bar{I}_{j,r,c,a}}} - (x_j - x_r)^2 - (y_j - y_r)^2, \quad (6.23)$$

Theorem interpretation: Theorem 6.1 emphasizes the fact that the optimal altitudes of UAV UEs will be a function of their objective function, locations of the ground BSs, network design parameters, and the interference level from other ground UEs and UAV UEs in the network. Hence, at every time instance t, UAV UE j must adjust its altitude depending on the values of $h_j^{\max}(v_j(t), z_j(t), z_{-j}(t))$ and $h_j^{\min}(v_j(t), z_j(t), z_{-j}(t))$, thus adapting to the network dynamics. The derived upper and lower bounds for the optimal altitude of the UAV UEs will hence allow reducing the action space in our game $\mathcal{G}$, thereby simplifying the process needed for the UAV UEs to find a solution (a so-called equilibrium) of the game. Next, the equilibrium point of the studied UAV UE path planning game $\mathcal{G}$ is analyzed.

6.3.2 Equilibrium of the UAV UE Path Planning Game

For the UAV UE path planning game $\mathcal{G}$, the subgame perfect Nash equilibrium (SPNE) in behavioral strategies must be studied. An SPNE is a profile of agent strategies that imposes a *Nash equilibrium* (NE) on every subgame of the original dynamic game. Here, we need to define the notion of a *behavioral strategy* that enables each UAV UE to assign independent probabilities to the set of actions at each network state that is independent across different network states. Note that, from the seminal result of Selten in [220], we know that, for any finite-horizon extensive game with perfect information, at least one SPNE will exist. We now define $\pi_j(v_j(t)) = (\pi_{j,z_1}(v_j(t)), \pi_{j,z_2}(v_j(t)), \cdots, \pi_{j,z_{|\mathcal{Z}_j|}}(v_j(t))) \in \Pi_j$ as the behavioral strategy of UAV j at state $v_j(t)$ and we define $\Delta(\mathcal{Z})$ as the set of all probability distributions over the action space $\mathcal{Z}$. Below, we can formally define the concept of an SPNE.

DEFINITION 6.2 *A behavioral strategy* $(\pi_1^*(v_j(t)), \cdots, \pi_J^*(v_j(t))) = (\pi_j^*(v_j(t)), \pi_{-j}^*(v_j(t)))$ *constitutes a* subgame perfect Nash equilibrium *if*, $\forall j \in \mathcal{J}$, $\forall t \in \mathcal{T}$ *and* $\forall \pi_j(v_j(t)) \in \Delta(\mathcal{Z})$, $\bar{u}_j(\pi_j^*(v_j(t)), \pi_{-j}^*(v_j(t))) \geq \bar{u}_j(\pi_j(v_j(t)), \pi_{-j}^*(v_j(t)))$.

Consequently, at every given state $v_j(t)$ and stage t, each UAV UE j will seek to maximize its expected sum of discounted rewards, computed as the summation of the immediate reward for a given state along with the expected discounted utility of the next states:

$$\bar{u}(v_j(t), \pi_j(v_j(t)), \pi_{-j}(v_j(t)))$$
$$= \mathbb{E}_{\pi_j(t)} \left\{ \sum_{l=0}^{\infty} \gamma^l u_j(v_j(t+l), z_j(t+l), z_{-j}(t+l)) | v_{j,0} = v_j \right\}$$
$$= \sum_{z \in \mathcal{Z}} \sum_{l=0}^{\infty} \gamma^l u_j(v_j(t+l), z_j(t+l), z_{-j}(t+l)) \prod_{j=1}^{J} \pi_{j, z_j}(v_j(t+l)), \quad (6.24)$$

where $\gamma^l \in (0,1)$ is a discount factor for delayed rewards and $\mathbb{E}_{\pi_j(v_j(t))}$ represents an expectation over trajectories of states and actions whereby actions are chosen based on $\pi_j(v_j(t))$. Note that u_j represents the short-term reward for being in state v_j and $\bar{u}_j$ represents the expected long-term total reward from state v_j onward.

We can now observe a clear coupling between the UAV UE's trajectory optimization, BS association vector, and power control levels. To find the SPNE, in a network with multiple UAV UEs, each UAV UE will have to acquire the entire knowledge of the future rewards (at each information set). Clearly, such knowledge is prohibitive in a wireless network because it requires each UAV UE to know all possible future actions of all other UAV UEs. This information-gathering process becomes even more challenging as the number of UAV UEs grows. We will overcome this challenge by resorting to tools from deep recurrent neural networks (RNNs) [221] whose dynamic temporal behavior and adaptive memory enables them to store key previous state information to predict future actions. In what follows, based on [219], a novel deep reinforcement learning (RL) algorithm that uses the RNN tools of ESNs (as also used in Chapter 5, Section 5.4.2) is investigated for solving the SPNE of the UAV UE path planning game $\mathcal{G}$. For clarity of our discussion, an introduction to deep ESN architecture (which expands on the shallow ESN studied in Chapter 5) is first provided to show how this tool can allow UAV UEs to store previous states whenever needed while learning future network states. Then, an RL algorithm based on the deep ESN architecture is studied and shown to be able to learn an SPNE of the game.

6.4 Deep Reinforcement Learning for Online Path Planning and Resource Management

We first begin by explaining what the deep ESN architecture is and how it differs from a canonical ESN.

6.4.1 Deep ESN Architecture

ESNs are RNNs that include feedback connections and belong to the family of reservoir computing (RC) [222]. An ESN is composed of an input weight matrix W_{in}, a recurrent matrix W, and an output weight matrix W_{out}. One key advantage of ESN is its quick and computationally efficient training, which stems from the fact that only the output weights are altered. Using this basic ESN architecture, we can stack multiple nonlinear reservoir layers to create a *deep ESN architecture*. Deep ESNs exploit the advantages of a hierarchical temporal feature representation at different levels of abstraction while preserving the RC training efficiency. They can learn data representations at different levels of abstraction, hence disentangling the difficulties in modeling complex tasks by representing them in terms of simpler ones hierarchically. We denote by $N_{j,R}^{(n)}$ the number of internal units of UAV UE j's reservoir at ESN layer n. Moreover, for UAV UE j, we also define $N_{j,U}$ as the external input dimension and $N_{j,L}$ as the number of stack layers. We can now introduce various ESN components:

- $v_j(t) \in \mathbb{R}^{N_{j,U}}$ the external input of UAV UE j at stage t capturing the current state of the network,
- $x_j^{(n)}(t) \in \mathbb{R}^{N_{j,R}^{(n)}}$ as the state of the reservoir, at layer n, for UAV UE j at stage t,
- $W_{j,\text{in}}^{(n)}$ as the input-to-reservoir matrix of UAV UE j at layer n, where $W_{j,\text{in}}^{(n)} \in \mathbb{R}^{N_{j,R}^{(n)} \times N_{j,U}}$ for $n = 1$, and $W_{j,\text{in}}^{(n)} \in \mathbb{R}^{N_{j,R}^{(n)} \times N_{j,R}^{(n-1)}}$ for $n > 1$,
- $W_j^{(n)} \in \mathbb{R}^{N_{j,R}^{(n)} \times N_{j,R}^{(n)}}$ as the recurrent reservoir weight matrix for UAV j, layer n, and
- $W_{j,\text{out}} \in \mathbb{R}^{|\mathcal{Z}_j| \times (N_{j,U} + \sum_n N_{j,R}^{(n)})}$ as the reservoir-to-output matrix of UAV j, layer n.

This deep ESN architecture can be essentially used to approximate a function $F_j = (F_j^1, F_j^2, \cdots, F_j^{N_{j,L}})$ for learning an SPNE for each UAV UE j at each stage t of our game. For each $n = 1, 2, \cdots, N_{j,L}$, function $F_j^{(n)}$ represents the evolution of the reservoir state at a layer n, i.e., $x_j^{(n)}(t) = F_j^{(n)}(v_j(t), x_j^{(n)}(t-1))$ for $n = 1$ and $x_j^{(n)}(t) = F_j^{(n)}(x_j^{(n-1)}(t), x_j^{(n)}(t-1))$ for $n > 1$. $W_{j,\text{out}}$ and $x_j^{(n)}(t)$ are initialized to zero while $W_{j,\text{in}}^{(n)}$ and $W_j^{(n)}$ are randomly generated. Even though the dynamic ESN reservoir is initially randomly generated, it is subsequently combined with the external input, $v_j(t)$, in order to store the network states and with the trained output matrix, $W_{j,\text{out}}$, so that it can approximate the reward function. Moreover, the spectral radius of $W_j^{(n)}$ (i.e., the largest eigenvalue in absolute value), $\rho_j^{(n)}$, must be strictly smaller than 1 to guarantee the stability of the reservoir [223]. In essence, the value of $\rho_j^{(n)}$ pertains to the variable memory length of the ESN reservoir, which enables the introduced deep ESN framework to store key previous state information. Naturally, higher values for $\rho_j^{(n)}$ imply a longer length for the memory.

A deep ESN architecture includes the input and reward functions. For each deep ESN at the level of a UAV UE j, we have two different input types: (a) external input, $v_j(t)$, which is fed to the first layer of the deep ESN and captures the current state of the network, and (b) input that is fed to all other layers for $n > 1$. For our deep ESN, the input to any layer $n > 1$ at stage t is nothing but the state of the previous layer, $x_j^{(n-1)}(t)$. At a game stage t, for UAV UE j, we define $\tilde{u}_j(v_j(t), z_j(t), z_{-j}(t)) = u_j(v_j(t), z_j(t), z_{-j}(t)) \prod_{j=1}^{J} \pi_{j,z_j}(v_j(t))$ to capture the expected value of the instantaneous utility function $u_j(v_j(t), z_j(t), z_{-j}(t))$ in (6.17). Hence, when UAV UE j takes an action z_j at a given network state $v_j(t)$, it will obtain the following payoff (reward):

6.4 Deep Reinforcement Learning for Online Path Planning

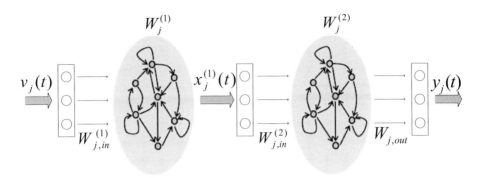

Figure 6.1 Studied deep ESN architecture.

$$r_j(v_j(t), z_j(t), z_{-j}(t)) = \begin{cases} \widetilde{u}_j(v_j(t), z_j(t), z_{-j}(t)), \text{ if UAV } j \text{ reaches } d_j, \\ \widetilde{u}_j(v_j(t), z_j(t), z_{-j}(t)) + \gamma \max_{z_j \in \mathcal{Z}_j} W_{j,\text{out}}(z_j(t+1), t+1) \\ [v'_j(t), x'^{(1)}_j(t), x'^{(2)}_j(t), \cdots, x'^{(n)}_j(t)], \text{ otherwise.} \end{cases} \quad (6.25)$$

In 6.25, $v'_j(t+1)$ and $x'^{(n)}_j(t)$, respectively, represent the next network state and reservoir state of layer (n), at stage $(t+1)$, when actions $z_j(t)$ and $z_{-j}(t)$ are taken at stage t. Figure 6.1 pictorially describes the studied reservoir architecture of a 2-layer deep ESN.

6.4.2 Deep ESN-Based UAV UE Update Rule

Each UAV UE will now use a deep ESN update phase in order to store and estimate the reward function of each path and the associated cell association/power allocation schemes at a game stage t. In this context, at any stage t, we adopt leaky integrator reservoir units [224] for updating the state transition functions $x^{(n)}_j(t)$. As a result, for the first layer $x^{(1)}_j(t)$, we can define the state transition function as follows:

$$x^{(1)}_j(t) = (1 - \omega^{(1)}_j)x^{(1)}_j(t-1) + \omega^{(1)}_j \tanh(W^{(1)}_{j,\text{in}} v_j(t) + W^{(1)}_j x^{(1)}_j(t-1)), \quad (6.26)$$

where $\omega^{(n)}_j \in [0, 1]$ is the leaking parameter at layer n for UAV UE j. This parameter is directly related to the speed of the reservoir dynamics in response to the input with larger values of $\omega^{(n)}_j$ resulting in a faster response of the corresponding n-th reservoir to the input. For $n > 1$, we can define the state transition $x^{(n)}_j(t)$ of UAV UE j as follows:

$$x^{(n)}_j(t) = (1 - \omega^{(n)}_j)x^{(n)}_j(t-1) + \omega^{(n)}_j \tanh(W^{(n)}_{j,\text{in}} x^{(n-1)}_j(t) + W^{(n)}_j x^{(n)}_j(t-1)), \quad (6.27)$$

At a stage t, the deep ESN output $y_j(t)$ can be used to estimate each UAV UE j's reward depending on this UAV UE's currently adopted action $z_j(t)$ and $z_{-j}(t)$ as well as the actions of other UAVs $(-j)$, respectively, for the current network state $v_j(t)$ after training $W_{j,\text{out}}$. This output, therefore, can be derived as follows:

$$y_j(v_j(t), z_j(t)) = W_{j,\text{out}}(z_j(t), t)[v_j(t), x^{(1)}_j(t), x^{(2)}_j(t), \cdots, x^{(n)}_j(t)]. \quad (6.28)$$

In order to train the deep ESN output matrix $W_{j,\text{out}}$, we employ a temporal difference RL scheme. In this context, we will use a linear gradient descent scheme that uses the reward error signal, as defined by the following update rule [225]:

$$W_{j,\text{out}}(z_j(t),t+1) = W_{j,\text{out}}(z_j(t),t) + \lambda_j(r_j(v_j(t), z_j(t), z_{-j}(t)) - y_j(v_j(t), z_j(t)))[v_j(t),$$
$$x_j^{(1)}(t), x_j^{(2)}(t), \cdots, x_j^{(n)}(t)]^T. \quad (6.29)$$

We now recall that the goal of each UAV UE will now be to minimize the value of the error function $e_j(v_j(t)) = |r_j(v_j(t), z_j(t), z_{-j}(t)) - y_j(v_j(t), z_j(t))|$.

6.4.3 Deep RL for Wireless-Aware Path Planning

We can now develop a multi-agent deep RL framework that leverages the introduced deep ESN architecture and update rule. This framework will be adopted by the UAV UEs in order to learn their path and resource allocation parameters at an SPNE of the path planning game $\mathcal{G}$. This RL algorithm will have two phases: *training* and *testing*. First, UAV UEs are trained offline. Subsequently, a testing phase will start, and it pertains to the actual execution of the RL algorithm during which the weights of $W_{j,\text{out}}, \forall j \in \mathcal{J}$ are optimized.

Training phase: For training, each UAV optimizes its output weight matrix $W_{j,\text{out}}$ to minimize the error function $e_j(v_j(t))$ at every stage t. The introduced training phase admits multiple iterations, each of which has multiple rounds, i.e., the number of steps required for all UAV UEs to reach their destinations d_j. In each round, the UAV UEs will face a tradeoff between playing the action associated with the highest expected utility and attempting to explore all their possible actions to enhance their reward function estimates in (6.25). This is nothing but the well-known RL exploration and exploitation tradeoff. Here, the UAV UEs must properly balance their RL process between exploring their network environment and exploiting the information that they acquire and accumulate from this environment exploration [226]. Here, we will employ the basic ϵ-greedy policy that enables the UAV UEs to select, with a probability of $1 - \epsilon + \frac{\epsilon}{|\mathcal{Z}_j|}$, the action that maximizes their utility value while exploring other possible actions with a probability of $\frac{\epsilon}{|\mathcal{A}_j|}$. As a result, we can now formally define each UAV UE j's strategy over its action space:

$$\pi_{j,z_j}(v_j(t)) = \begin{cases} 1 - \epsilon + \frac{\epsilon}{|\mathcal{Z}_j|}, & \text{argmax}_{z_j \in \mathcal{Z}_j} y_j(v_j(t), z_j(t)), \\ \frac{\epsilon}{|\mathcal{Z}_j|}, & \text{otherwise.} \end{cases} \quad (6.30)$$

Based on the chosen action $z_j(t)$, each UAV UE j updates its location, transmit power level, and cell association choice, and, then, it calculates its reward function using (6.25). To identify the next state of the network, each UAV UE j will broadcast its chosen action to all other UAV UEs. Subsequently, every UAV UE j will update its state transition vector $x_j^{(n)}(t)$ for each deep ESN layer (n) by using (6.26) and (6.27). At any stage t, the output y_j can now be updated using (6.28). Finally, for each UAV UE j, the output matrix $W_{j,\text{out}}$ weights can be updated based on the linear gradient descent rule of (6.29). We summarize this introduced training phase in Algorithm 2.

6.4 Deep Reinforcement Learning for Online Path Planning

Algorithm 2 Training phase of our deep RL algorithm

Initialization:
$\pi_{j,z_j}(v_j(t)) = \frac{1}{|\mathcal{A}_j|} \forall t \in T, z_j \in \mathcal{Z}_j, y_j(v_j(t), z_j(t)) = 0, W_{j,\text{in}}^{(n)}, W_j^{(n)}, W_{j,\text{out}}$.

for Total number of iterations for training **do**
    while At least one UAV UE j has not reached its destination d_j, **do**
        for all UAVs j (in a parallel manner) **do**
        Input: Each UAV UE j obtains an input $v_j(t)$ based on (6.16).
        Step 1: Action selection
Each UAV UE j picks a random action $z_j(t)$ with probability ϵ,
Otherwise, UAV j selects $z_j(t) = \text{argmax}_{z_j \in \mathcal{Z}_j} y_j(v_j(t), z_j(t))$.
        Step 2: Location, cell association, and transmit power update
Each UAV UE j updates its location, cell association and transmission power level based on the selected action $z_j(t)$.

        Step 3: Reward computation
Each UAV UE j calculates the values of its reward using (6.25).

        Step 4: Action broadcast
Each UAV UE j broadcasts its selected action $z_j(t)$ to all other UAV UEs.

        Step 5: Deep ESN update
- Each UAV UE j updates the state transition vector $x_j^{(n)}(t)$ for each layer (n) of the deep ESN architecture based on (6.26) and (6.27).
- Each UAV UE j finds its output $y_j(v_j(t), z_j(t))$ using (6.28).
- The weights of the output matrix $W_{j,\text{out}}$ of each UAV UE j are updated using the linear gradient descent update rule defined in (6.29).

        end for
    end while
end for

Testing phase: The testing phase is summarized in Algorithm 3. The testing phase pertains to the actual execution of the RL process in the network. During the testing phase, each UAV UE will pick its action in a greedy way (for every state $v_j(t)$). It will then update its location, cell association, and transmit power level accordingly. Each UAV UE will subsequently broadcast its chosen action and update its state transition vector $x_j^{(n)}(t)$ for each deep ESN layer n by using (6.26) and (6.27).

Convergence: Guaranteeing the convergence of the studied deep RL scheme is challenging as it is highly dependent on the hyperparameters used during training. For instance, using too few neurons in the hidden layers results in underfitting that undermines the neural network to detect the signals in a complicated data set. On the other hand, using too many neurons in the hidden layers can yield overfitting or an increase in the training time that hampers the training procedure. However, as shown in [219], if the algorithm converges, it is guaranteed to find an SPNE of the formulated game. Moreover, the simulations conducted in [219] did not observe major cases of non-convergence. Naturally, if a cycling behavior is observed due to non-convergence, the UAV UEs can simply stop their learning process at any suboptimal solution. While that solution may

---

Algorithm 3 Testing phase of our deep RL algorithm

---

while At least one UAV UE j has not reached its destination d_j, **do**
    for all UAV UEs j (in parallel) **do**
        Input: Each UAV UE j obtains an input $v_j(t)$ using (6.16).
        Step 1: Action selection
Each UAV UE j chooses an action $z_j(t) = \mathrm{argmax}_{z_j \in \mathcal{Z}_j} y_j\left(v_j(t), z_j(t)\right)$.
        Step 2: Location, cell association, and transmit power update
Each UAV UE j updates its location, cell association, and transmission power level based on the selected action $z_j(t)$.

        Step 3: Action broadcast
Each UAV j broadcasts its selected action $z_j(t)$ to all other UAV UEs.

        Step 4: State transition vector update
Each UAV UE j updates the state transition vector $x_j^{(n)}(t)$ for each deep ESN layer (n) using (6.26) and (6.27).

    end for
end while

---

not be an SPNE, it can still provide a useful suboptimal outcome for a complicated data set. Naturally, one can consider future extensions to the architecture of the proposed RL algorithm whereby convergence is always guaranteed.

6.5 Representative Simulation Results

We simulate an 800 m × 800 m square area that we divide into 40 m × 40 m grid areas. In this area, we randomly and uniformly deploy 15 BSs and we use an uncorrelated Rician channel with $\widehat{K} = 1.59$ [227]. We discretize the maximum transmit power of each UAV into 5 equally separated levels. The external input of our deep ESN, $v_j(t)$, is a function of the number of UAV UEs; hence, the number of hidden nodes per layer, $N_{j,R}^{(n)}$, will change with the number of UAV UEs. Our key simulation parameters are summarized in Table 6.1.

Figure 6.5 presents the upper bound for the optimal altitude of UAV UE j as the SINR threshold value, $\bar{\Gamma}$ changes, for various values of the transmit power. Meanwhile, in Figure 6.5, we present the lower bound for the optimal UAV UE altitude as the interference threshold varies, for different transmit power levels. From these figures, we can observe that the upper bound on the UAV UE's optimal altitude decreases as $\bar{\Gamma}$ increases while its lower bound decreases as the interference threshold $\sum_{c=1}^{C_{j,s}(t)} \bar{I}_{j,r,c,a}$ increases. We also know that the maximum altitude of a UAV UE will be smaller as the ground network gets denser while its lower bound increases as the ground network data requirements increase. As a result, in such scenarios, a UAV UE is better off operating at higher altitudes.

6.5 Representative Simulation Results

Table 6.1 System parameters.

Parameters	Values	Parameters	Values
Maximum transmit power ($\overline{P}_j$) for a UAV UE	20 dBm	SINR threshold ($\overline{\Gamma}_j$)	−3 dB
Transmit power for ground ($\widehat{P}_q$) for a ground UE	20 dBm	Learning rate (λ_j)	0.01
Noise power spectral density (N_0)	−174 dBm/Hz	RB bandwidth (B_c)	180 kHz
Total bandwidth (B)	20 MHz	# of interferers (L)	2
Packet arrival rate ($\lambda_{j,s}$)	(0,1)	Packet size (ν)	2000 bits
Carrier frequency ($\hat{f}$)	2 GHz	Discount factor (γ)	0.7
Number of hidden layers	2	Step size ($\widetilde{a}_j$)	40 m
Leaky parameter/layer ($\omega_j^{(n)}$)	0.99, 0.99	ϵ	0.3

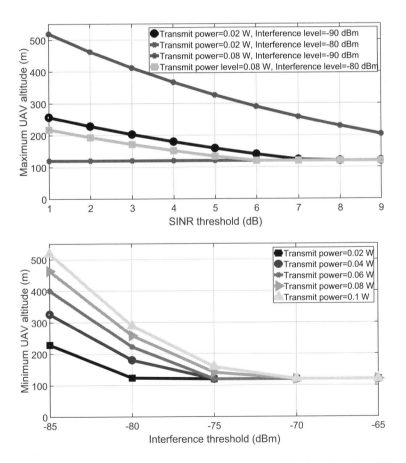

Figure 6.2 The (a) upper bound for the optimal UAV UE altitude as function of the SINR threshold ($\overline{\Gamma}$) for various transmit power levels and ground network density and (b) lower bound for the optimal UAV UE altitude as function of the interference threshold ($\sum_{c=1}^{C_{j,s}(t)} \overline{I}_{j,r,c,a}$), for various transmit powers.

Table 6.2 Performance assessment for one UAV.

	# of steps	delay (ms)	average rate per UE (Mbps)
Wireless-aware approach	32	6.5	0.95
Shortest path	32	12.2	0.76

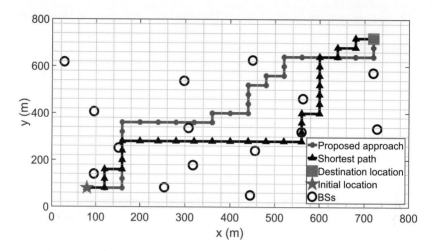

Figure 6.3 Illustrative snapshot showing the path of a single UAV UE resulting from our wireless-aware approach and from a shortest path scheme.

Figure 6.3 shows a snapshot of a representative path that a single UAV adopts. In this snapshot, we show the path resulting from our approach, which is wireless-aware, and we compare it with the path resulting from a baseline shortest path scheme whose goal is to minimize the UAV UE mission time. Indeed, in contrast to our studied scheme, which explicitly factors in wireless metrics during path planning, in the shortest path baseline, the goal of the UAV UEs is to reach their destinations with the minimum number of steps. Table 6.2 provides the performance results for the UAV UE's paths in Figure 6.3. In our studied RL approach, the UAV UE selects a path away from the densely deployed ground network area while still maintaining proximity to its serving ground BS. Therefore, in our approach, the UAV UE clearly balances the goals of optimizing its network performance and minimizing the time steps needed to complete its mission and reach its destination. This path clearly minimizes the interference level that the UAV UE induces on the ground UEs as well as the UAV UE's wireless latency (see Table 6.2). Therein the proposed approach achieves a 25% increase in the average rate per ground UE and a 47% decrease in the wireless latency compared to the shortest path, while requiring the same number of steps to reach the destination.

Figure 6.4 compares the average wireless latency per UAV UE and the average ground UE rate that result from our wireless-aware approach and the shortest path baseline. In addition, we use Table 6.3 to compare the number of steps required by all UAV UEs to arrive at their missions' destinations. From Figure 6.4 and Table 6.3, we can clearly observe that our approach achieves a lower wireless latency per UAV UE and a higher

6.5 Representative Simulation Results

Table 6.3 The required number of steps for all UAV UEs to reach their corresponding destinations based on our approach and that of the shortest path scheme for different numbers of UAV UEs.

# of steps	1 UAV	2 UAVs	3 UAVs	4 UAVs	5 UAVs
Wireless-aware approach	4	4	6	7	8
Shortest path	4	4	6	6	7

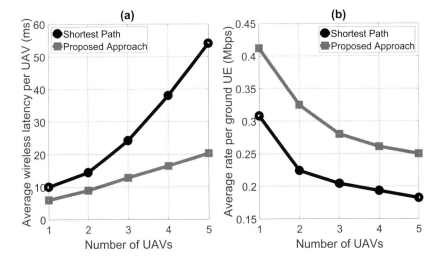

Figure 6.4 Performance evaluation of our approach in terms of average (a) wireless latency per UAV and (b) rate per ground UE as compared to the shortest path approach, for different network sizes.

rate per ground UE, compared to the shortest path baseline. This advantage is seen for different network sizes. We can also see that our wireless-aware solution needs a number of steps that is comparable to the shortest path case. In fact, the wireless-aware scheme provides a better tradeoff between UAV UE latency and ground UE data rate compared to the shortest path. For example, taking the case of 5 UAV UEs, we can observe that the introduced solution achieves a 37% increase in the average ground UE data rate and a 62% decrease in the average (per UAV UE) wireless latency. Indeed, one can adjust the multi-objective weights of our utility function based on several parameters, such as the rate requirements of the ground network, the power limitation of the UAV UEs, and the maximum tolerable wireless latency of the UAV UEs. Figure 6.4 also demonstrates that an increase in the number of UAV UEs will lead to an increase in the average delay per UAV UE and a decrease in the average rate per ground UE, for all schemes. This stems from the increase in the interference level on the ground UEs and other UAV UEs owing to the LOS links between UAV UEs and their serving ground BSs.

Figure 6.5 examines the impact of the altitude of the UAV UEs on the average, per UAV UE, wireless latency and the average data rate per ground UE for different utility functions. Clearly, higher UAV UEs' altitudes lead to an increase in the average wireless latency (for all utility functions) due to the associated increase in the distance

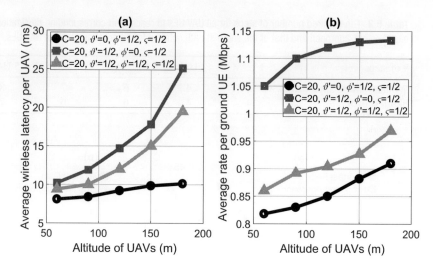

Figure 6.5 Performance evaluation of our approach in terms of average (a) wireless latency per UAV UE and (b) rate per ground UE for different utility functions and for different UAV UE altitudes.

between the UAV UEs and their serving BSs, which accentuates the path loss effect. Meanwhile, higher UAV UE altitudes will yield a higher average data rate for the ground UEs because the path loss effect will now reduce the air-to-ground interference. Clearly, we can now observe an interesting tradeoff between minimizing the average delay per UAV UE and maximizing the average ground UE data rate. Hence, alongside the multi-objective weights, the altitude of the UAV UEs can be modified so as to meet the rate requirements of the ground UEs while also minimizing the communication latency for each UAV UE depending on its mission goal.

In Figure 6.6, we show the average transmit power per UAV UE along its path as function of the number of ground BSs under two utility functions: one that focuses on minimizing the average latency for each UAV UE and another that focuses on minimizing the interference on the ground UEs. Figure 6.6 shows that adding ground BSs will have a direct effect on the power levels of the UAV UEs. For instance, for denser networks, an increase in the transmit power level occurs because of the increase in the interference from the ground UEs. Hence, the UAV UEs will use a larger transmission power in order to meet their wireless delay requirements. We can also observe that the average transmit power level per UAV UE decreases from 36 mW to 29 mW in the case of minimizing the interference level caused on neighboring ground BSs. This stems from the fact that, as the number of ground BSs increases, the interference caused by each UAV UE on the ground network will also increase. This, in turn, will force each UAV UE to reduce its transmit power. We note that, when minimizing the wireless latency, the average transmit power per UAV UE will be larger than the case of minimizing the interference level, irrespective of the ground network size. Hence, the UAV UEs' transmit power level is a function of their mission objective and the number of ground BSs.

6.5 Representative Simulation Results

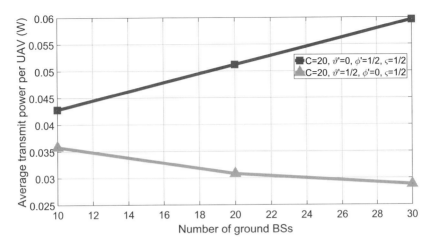

Figure 6.6 Impact of densifying the ground network on the average transmit power of the UAV UEs along their paths.

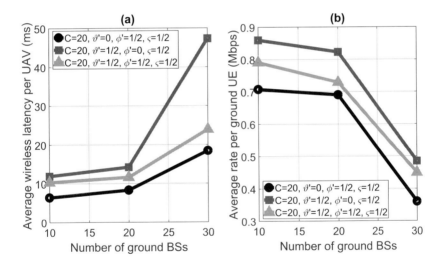

Figure 6.7 Impact of the ground network size on the average (a) wireless latency per UAV UE and (b) rate per ground UE for different utility functions and for a constant UAV UE altitude of 120 m.

In Figure 6.7, we show the (a) wireless latency per UAV UE and (b) rate per ground UE for different utilities as a function of the number of BSs and for a fixed UAV UE altitude of 120 m. For a denser ground network, we can see that the average wireless latency per UAV UE will increase while the average per UE rate will decrease. For instance, when the objective is to minimize the interference level along with energy efficiency, the average wireless latency per UAV UE increases from 13 ms to 47 ms, and the average rate per ground UE decreases from 0.86 Mbps to 0.48 Mbps as the number of BSs increases from 10 to 30.

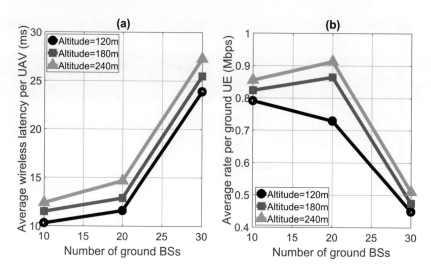

Figure 6.8 Average (a) wireless latency per UAV and (b) rate per ground UE as the ground network size changes, for various utility functions and for different UAV UE altitudes.

Figure 6.8 shows how the (a) wireless latency per UAV UE and (b) rate per ground UE will vary for different UAV UE altitudes and different ground network size (number of BSs). As the UAV altitude increases and/or the ground network becomes denser, the average wireless latency per UAV UE becomes higher. For example, we can observe a delay increase of 27% as the UAV UEs' altitude doubles for 20 BSs. Meanwhile, we observe an increase of 120% as the number of BSs increases from 10 to 30 for a fixed UAV UE altitude of 180 m. The results in Figure 6.5 demonstrate that the UAV UE's maximum altitude will decrease for denser ground networks, and, hence, it is desirable to operate the UAVs at a lower altitude when the number of BSs increases from 10 to 30. From Figure 6.8, we can also see that a denser ground network leads to a decrease in the average data rate per ground UE because of the increase in the interference. Meanwhile, the average data rate per ground UE will increase as the UAV UEs' altitude becomes higher. Clearly, the overall network performance tightly depends on both the UAV UEs' altitude and the number of BSs. In a dense ground network, UAV UEs must fly at a lower altitude for applications in which the wireless transmission latency is more critical and at a higher altitude when a minimum achievable data rate for the ground UEs is required.

In Figure 6.9, we study the impact of changing the number of nearest BSs (L_j) in the observed network state, $v_j(t)$, of a UAV UE j on the average rate per ground UE for different utility functions. An improvement can be observed in the average rate per ground UE as the number of nearest BSs in the state definition increases. For instance, when UAV UEs minimize the interference caused on the ground network, the average rate per ground UE increases by 28% as the number of BSs in the state definition increases from 1 to 5. This gain results from the fact that, as L_j increases, the UAV UEs get a better sense of their surrounding environment and, hence, they can more properly

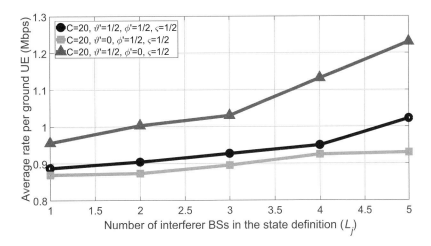

Figure 6.9 Average data rate per ground UE when the number of interferer BSs in the state definition (L_j) changes.

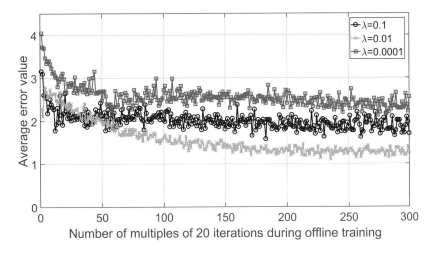

Figure 6.10 Convergence of the offline training phase as function of the learning rate.

choose their next location in a way to minimize the interference level they induce on the ground network. Another important observation here is that, as L_j increases, the size of the external input ($\boldsymbol{v}_j$) will be larger, which requires more neurons per layer. This will then increase the number of iterations needed to converge, which highlights an interesting tradeoff between improving the ground UEs' performance and maintaining low algorithmic complexity.

Figure 6.10 shows how the learning rates, λ, affect the average training error $e_j(\boldsymbol{v}_j(t))$. In essence, the choice of a learning rate directly determines the step size that the RL process must take to reach the SPNE. In this regard, a choice of small values for the learning rate, e.g., $\lambda = 0.0001$, will yield slower convergence. In contrast, choosing large learning rate values (e.g., $\lambda = 0.1$) reduces the error function for the first few iterations but that effect levels off thereafter. Figure 6.10 shows that $\lambda = 0.1$ will not lead to convergence during the testing phase. In contrast, $\lambda = 0.0001$ and $\lambda = 0.01$ will result in convergence. This is because large initial learning rates will decay the loss function faster and thus make the model get stuck at a particular region of the optimization space instead of better exploring it. Clearly, a better performance is achieved for $\lambda = 0.01$, as compared to smaller and larger values of the learning rate. To overcome this issue, we adopt the early stopping technique to avoid overfitting.

6.6 Chapter Summary

In this chapter, a novel interference-aware path planning scheme that allows cellular-connected UAV UEs to minimize interference caused on a ground network was investigated. The problem has been formulated as a noncooperative game in which the UAV UEs are the players. To solve the game, a deep RL algorithm based on ESNs was investigated and shown to find an SPNE of the formulated path planning game, if it converges. The introduced algorithm enables each UAV to decide on its next location, transmission power level, and cell association vector in an autonomous manner, thus adapting to the changes in the network. Simulation results have shown that the proposed approach achieves better wireless latency per UAV and rate per ground UE while requiring a number of steps comparable to the shortest path scheme. The results show that a UAV UE's altitude plays a vital role in minimizing the interference level on the ground UEs as well as the wireless transmission delay of the UAV UE. In particular, the altitude of the UAV is a function of the ground network density, the UAV UE's objective, and the actions of other UAV UEs in the network. Finally, we note that the model that we designed in this chapter can be to many of the UAV UE use cases that we discussed in Chapter 2.

7 Resource Management for UAV Networks

The deployment of UAVs and their mobility require network operators to revisit the way in which the network resources, such as spatial resources (e.g., cell association) and spectrum resources, are managed. In particular, the three-dimensional nature of UAV networks and their mobility bring forward new challenges for resource management. In this chapter, we primarily focus on how wireless communication resources can be optimized and managed in wireless networks that support UAVs. In Section 7.1, we start by analyzing a very unique problem related to wireless networks supported by hovering UAV BSs: Cell association in the presence of explicit hover time constraints. Naturally, the presence of hover times for UAVs will drastically change the way in which cell association is performed. Then, in Section 7.2, we generalize the problem of cell association to a fully fledged 3D cellular system that integrates both UAV BSs and UAV UEs. Subsequently, in Section 7.3, we investigate the problem of spectrum and cache management in a wireless network supported by UAV BSs that are able to access both licensed and unlicensed spectrum resources. We conclude this chapter with insights on the problems of resource management in UAV networks.

7.1 Cell Association in UAV-Assisted Wireless Networks under Hover Times Constraints

In this section, we study the problem of cell association in a wireless network that uses UAV BSs to provide downlink connectivity to ground users based on our work in [228]. In particular, we study the potential of using hovering UAVs that can stay relatively static over certain geographical areas to provide wireless communication and connectivity. As discussed in previous chapters, UAV BSs can indeed complement existing wireless systems by delivering connectivity to congested (e.g., hotspot) areas, to disaster-affected areas, as well as to temporary events such as in a stadium or an open-air theater. However, to provide such connectivity, LAP UAVs such as quadrotor drones, must be able to hover over a given area for a specified period of time. In this context, the hover time that a UAV BS can spend over a given geographical area is limited by the battery, energy, and other hardware limitations of UAVs. As a result, the presence of hover time constraints imposes a number of limitations on the quality of the data service that UAV BSs can provide. Here, our goal is to study how such hover time constraints can strongly impact the performance and connectivity provided by UAV BSs. In particular, we focus on how cell association can be optimally managed to meet the data service requirements of users, while being cognizant of the hover time constraints of the UAV BSs.

It is noteworthy to mention that problems of cell association for wireless networks with UAV BSs have been studied in some prior works, such as [229] and [230]. Although these work show how the area-to-UAV assignment can enhance the wireless capacity of a cellular system, they do not take into account the fact that ground users in a UAV-assisted wireless network can be spatially distributed in a rather arbitrary manner. Moreover, these works do not explicitly factor in the impact of hover times; which is necessary to design proper resource and cell association mechanisms. Indeed, the *flight/hover time duration of UAV BSs* will bring forward important technical challenges that are unique to UAV-assisted wireless systems [231] and [232]. In this context, the *hover time* of a UAV BS, defined as the flight time during which a UAV BS can remain relatively stationary in the sky over a certain geographical area to wirelessly communicate with ground users, affects the performance of the system in various ways. For instance, a longer UAV hover time can enable the UAV BS to service ground wireless users for longer periods of time which, in turn, allows it to sustain higher load requirements and service a larger geographical area. In contrast, a lower hover time (as constrained by the onboard battery of a UAV as well as some of the flight regulations discussed in Chapter 1) can limit the amount of data that can be delivered by this UAV BS. Therefore, to analyze cell association and resource management in a UAV-assisted wireless network, the hover time constraints must be explicitly accounted for. In this context, most of the prior UAV studies related to resource management, such as [25, 42, 157, 169, 171, 229, 233–237], have not considered the hover time constraints, which motivates the study provided in this section.

7.1.1 System Model

We focus on a $\mathcal{D} \subset \mathbb{R}^2$ geographical area in which multiple wireless users are deployed. In the two-dimensional spatial plane, the users are distributed according to a given, generic distribution $f(x, y)$. The network deploys a set $\mathcal{M}$ of M UAVs that will be used as aerial BSs to provide connectivity for the users on the ground[1]. For each UAV $i \in \mathcal{M}$ located at an altitude h_i, we define $s_i = (x_i, y_i, h_i)$ as its 3D coordinate. We consider the downlink of the wireless network, and we use a frequency division multiple access scheme for the UAV BSs. For each UAV BS i, we define P_i as the maximum transmit power and B_i as the total available bandwidth.

Since we are interested in cell association, we will partition the geographical area into different subareas, as shown in Figure 8.8. For each UAV BS i, we define $\mathcal{A}_i$ as the partition of the geographical area that will be serviced by UAV BS i. In other words, all users that are positioned in the area $\mathcal{A}_i$ will be associated with UAV BS i for wireless connectivity. Given that we have M UAV BSs, then, our area will be divided into a total of M disjoint partitions (one partition per UAV). Each UAV i will have a *hover time* τ_i, which represents the time duration used by UAV i to hover (i.e., relatively stop) over a corresponding cell area partition to serve ground wireless users. Within

---

[1] To provide backhaul connectivity for this UAV-enabled network, for model simplicity, we consider that readily available solutions, such as satellites or WiFi, are used [238].

7.1 Cell Association in UAV-Assisted Wireless Networks

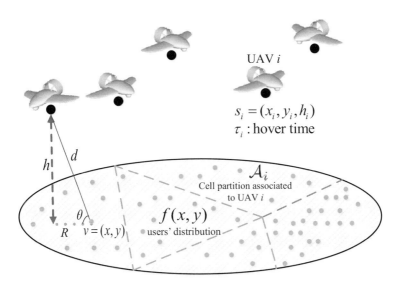

Figure 7.1 System model.

its hover time duration, each UAV BS will connect to the cellular ground users, perform control and computations, and transmit the needed data to the devices. Hence, we can define a variable T_i that captures the effective data transmission duration within which a given UAV i will be serving its associated users. Naturally, T_i will be smaller than the total hover time, because the UAV will use its hover time not only for wireless transmission purposes but also to perform control and computing functions. Therefore, we define a *UAV control time* function $g_i(.)$ that depends on the total number of users in $\mathcal{A}_i$. $g_i(.)$ captures the fraction of the hover time that is not used for effective wireless data transmission. This control time translates into the total duration that a UAV BS i must spend for control and computation functions, setting up connections, and performing signaling. Intuitively, the control time increases with the total number of users in a given cell.

Hereinafter, the total amount of data (i.e., in bits) that each UAV BS sends to a given ground user will be referred to as the *data service*. This data service is impacted by a number of key factors such as the effective data transmission time (stemming from the flight time) and the network bandwidth, which itself depends on the hover time and the control time. Clearly, this data service will be impacted by the hover time. In particular, for a wireless network with UAV BSs, we can consider two types of resources: (a) bandwidth (as is done in a classical cellular network), and (b) effective data transmission times that depend on the hover time (a unique feature of UAV networks).

Now that the model is setup, one can consider two important resource management scenarios. **Scenario 1**, which we call *UAV communications under hover time constraints*, is a case in which, given the maximum imposed hover time durations (dictated by each UAV BS's flight and energy restrictions), the goal is to find an optimal cell partitioning for the area that maximizes the average amount of data transmitted to ground users in a fair manner. In particular, Scenario 1 involves partitioning the area in an optimal way based on both the hover time constraints and the spatial distribution of the wireless users. In short, Scenario 1 asks the following question: Under a maximum, per UAV hover time constraint, how can the network maximize the total data service while including fairness (e.g., load balancing) constraints? Scenario 1 mainly pertains to resource-limited wireless transmission scenarios in which the amount of resources (for our case: hover times and bandwidth resources) is not sufficient to completely meet the needs of the wireless users. An illustrative example of Scenario 1 is one in which battery-limited UAV BSs are deployed in geographical hotspots that have a high load of users and associated demands for wireless networking.

Alternatively, one can consider a second scenario, **Scenario 2**, called *UAV communication under load constraints*. Here, the objective is to optimize the hover times of the UAV BSs in a way to completely meet the data service demands (in bits) required by the ground users. Scenario 2, therefore, poses the following problem: Under a known load requirement for each ground user (at each location), how can the network minimize the average hover times of the UAV BSs? Answering this question allows a wireless network to satisfy the demands of its users with a minimum hover time from the UAV BSs. Minimizing the hover time will, in turn, lead to indirectly minimizing the energy consumption of the UAVs. The analysis of Scenario 2 is particularly apropos for public safety and emergency situations in which all ground users must be quickly and efficiently served by aerial UAV BSs.

In the scope of this chapter, we will provide an analytical exposition on the solution of Scenario 1. For Scenario 2, given that the approach is somewhat analogous to Scenario 1, we will restrict our attention to a few insightful numerical results.

To model our network, we need to use a proper AG channel model. As discussed in Chapter 3, several AG channel models exist for UAV networks. Within the scope of this section, we adopt the widely used probabilistic path loss model discussed in Chapter 3 and provided by ITU [239] and the work in [157]. This model includes both an LOS and NLOS component. For notational convenience and coherence, we redefine this model here. For instance, the path loss between UAV BS i and a given user at location (x, y) will be:

$$\Lambda_i(x, y) = \begin{cases} \left(\frac{4\pi f_c d_o}{c}\right)^2 \left(d_i(x,y)/d_o\right)^2 \mu_{\text{LOS}}, & \text{LOS link,} \\ \left(\frac{4\pi f_c d_o}{c}\right)^2 \left(d_i(x,y)/d_o\right)^2 \mu_{\text{NLOS}}, & \text{NLOS link,} \end{cases} \quad (7.1)$$

where μ_{LOS} and μ_{NLOS} are different attenuation factors for LOS and NLOS connections. Meanwhile, the variable f_c represents the carrier frequency, d_o represents the free-space reference distance, and c is the speed of light. $d_i(x, y) = \sqrt{(x - x_i)^2 + (y - y_i)^2 + h_i^2}$ represents the distance between a UAV i and an arbitrary user at location (x, y). We then define the LOS probability for any UAV BS to user link, as follows:

$$P_{\text{LOS},i} = b_1 \left(\frac{180}{\pi}\theta_i - 15\right)^{b_2}. \quad (7.2)$$

In (7.2), $\theta_i = \sin^{-1}(\frac{h_i}{d_i(x,y)})$ represents the elevation angle (in radians) between the UAV BS and the ground user while b_1 and b_2 are constants that capture the environment impact. Recall that the NLOS probability will be given by: $P_{\text{NLOS},i} = 1 - P_{\text{LOS},i}$. Now, given that $d_o = 1$ m and $K_o = \left(\frac{4\pi f_c}{c}\right)^2$, we can define the average path loss as

$K_o d_i^2(x,y)[P_{\text{LOS},i}\mu_{\text{LOS}} + P_{\text{NLOS},i}\mu_{\text{NLOS}}]$. Then, we can find the power of the received signal from a UAV i:

$$\bar{P}_{r,i}(x,y) = \frac{P_i}{K_o d_i^2(x,y)\left[P_{\text{LOS},i}\mu_{\text{LOS}} + P_{\text{NLOS},i}\mu_{\text{NLOS}}\right]}, \qquad (7.3)$$

with P_i being the transmit power of UAV BS i. We can write the SINR received by any ground user at coordinate (x,y) and served by UAV BS i:

$$\gamma_i(x,y) = \frac{\bar{P}_{r,i}(x,y)}{I_i(x,y) + \sigma^2}, \qquad (7.4)$$

where the term $I_i(x,y) = \beta \sum_{j \neq i} \bar{P}_{r,j}(x,y)$ captured the interference experienced by a user at location (x,y) and stemming from all UAV BSs other than UAV BS i. We also define $0 \leq \beta \leq 1$ as a weight factor to control the amount of interference (e.g., as a term that can be used to allow some form of interference mitigation). $\beta = 1$ and $\beta = 0$, respectively, represent two extreme scenarios: a full interference and an interference-free scenario.

Next, for a given user at coordinate (x,y) and served by UAV BS i, we can define the data rate:

$$C_i(x,y) = W(x,y)\log_2\left(1 + \gamma_i(x,y)\right), \qquad (7.5)$$

where $W(x,y)$ is the bandwidth allocated to the user at (x,y). As a result, the data service provided by UAV BS i to the user at (x,y) will be given by:

$$L_i(x,y) = T_i C_i(x,y), \qquad (7.6)$$

where T_i is UAV BS i's effective transmission time. Here, $L_i(x,y)$ is essentially the total number of bits sent to a ground user located at coordinate (x,y). This data service term will be a function of several parameters such as the user's location, the allocated bandwidth, the location of the serving UAV, as well as T_i, the effective data transmission time of UAV BS i. As discussed earlier, now, each UAV has two key resources to allocate to its ground users: the effective data transmission times and the bandwidth. Naturally, the amount of resources that each user can obtain is dependent on various network parameters that include the partitioning of the cells, the bandwidth and hover time of the UAVs, and the total number of users. Having defined our general model, we can now analyze cell association and partitioning for Scenario 1.

7.1.2 Optimal and Fair Cell Partitioning for Data Service Maximization under Hover Time Constraints

As discussed previously, in Scenario 1, the goal is to derive the optimal cell partitions that can maximize the average data service to the wireless users while taking into account the restrictions on the hover times of the UAV BSs and the current spatial distribution of the ground users. We assume that each cell will be assigned to a single UAV BS. Meanwhile, all users within a given cell will be serviced by the UAV BS that is assigned to that cell. To perform cell partitioning, it is customary to use classical partitioning approaches, such as Voronoi and weighted Voronoi diagrams [240]. However, these known techniques do not explicitly take into account the spatial distribution of users, which can lead to imbalanced cell partitions and loads across cells. In short, conventional Voronoi-based cell partitioning approaches can lead to a highly unfair data service for the users. In contrast, while maximizing the total data service, our approach will ensure that resources are equally shared among all users. Hence, our approach avoids creating unbalanced cell partitions, thereby leading to better fairness compared to classical Voronoi solutions.

The hover time τ_i that UAV BS i uses to serve the users located in its cell $\mathcal{A}_i$ will encompass the effective data transmission time and the control time. To ensure fairness, we impose the following condition:

$$T_i = \tau_i - g_i \left(\int_{\mathcal{A}_i} f(x,y) \mathrm{d}x \mathrm{d}y \right), \quad \forall i \in \mathcal{M}, \tag{7.7}$$

where g_i is the control time, which depends on the number of the users in $\mathcal{A}_i$. Here, we can note that, given the spatial distribution of users, $f(x,y)$, and the total number of users, N, the average number of users in partition $\mathcal{A}_i$ can be defined as $N \int_{\mathcal{A}_i} f(x,y) \mathrm{d}x \mathrm{d}y$ [241].

From (7.5) and (7.6), we observe that the term $T_i B_i$ can be viewed as the resources that UAV BS i employs to serve users in $\mathcal{A}_i$. Therefore, in order to ensure a fair resource allocation policy, we should have:

$$\frac{T_i B_i}{\int_{\mathcal{A}_i} f(x,y) \mathrm{d}x \mathrm{d}y} = \frac{T_j B_j}{\int_{\mathcal{A}_j} f(x,y) \mathrm{d}x \mathrm{d}y}, \quad \forall i \neq j \in \mathcal{M},^2 \tag{7.8}$$

where (7.8) guarantees that a UAV BS that has more bandwidth and a longer hover time will provide service to more users.

Given (7.8) and $\int_{\mathcal{D}} f(x,y) \mathrm{d}x \mathrm{d}y = \sum_{k=1}^{M} \int_{\mathcal{A}_k} f(x,y) \mathrm{d}x \mathrm{d}y = 1$, we can derive the following constraint on the number of users in each partition:

$$\int_{\mathcal{A}_i} f(x,y) \mathrm{d}x \mathrm{d}y = \frac{B_i T_i}{\sum_{k=1}^{M} B_k T_k}, \quad \forall i \in \mathcal{M}. \tag{7.9}$$

By inspecting (7.9), we can observe that the number of users in each generated optimal partition depends on the resources of the UAV BSs. For instance, if the UAV BSs have the same bandwidths and hover times, then (7.7)–(7.9) lead to $\int_{\mathcal{A}_i} f(x,y) \mathrm{d}x \mathrm{d}y = \frac{1}{M}$, $\forall i \in \mathcal{M}$. In other words, identical UAV BSs will serve equally loaded cells.

Given (7.5), (7.6), and (7.9), we can define the average data service at location $(x,y) \in \mathcal{A}_i$:

$$L_i(x,y) = \frac{T_i B_i}{N \int_{\mathcal{A}_i} f(x,y) \mathrm{d}x \mathrm{d}y} \log_2(1 + \gamma_i(x,y)) = \left(\frac{1}{N} \sum_{k=1}^{M} B_k T_k \right) \log_2(1 + \gamma_i(x,y)). \tag{7.10}$$

---

[2] Note that, given hover times of the UAVs, τ_i, $\forall i \in \mathcal{M}$, we can compute T_i, $\forall i \in \mathcal{M}$ by solving the system of equations in (7.7) and (7.8).

We can now formally pose an optimization problem whose goal is to optimally partition the service area of the UAV BSs so as to maximize the average data service, as follows:

$$\max_{\mathcal{A}_i, i \in \mathcal{M}} \sum_{i=1}^{M} \int_{\mathcal{A}_i} \left(\frac{1}{N} \sum_{k=1}^{M} B_k T_k \right) \log_2 (1 + \gamma_i(x,y)) f(x,y) \mathrm{d}x \mathrm{d}y, \quad (7.11)$$

$$\text{s.t.} \int_{\mathcal{A}_i} f(x,y) \mathrm{d}x \mathrm{d}y = \frac{B_i T_i}{\sum_{k=1}^{M} B_k T_k}, \quad \forall i \in \mathcal{M}, \quad (7.12)$$

$$\gamma_i(x,y) \geq \gamma_{th}, \text{ if } (x,y) \in \mathcal{A}_i, \quad \forall i \in \mathcal{M}, \quad (7.13)$$

$$\mathcal{A}_l \cap \mathcal{A}_m = \emptyset, \quad \forall l \neq m \in \mathcal{M}, \quad (7.14)$$

$$\bigcup_{i \in \mathcal{M}} \mathcal{A}_i = \mathcal{D}, \quad (7.15)$$

where (7.12) is a constraint on the load of each cell while (7.13) is the necessary condition for associating each user to a UAV i. (7.30) and (7.31) guarantee disjoint cell partitions whose union covers the entire considered area $\mathcal{D}$.

Given (7.13), we introduce a function $q_i(x,y) = \left(\frac{\gamma_i(x,y)}{\gamma_{th}} \right)^{-n}$ with n being a large number (i.e., tends to $+\infty$), and, then, we subtract $q_i(x,y)$ from the objective function in (7.11). Now, we can observe that, whenever constraint (7.13) is violated, $q_i(x,y)$ goes $+\infty$ and, thus, point (x,y) will not be assigned to UAV i or equivalently $(x,y) \notin \mathcal{A}_i$. Hence, whenever the problem is feasible, we can omit (7.13) while penalizing the objective function in (7.11) by $q_i(x,y)$. We now let $\lambda = \frac{1}{N} \sum_{k=1}^{M} B_k T_k$, and $\omega_i = \frac{B_i T_i}{\sum_{k=1}^{M} B_k T_k}$. Then, the optimization in (7.11) can be cast as the following minimization problem:

$$\min_{\mathcal{A}_i, i \in \mathcal{M}} \sum_{i=1}^{M} \int_{\mathcal{A}_i} -\left(\lambda \log_2(1 + \gamma_i(x,y)) - q_i(x,y) \right) f(x,y) \mathrm{d}x \mathrm{d}y, \quad (7.16)$$

$$\text{s.t.} \int_{\mathcal{A}_i} f(x,y) \mathrm{d}x \mathrm{d}y = \omega_i, \quad \forall i \in \mathcal{M}, \quad (7.17)$$

$$\mathcal{A}_l \cap \mathcal{A}_m = \emptyset, \quad \forall l \neq m \in \mathcal{M}, \quad (7.18)$$

$$\bigcup_{i \in \mathcal{M}} \mathcal{A}_i = \mathcal{D}. \quad (7.19)$$

There are many challenges that must be overcome in order to solve (7.16). These challenges include the continuity of the optimization variables $\mathcal{A}_i, \forall i \in \mathcal{M}$, the fact that $f(x,y)$ is a generic function of x and y, and the presence of complex constraints in (7.17). We will therefore use mathematical tools from the field of *optimal transport theory* [242] to overcome these challenges and find the optimal cell partitions, $\mathcal{A}_i$, for which the average total data service is maximized. Optimal transport theory essentially studies matching problems between two continuous or discrete sets using a so-called *transport map T*, which is used to map one set to another. In our considered scenario, we have a continuous distribution of users that must be matched to a discrete set of

UAV BS locations. In general, the optimal cell partitions can be obtained by optimally mapping the users to the UAV BSs.

Given (7.16), the cell partitions are related to the concept of a transport map by [243]:

$$\left\{ T(v) = \sum_{i \in \mathcal{M}} s_i \mathbb{1}_{\mathcal{A}_i}(v); \int_{\mathcal{A}_i} f(x,y) \mathrm{d}x\mathrm{d}y = \omega_i \right\}, \quad (7.20)$$

where $\omega_i = \frac{B_i T_i}{\sum_{k=1}^{M} B_k T_k}$, as given in (7.17), is directly related to the hover time and the bandwidth of the UAV BSs. Also, $\mathbb{1}_{\mathcal{A}_i}(v)$ is an indicator function that will be 1 if $v \in \mathcal{A}_i$, and 0 otherwise. Given this notation, we can cast (7.16) within the optimal transport framework as follows. Given a continuous probability measure f of users, and a discrete probability measure $\Gamma = \sum_{i \in \mathcal{M}} \omega_i \delta_{s_i}$ corresponding to the UAVs, we must find the optimal transport map for which $\int_{\mathcal{D}} J(v, T(v)) f(x,y) \mathrm{d}x\mathrm{d}y$ is minimized. In this case, δ_{s_i} is the Dirac function, and J is the transportation cost function, which is used in (7.16) and is given by:

$$J(v, s_i) = J(x, y, s_i) = q_i(x,y) - \lambda \log_2 (1 + \gamma_i(x,y)). \quad (7.21)$$

We can now see that our cost function, J, and the source distribution, f, are continuous. In this case, by using the so-called Monge-Kantorovich problem from optimal transport, we can state the following theorem (whose proof can be found in [228]):

THEOREM 7.1 *The optimization problem in (7.16) is equivalent to the following unconstrained maximization problem:*

$$\max_{\psi_i, i \in \mathcal{M}} \left\{ F(\boldsymbol{\psi}^T) = \sum_{i=1}^{M} \psi_i \omega_i + \int_{\mathcal{D}} \psi^c(x,y) f(x,y) \mathrm{d}x\mathrm{d}y \right\}, \quad (7.22)$$

where $\boldsymbol{\psi}^T$ is a vector of variables $\psi_i, \forall i \in M$, and $\psi^c(x,y) = \inf_i J(x,y,s_i) - \psi_i$.

Theorem 8.2 shows that the complex optimal cell partitioning problem in (7.16) can be transformed to a tractable optimization problem with M variables. Hence, by solving (7.22), we can obtain the optimal values of $\psi_i, \forall i \in \mathcal{M}$ that can then be used to derive the optimal cell partitions. This solution, in fact, can be completely characterized by the following theorem [228]:

THEOREM 7.2 *Given (7.22), F is a concave function of variables $\psi_i, i \in \mathcal{M}$. We also have:*

$$\frac{\partial F}{\partial \psi_i} = \omega_i - \int_{\mathcal{D}_i} f(x,y) \mathrm{d}x\mathrm{d}y, \quad (7.23)$$

where $\mathcal{D}_i = \{(x,y) | J(x,y,s_i) - \psi_i \leq J(x,y,s_j) - \psi_j, \forall j \neq i\}$.

Theorem 7.2 shows the concavity of F as a function of $\boldsymbol{\psi}^T$. As a result, by maximizing F, we can derive the optimal values for $\psi_i, \forall i \in \mathcal{M}$. Subsequently, given the optimal $\psi_i, \forall i \in \mathcal{M}$, (7.20) can be employed to derive the optimal cell partitions corresponding to (7.16). In particular, as shown in [228], a gradient-descent-based approach can ultimately be used to find the optimal partition by leveraging the result of Theorem 7.2.

Table 7.1 Simulation parameters.

Parameter	Description	Value
f_c	Carrier frequency	2 GHz
P_i	Transmit power of each UAV BS	0.5 W
N_o	Noise power spectral density	-170 dBm/Hz
N	Number of ground users	300
μ_{LOS}	Additional path loss to free space for LOS	3 dB
μ_{NLOS}	Additional path loss to free space for NLOS	23 dB
B	Bandwidth	1 MHz
α	Control time factor	0.01
h	Altitude of a UAV BS	200 m
u	Load per user	100 Mb
μ_x, μ_y	Mean of the truncated Gaussian distribution	250 m, 330 m
b_1, b_2	Environmental parameters (dense urban)	0.36, 0.21 [157]

As proven in [228], one can follow a similar approach to address the cell partitioning problem under Scenario 2. Next, we will provide a set of simulation results that showcase the impact of hover time on the overall operation of a wireless network that uses UAV BSs. The results will include insights from both Scenario 1 and Scenario 2.

7.1.3 Extensive Simulations and Numerical Results

We evaluate the developed framework by using extensive simulations. We use a two-dimensional truncated Gaussian distribution to deploy wireless ground users within a rectangular area of size 1000 m × 1000 m. This spatial distribution is chosen since it is an accurate representation of a hotspot area. We use a grid-based deployment for the UAVs, and we deploy them at an altitude of 200 m. Unless stated otherwise, we consider a full interference case with $\beta = 1$. We use $g_i(Na_i) = \alpha(Na_i)^2$ for the control time with α being an arbitrary constant. Other parameters are provided in Table 7.1. We compare our results, obtained based on the developed optimal cell partitioning approach, with the classical weighted Voronoi diagram baseline. All statistical results are averaged over a large number of independent runs.

Representative Results for Scenario 1

In Figures 7.2 and 7.3, we show an illustrative, comparative example of the partitions resulting from our studied approach and a classical weighted Voronoi diagram. Here, an illustration of UAV partitions under a nonuniform user distribution is presented. This example includes a total of five UAV BSs that are servicing ground users distributed according to a nonuniform, truncated Gaussian spatial distribution. Here, we set the maximum hover time to 30 minutes, a typical value for quadcopter UAV BSs [244]. These figures use a darker color for areas that have a higher user density. Figure 7.3 shows that the cell partitions related to UAV BSs 4 and 5 have significantly more users compared to partition 1. Hence, under hover time restrictions, ground users that are positioned in cell partitions 4 and 5 cannot be fairly served by UAV BSs. However,

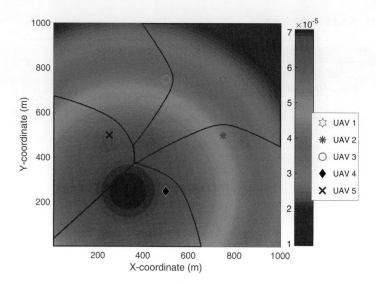

Figure 7.2 Optimal transport-based cell partitions.

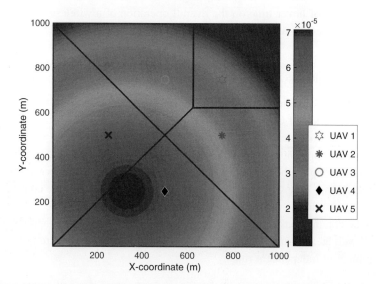

Figure 7.3 Weighted Voronoi diagram.

the optimal cell partitions resulting from our discussed approach are derived in a way that the average data service under a fairness constraint is maximized. For instance, Figure 7.2 demonstrates that the size of cell partitions 4 and 5 decreases compared to the weighted Voronoi diagram. Hence, the developed solution leads to better fairness among the users compared to the weighted Voronoi case.

The fairness of the studied scheme is further evaluated in Figure 7.4 using the popular Jain's fairness index for different values of σ_o, which is a parameter that determines how uniform the truncated Gaussian distribution will be. Here, we note that larger values of σ_o imply a more uniform spatial distribution. Figure 7.4 clearly demonstrates

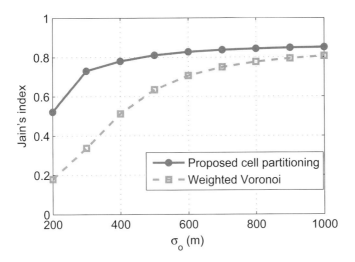

Figure 7.4 Jain's index for fairness (in terms of average data service) of the developed solution compared to Voronoi.

that the studied, hover-time cognizant solution yields better fairness as exemplified by the minimum Jain's index resulting from both solutions. In particular, the developed framework is much more fair in scenarios with highly non-uniform spatial distributions (i.e., practical, real-worldistic hotspot scenarios). Clearly, whenever the distribution becomes more uniform, the developed approach will tend toward the Voronoi case.

In Figure 7.5, we show how the interference factor β affects the average total data service. Figure 7.5 first corroborates the intuition that a lower interference will lead to a higher data service for the users. For instance, for a scenario with 5 UAV BSs, the data service can be tripled by reducing β from 1 to 0.1. From this figure, we can also observe that using more UAV BSs is only beneficial if proper interference mitigation (i.e., a low β) is done. For example, Figure 7.5 shows that doubling the number of UAV BSs from 5 to 10 provides a substantial gain (about 56%) if $\beta = 0.1$. However, this gain is only 5% for $\beta = 1$.

Figure 7.6 shows the impact of the maximum hover time on the data service. One interesting result from Figure 7.6 is the fact that the use of a small number of UAV BSs (i.e., 5 UAV BSs) with a relatively large hover time (40 minutes) yields a better performance compared to a case with a double number of UAV BSs (10 UAV BSs) with a maximum hover time of 30 minutes. Therefore, an additional 10 minutes of hover time can lead a better performance than doubling the number of UAV BSs. Naturally, these gains stem not only from the hover time but also from the fact that adding UAV BSs can increase interference. Nonetheless, this result clearly showcases the importance of the hover time in resource management for wireless networks with UAV BSs. For system operators, in many scenarios, it can be more efficient to deploy more capable UAV BSs (with more energy to fly/hover as discussed in Chapter 1) than to deploy a larger number of UAV BSs with shorter flying times.

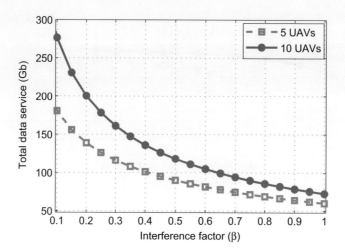

Figure 7.5 Average data service resulting from the studied solution as a function of the interference factor.

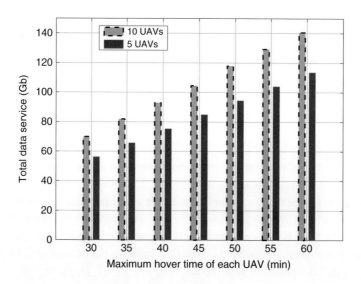

Figure 7.6 Impact of the maximum hover time on the network performance.

Representative Results for Scenario 2

Next, we present illustrative results for Scenario 2. Recall that, in this scenario, the objective is to minimize the hover time under a data service requirement for each ground user. In the numerical results, we set this data service requirement to a value of 10 Mb.

First, in Figure 7.7, we show how the network bandwidth impacts the total hover time needed by the UAVs. In this figure, we compare two bandwidth allocation schemes: an optimal bandwidth allocation that minimizes the hover time (derived in [228]), and a baseline equal bandwidth allocation. From this figure, we can first see that a larger bandwidth leads to a smaller hover time. Therefore, having more bandwidth can enable

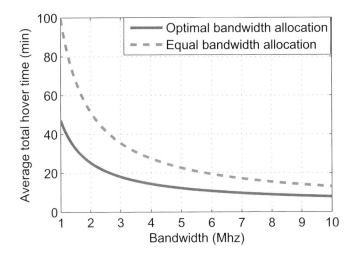

Figure 7.7 Impact of bandwidth and bandwidth allocation on the average (per UAV) hover time.

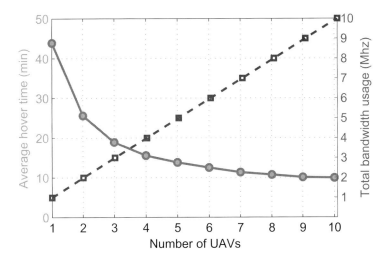

Figure 7.8 Average hover time as function of the network size (number of UAVs) and bandwidth.

the UAV BSs to serve the ground user (at a given target rate) more quickly. Moreover, Figure 7.7 also shows that optimizing the bandwidth while being aware of the hover time can reduce this hover time by up to 51% compared to a hover time-agnostic equal bandwidth allocation.

In Figure 7.8, we present the average total UAV BS hover time as a function of the number of UAV BSs for an interference-free scenario. In this case, we can clearly see that the total bandwidth usage linearly increases with the network size. Figure 7.8 also shows that using a larger number of UAV BSs can potentially reduce the total hover time for an interference-free scenario. For example, from Figure 7.8, we can see that tripling the number of UAV BSs increases from 2 to 6 leads to

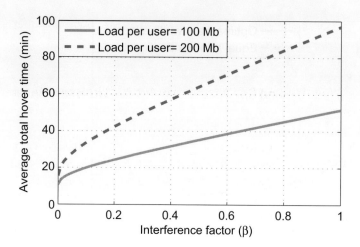

Figure 7.9 Impact of interference on the average hover time of UAV BSs.

a 53% decrease in the total hover time. Nevertheless, deploying more UAVs in an interference-free scenario will require more bandwidth. Therefore, the results of this figure uncover a very clear fundamental tradeoff between bandwidth and hover time: A network designer can operate its network in a more spectrally efficient manner (i.e., use less bandwidth), if it can deploy UAV BSs that can hover for longer periods of time.

Finally, in Figure 7.9, we assess the impact of interference on the hover time. Naturally, a higher interference will require longer hover times from the UAV BSs because of the lower associated transmission rate. For lower transmission rates, the UAV BSs need to hover for a longer time duration to meet the service demands of the ground users. From this figure, we can see that, for a full interference case (at $\beta = 1$), the average hover time is more than four times larger than the one resulting from the interference-free case (at $\beta = 0$). Hence, to account for the hover time capabilities of the UAV BSs, one must properly control the way in which interference is managed in the network via scheduling and interference mitigation techniques.

7.1.4 Summary

In this section, we have analyzed the impact of the hover time of UAV BSs on the performance that they can achieve in a network. In particular, we have studied a comprehensive solution for optimizing cell association and partitioning, for UAV-assisted wireless networks, in the presence of hover time constraints. We have first shown that a complete characterization of the solution is possible using tools from optimal transport theory. Then, we have shed light on how various parameters, such as bandwidth, interference, and the users' spatial distribution, can impact the hover time and data service performance of a network with UAV BSs. The results of this section can serve as a basis for studying more complex resource management and cell association mechanisms in a UAV-assisted wireless network.

7.2 Resource Planning and Cell Association for 3D Wireless Cellular Networks

In the previous section, we focused on wireless networks in which UAVs play the sole role of UAV BSs that support ground users. However, in future wireless networks, particularly in 5G and 6G networks, we envision that two types of UAVs will be deployed: UAV BSs and UAV UEs, as articulated in Chapters 1 and 2 of this book. In such networks, UAV BSs may be used to support not only ground users but also flying UAV UEs. To this end, in this section, building on the work in [245], we go beyond the two-dimensional model of Section 7.1 to design a fully fledged 3D cellular network that encompasses both UAV BSs and UAV UEs. In such a scenario, we focus on the interactions between these two types of UAVs, and we study how one can generalize classical two-dimensional network models (e.g., hexagonal cells) to 3D space. Such a generalization will entail proper planning of spatial reuses (i.e., how to perform frequency planning in 3D) and will require new ways to perform cell association in 3D space. The developed model will also incorporate HAPs for providing backhaul connectivity.

Here, we note that, in the current state of art on UAV communications, most model have focused on a single use case: UAV BS or UAV UE. For instance, the majority of works on deployment such as [169, 174, 246] and resource management [247–249] consider only UAV BSs with ground users and no flying UEs. Meanwhile, the prior works on UAV UEs [219, 250, 251] have only focused on how such UAV UEs can make use of ground BSs and did not focus on the possibility of UAV-to-UAV communication between UAV BSs and UAV UEs. In general, given the high potential of having both UAV UEs and UAV BSs deployed in future networks, as motivated by the various applications of Chapter 2, it is imperative to study a fully fledged 3D wireless cellular network that integrates both UAV use cases. Such a study will be done in this section with a focus on resource planning and management.

7.2.1 A Rigorous Model for 3D Cellular Networks

We consider the 3D wireless cellular network illustrated in Figure 8.8 that encompasses a set $\mathcal{N}$ of N UAV BSs (LAPs), a set $\mathcal{L}$ of L UAV UEs, and several HAP UAVs. In the considered aerial 3D network, UAV BSs serve UAV UEs over the downlink while HAPs are used to provide backhaul connectivity [252] for UAV BSs. We consider each UAV BS to be equipped with omnidirectional antennas for full 3D connectivity. HAPs can provide a suitable backhaul solution for our 3D network, since they can establish LOS backhaul links while also adjusting their location with respect to the positions of the UAV BSs. While it is possible to use different backhaul types for the considered 3D network [253], we employ HAP UAVs that can connect via free-space optical (FSO) communications backhaul links to the UAV BSs. Such a choice is done to enhance the reliability and latency of the backhaul link, compared to the use of terrestrial wireless connections. We also assume that each UAV BS will connect to the closest HAP (providing the highest rate) for backhauling. We use C_n to represent UAV BS n's transmission rate over the backhaul. This rate is assumed to be constant and predetermined for the considered model. Each UAV BS n will have a transmit power P_n and a bandwidth B_n. The spatial distribution (in 3D space) of the UAV UEs is given by a generic function $f(x, y, z)$. This function captures the probability with which each UAV UE can be present around a 3D location (x, y, z). To estimate this distribution of the UAV UEs, UAV BSs can use machine learning techniques (e.g., see [245]) without requiring a continuous tracking of UAV UEs. Analogously to the previous section, we are interested in partitioning the space to find the cells associated with each UAV BS. Here, we focus on a 3D space and, therefore, we divide our space into N 3D cells. Each cell represents a spatial volume that must be served by a single UAV BS. We define the set $\mathcal{V}_n$ to represent a 3D cell that is associated with a UAV BS n that is serving UAV UEs located within this 3D cell. As a result, we can compute the average number of UAV UEs within cell $\mathcal{V}_n$:

$$K_n = L \int_{\mathcal{V}_n} f(x, y, z) \mathrm{d}x \mathrm{d}y \mathrm{d}z. \tag{7.24}$$

In the considered model, we use the FDMA scheme for the UAV BSs. As a result, the average downlink data rate from a UAV BS n to a UAV UE at coordinate (x, y, z) will be:

$$R_n(x, y, z) = \frac{B_n}{K_n} \log_2 \left(1 + \gamma_n(x, y, z)\right), \tag{7.25}$$

where $\frac{B_n}{K_n}$ represents the amount of bandwidth used to service each UAV UE within cell $\mathcal{V}_n$. This bandwidth is determined by sharing the total bandwidth among the UAV UEs. In (7.25), $\gamma_n(x, y, z)$ represents the SINR experienced by a UAV UE that is positioned at coordinate (x, y, z) and serviced UAV UE $n \in \mathcal{N}$.

To quantify the performance of UAV UEs, we use the average latency as a key metric. We consider three types of latency measures: (a) transmission latency from UAV BSs to UAV UEs, (b) computation latency that UAV BSs use to serve UAV UEs, and (c) backhaul latency for the UAV BS to HAP links. For a UAV BS $n \in \mathcal{N}$ that is transmitting data to a UAV UE at coordinate (x, y, z), the transmission latency will be:

$$\tau_n^{\mathrm{Tr}}(x, y, z, K_n) = \frac{\beta}{R_n(x, y, z)}, \tag{7.26}$$

where β represents the size (in bits) of each packet transmitted to each drone-UE.

Next, we can easily define the backhaul transmission latency, which depends on the load of the UAV BSs and the backhaul data rate. In particular, the average backhaul latency for the link between a UAV BS $n \in \mathcal{N}$ and its backhaul-serving HAP, will be:

$$\tau_n^{\mathrm{B}}(K_n) = \frac{\beta L \int_{\mathcal{V}_n} f(x, y, z) \mathrm{d}x \mathrm{d}y \mathrm{d}z}{C_n} = \frac{\beta K_n}{C_n}, \tag{7.27}$$

where $\beta L \int_{\mathcal{V}_n} f(x, y, z) \mathrm{d}x\mathrm{d}y\mathrm{d}z$ is the average load on UAV BS n and C_n is the maximum backhaul data rate between UAV BS n and its serving HAP.

Next, to define the computation time, we first observe that it depends on two factors: (a) the processing speed of a UAV BS, and (b) the size of the data (i.e., load) processed at each UAV BS. To this end, we introduce a function $g_n(\beta K_n)$ to represent the computation latency at UAV BS n. Here, βK_n represents the total data size that UAV BS n must process. Consequently, we can now define the total latency experienced by any UAV UE located at a coordinate (x, y, z) and served by UAV BS n:

$$\tau_n^{\mathrm{tot}}(x, y, z, K_n) = \tau_n^{\mathrm{Tr}}(x, y, z, K_n) + \tau_n^{\mathrm{B}}(K_n) + g_n(\beta K_n). \tag{7.28}$$

Having provided a concrete model for our 3D network, we can now define our objective, which is to minimize the average latency of the UAV UEs by finding an optimal 3D cell association between UAV BSs and UAV UEs. To do so, we must first determine how the UAV BSs can be deployed in a 3D cellular structure. Then, given such a deployment and given an estimation of the spatial distribution of the UAV UEs, we can determine the optimal 3D cell partitions $\mathcal{V}_n$, $\forall n \in \mathcal{N}$ that lead to a minimum average latency for UAV UEs. This problem can be posed formally as follows:

$$\min_{\mathcal{V}_1,\ldots,\mathcal{V}_N} \sum_{n=1}^{N} \bigg[\int_{\mathcal{V}_n} \tau_n^{\mathrm{Tr}}(x, y, z, K_n) f(x, y, z) \mathrm{d}x\mathrm{d}y\mathrm{d}z$$

$$+ \tau_n^{\mathrm{B}}(K_n) + g_n(\beta K_n) \bigg], \tag{7.29}$$

$$\text{s.t. } \mathcal{V}_l \cap \mathcal{V}_m = \emptyset, \quad \forall l \neq m \in \mathcal{N}, \tag{7.30}$$

$$\bigcup_{n \in \mathcal{N}} \mathcal{V}_n = \mathcal{V}, \tag{7.31}$$

where $K_n = L \int_{\mathcal{V}_n} f(x, y, z) \mathrm{d}x\mathrm{d}y\mathrm{d}z$ is the average number of UAV UEs in $\mathcal{V}_n$, which depends on the 3D cell association and $\mathcal{V}$ is the entire considered space in which UAV UEs can fly. (7.30) and (7.31) are constraints used to guarantee that the derived 3D partitioning will lead to disjoint spaces whose union covers the entire 3D region considered $\mathcal{V}$.

7.2.2 3D Deployment of a Cellular Network with UAV BSs: A Truncated Octahedron Structure

Prior to solving our cell association problem, we must properly deploy and plan our network in 3D space. Given that a 3D cellular network fundamentally differs from a classical, two-dimensional hexagonal network, it is imperative to first develop a new approach for 3D deployment of UAV BSs with an associated frequency planning mechanism in 3D space. Inspired by the way in which hexagons were used for the two-dimensional case, for 3D, we adopt the notion of a truncated octahedron structure to deploy the UAV BSs in 3D and to derive feasible integer frequency factors that allow a characterization of the co-channel interference among UAV BSs.

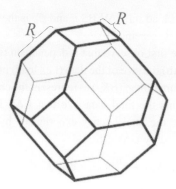

Figure 7.10 An illustration of a 3D truncated octahedron.

Indeed, in a classical ground cellular network, hexagonal cells are used for BS planning. The hexagonal shape was used because non-overlapping hexagons can fully cover (with no gaps) a two-dimensional space. This shape can also properly approximate the circular radiation pattern of an omni-directional BS antenna. Inspired from this approach, in this section, we consider a similar problem in 3D. For instance, in 3D, regular polyhedron geometric shapes that can tessellate a given space (i.e., fill it entirely with no gaps) include hexagonal prism, cube, rhombic dodecahedron, and truncated octahedron [254]. From this set, one can see that the closest approximation of a sphere can be done with the truncated octahedron. In addition, the truncated octahedron [254] minimizes the number of polyhedrons needed to completely cover a 3D space. The truncated octahedron is a polyhedron in three dimensions with regular polygons faces. As we can see from Figure 7.10, the truncated octahedron has 14 faces with 8 regular hexagonal and 6 square, 24 vertices, and 36 edges [255]. As already mentioned, a truncated octahedron can completely fill and tessellate a 3D Euclidean space without overlap among the different cells. Due to these features, we adopt a truncated octahedron-based cell structure for modeling and deploying a 3D cellular network.

To form a 3D wireless network, we will deploy the UAV BSs in a way to cover the entire desired space. Hence, we will first introduce an arrangement of multiple truncated octahedron cells that completely cover a given space. Then, the UAV BSs will be placed at the center of each truncated octahedron, as illustrated in Figure 7.11. This deployment approach provides full coverage for a 3D space, and, as will be evident from the rest of this section, it will also provide a tractable way to analyze a 3D wireless network. In addition, this approach will also facilitate tractable frequency planning in 3D space. Given this approach, we can now determine the exact locations of the UAV BSs and their associated truncated octahedron cells, from the following theorem:

THEOREM 7.3 The three-dimensional locations of drone-BSs in the studied 3D cellular network are given by:

$$P_{\{a,b,c\}} = [x_o, y_o, z_o] + \sqrt{2}R[a+b-c, -a+b+c, a-b+c], \tag{7.32}$$

7.2 Resource Planning and Cell Association for 3D Wireless Cellular Networks 163

Figure 7.11 Using truncated octahedron cells to deploy a 3D cellular network based on UAV BSs.

where a, b, c are integers chosen from set $\{..., -2, -1, 0, 1, 2, ...\}$, and R is the edge length of the considered truncated octahedrons. $[x_o, y_o, z_o]$ is the Cartesian coordinates of a given reference location (e.g., center of a specified space).

The proof of this theorem is found in [245]. Theorem 7.3 can be used to find the exact 3D coordinates of UAV BSs that are placed at the centers of truncated octahedrons. By using Theorem 7.3, as shown in [245], one can also determine the frequency reuse factor as well as interfering UAV BSs, using the following result:

THEOREM 7.4 In the considered 3D cellular network, any feasible integer frequency reuse factors can be determined by solving the following equations:

$$\begin{cases} q = \sqrt{\dfrac{\left[3(n_1^2 + n_2^2 + n_3^2) - 2(n_1 n_2 + n_1 n_3 + n_2 n_3)\right]^3}{27}}, \\ q = \sqrt{\dfrac{\left[3(m_1^2 + m_2^2 + m_3^2) - 2(m_1 m_2 + m_1 m_3 + m_2 m_3)\right]^3}{64}}, \end{cases} \quad (7.33)$$

where q is a positive integer that represents the frequency reuse factor. $n_1, n_2, n_3, m_1, m_2,$ and m_3 are integers that satisfy (7.33) by generating feasible frequency reuse factors.

Theorem 7.4 allows a network operator to determine feasible frequency reuse factors in a 3D wireless network. This theorem will also allow the operator to determine the 3D locations of co-channel UAV BSs during frequency planning. For example, for $(n_1, n_2, n_3) = (1, 0, 0)$, and $(m_1, m_2, m_3) = (1, 1, 0)$, we can obtain a frequency reuse factor of 1. Indeed, $q = 1$ pertains to a worst-case scenario where all UAV BSs interfere with one another. For this worst-case scenario, we can determine the locations of co-channel interfering UAV BSs corresponding to a reference cell with an edge length R and center $(0, 0, 0)$ by using the columns of the following matrix:

$$\boldsymbol{H} = \sqrt{2}R\begin{bmatrix} \boldsymbol{H_1} & \boldsymbol{H_2} \end{bmatrix}_{3 \times 16}, \quad (7.34)$$

where

$$H_1 = \begin{pmatrix} 1 & 1 & -1 & 1 & 1 & -1 & -1 & -1 \\ -1 & 1 & 1 & 1 & 1 & -1 & 1 & -1 \\ 1 & -1 & 1 & -1 & 1 & -1 & -1 & 1 \end{pmatrix},$$

$$H_2 = \begin{pmatrix} 1 & -1 & 2 & 0 & 0 & -2 & 0 & 0 \\ -1 & -1 & 0 & 2 & 0 & 0 & -2 & 0 \\ -1 & 1 & 0 & 0 & 2 & 0 & 0 & -2 \end{pmatrix}.$$

Each column of matrix H represents a 3D location of one co-channel UAV BS. Having determined the 3D planning of the wireless network and given an estimation of the spatial distribution, we can next solve the posed cell association problem.

7.2.3 Latency-Minimal 3D Cell Association

We can now rewrite our 3D cell association problem as follows:

$$\min_{\mathcal{V}_1,\dots,\mathcal{V}_N} \sum_{n=1}^{N} \left[\int_{\mathcal{V}_n} \frac{\beta K_n}{B_n \log_2\left(1 + \gamma_n(x,y,z)\right)} \hat{f}(x,y,z) \mathrm{d}x\mathrm{d}y\mathrm{d}z \right.$$
$$\left. + \frac{\beta K_n}{C_n} + g_n(\beta K_n) \right], \tag{7.35}$$

$$\text{s.t.} \quad K_n = L \int_{\mathcal{V}_n} \hat{f}(x,y,z) \mathrm{d}x\mathrm{d}y\mathrm{d}z, \tag{7.36}$$

$$\mathcal{V}_l \cap \mathcal{V}_m = \emptyset, \quad \forall l \neq m \in \mathcal{N}, \tag{7.37}$$

$$\bigcup_{n \in \mathcal{N}} \mathcal{V}_n = \mathcal{V}, \tag{7.38}$$

where $\gamma_n(x,y,z)$ represents the downlink SINR of a UAV UE at coordinate (x,y,z) serviced by a UAV BS $n \in \mathcal{N}$. For air-to-air communications, we consider a practical bounded path loss model (e.g., see [256] and our discussion in Chapter 3). As a result, we can write the SINR as follows:

$$\gamma_n(x,y,z) = \frac{\eta \kappa_n(x,y,z) P_n [1 + d_n(x,y,z)]^{-\alpha}}{\sum_{u \in \mathcal{I}_{\text{int}}} \eta \kappa_u(x,y,z) P_u [1 + d_u(x,y,z)]^{-\alpha} + N_o B_n}, \tag{7.39}$$

$$d_n(x,y,z) = \sqrt{(x-x_n)^2 + (y-y_n)^2 + (z-z_n)^2}, \tag{7.40}$$

$$d_u(x,y,z) = \sqrt{(x-x_u)^2 + (y-y_u)^2 + (z-z_u)^2}, \quad u \in \mathcal{I}_{\text{int}}. \tag{7.41}$$

Here, we define $\kappa_n(x,y,z)$ as a channel gain factor between UAV BS n and a UAV UE at location (x,y,z). This factor will depend on the environment and the positions of UAV UEs and UAV BSs. For example, $\kappa_n(x,y,z) = 1$ captures an LOS AA communication while $0 < \kappa_n(x,y,z) < 1$ represents NLOS conditions. Moreover, in 7.39, the parameter α represents the path loss exponent, N_o is the noise power spectral density, η is the path

7.2 Resource Planning and Cell Association for 3D Wireless Cellular Networks

loss constant, and (x_n, y_n, z_n) is UAV BS n's 3D position. $d_n(x, y, z)$ and $d_u(x, y, z)$ are, respectively, the distance between UAV BSs n and u and a UAV UE located at (x, y, z). $\mathcal{I}_{\text{int}}$ represents the set of co-channel interfering UAV BSs that operate over the same frequency as UAV BS n.

As observed in the cell association problem of Section 7.1, solving a cell association problem such as (7.35) is challenging due to the complexity of partitioning a geographical space. This challenge is exacerbated in the 3D case by the fact that the optimization variables $\mathcal{V}_n, \forall n \in \mathcal{N}$, are continuous and mutually dependent 3D association spaces that are unknown a priori. Also, we can note that the objective function in (7.35) cannot be expressed in closed form, and, thus, the problem becomes intractable. As done in Section 7.1, we will overcome these challenges by resorting to tools from optimal transport theory. However, the 3D aspect of the studied problem here will require new results to characterize the optimal cell association.

Again, we can use the semi-discrete optimal transport framework to solve our cell association problem. In particular, we deal with a mapping between a continuous 3D distribution of UAV UEs and a discrete set of UAV BSs. We can then use optimal transport tools to characterize the solution [245]:

THEOREM 7.5 For a UAV BS l, the optimal 3D cell association that minimizes the average latency in (7.35) will be given by:

$$\mathcal{V}_l^* = \left\{ (x, y, z) \middle| \alpha_l + \frac{K_l}{L} h_l(x, y, z) + \frac{\beta}{C_l} + g_l'(\beta K_l) \right. \\ \left. \leq \alpha_m + \frac{K_m}{L} h_m(x, y, z) + \frac{\beta}{C_m} + g_m'(\beta K_m), \forall l \neq m \right\}, \tag{7.42}$$

where $h_l(x, y, z) \triangleq \frac{\beta}{B_l \log_2\left(1 + \gamma_l(x,y,z)\right)}$, and $\alpha_l \triangleq \int_{\mathcal{V}_l} h_l(x, y, x) \hat{f}(x, y, z) \mathrm{d}x \mathrm{d}y \mathrm{d}z$.

Theorem 7.5 can be used to completely determine the optimal 3D cell partitions that allows each UAV BS to minimize the average latency of its transmissions to UAV UEs. Clearly, from (7.42), we observe that this optimal 3D cell association will be a function of the different wireless network parameters such as the UAV UEs' spatial distribution, the UAV BSs' locations, the backhaul rate, the network load, and the computing speed at UAV BSs. Given such parameters, we can use Theorem 7.5 to optimally partition a 3D space of interest and determine a minimum latency 3D cell association scheme. Naturally, to minimize the average latency, a UAV BS that is experiencing a better backhaul link and that possesses a higher computational capabilities or higher bandwidth and transmit power will serve more UAV UEs. Using this theorem coupled with known optimization algorithms such as those proposed in [245] and [243], one can design efficient iterative algorithms to find the 3D cell partitioning for any wireless network with UAV BSs and UAV UEs.

Table 7.2 Typical parameters used in our 3D network simulations.

Parameter	Description	Value
f_c	Carrier frequency	2 GHz
P_n	Transmit power of UAV BS	0.5 W
N_o	Power spectral density of the noise	-170 dBm/Hz
L	Number of UAV UEs	200
B_n	Bandwidth for every UAV BS	10 MHz
α	Path loss exponent	2
η	Path loss constant	1.42×10^{-4}
β	UAV UE packet size	10 kb
q	Frequency reuse factor	1
C_n	UAV BS n's backhaul rate	$(100+n)$ Mb/s
ω_n	Computation speed for each UAV BS	10^2 Tb/s
μ_x, μ_y, μ_z	Mean of the truncated Gaussian distribution in the x, y, and z directions	1000 m, 1000 m, 1000 m
$\sigma_x, \sigma_y, \sigma_z$	Standard deviation of the spatial distribution in the x, y, and z directions	600 m, 600 m, 600 m
κ_n	Channel gain factor	1

7.2.4 Representative Simulation Results

To simulate the studied system, we consider a 3D cubic space of size 3 km × 3 km × 3 km. We then deploy a total of 18 UAV BSs using the developed truncated octahedron solution. The positions of the UAV BSs are found from (7.32) with $a \in \{-1, 0, 1\}, b \in \{-1, 0, 1\}, c \in \{0, 1\}$, and $R = 400$ m. We generate a random realization of a continuous spatial distribution for the locations of the UAV UEs, using a 3D truncated Gaussian distribution with a given mean and variance. The samples are used for estimating the spatial distribution of the UAV UEs. We consider a quadratic function of data size (i.e., load of each UAV BS) to represent the computation time and, thus, for any given UAV BS n, we define the computation time as $g_n(\beta K_n) = \frac{(\beta K_n)^2}{\omega_n}$ where ω_n is the processing speed of UAV BS n. Unless stated otherwise, we use the parameters of Table 7.2. We compare our developed 3D cell association with a conventional SINR-based cell association (i.e., weighted Voronoi diagram).

In Figure 7.12, we present the average total latency resulting from the studied approach and the SINR-based scheme, as the number of UAV UEs varies in the network. Figure 7.12 shows that an increase in the total number of UAV UEs will lead to a higher total latency. This stems from the fact that a larger network of UAV UEs will lead to a higher load on the UAV BSs, and, thus, it will increase transmission, backhaul, and computational latencies. From Figure 7.12, we can observe that an increase in the number of UAV UEs from 200 to 300, leads to, respectively, a 56% and a 42% increase in the total latency for the SINR-based association and our approach. Moreover, Figure 7.12 clearly demonstrates that the developed solution yields significant latency reductions compared to the baseline SINR-based approach. This is because the developed solution explicitly accounts not only for the SINR but also for the impact of network congestion on the

7.2 Resource Planning and Cell Association for 3D Wireless Cellular Networks

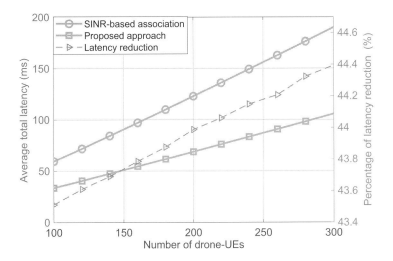

Figure 7.12 Variation in the average total latency as the number of UAV UEs changes.

transmission, backhaul, and computational latencies. As a result, the studied solution can help the network avoid the introduction of highly loaded 3D cells that experience an excessively high latency. Finally, from Figure 7.12, we can observe that the developed framework can lead to an average of 43.9% reduction in the average total latency compared to the baseline.

Next, in Figure 7.13, we evaluate the impact of the transmission bandwidth on the latency. For instance, a higher bandwidth can lead to a lower transmission latency due to the associated improvements in the data rates. Consequently, from Figure 7.13, we can observe how the developed solution can yield significant enhancements in spectrum efficiency compared to the SINR-based baseline. The developed solution can achieve a similar performance to the SINR association case, while using less bandwidth. For example, as per Figure 7.13, to guarantee a maximum total latency of 70 ms, the developed solution will use 57% less bandwidth compared to the SINR-based baseline. Figure 7.13 also shows that the rate with which latency is reduced becomes smaller for larger bandwidth. This is due to the fact that, for a network a larger bandwidth, the transmission latency will be much smaller than the other two latency components (backhaul and computation).

Next, in Figure 7.14, we study how the transmission, computation, and backhaul latencies will be affected by a change in the UAV UE load. First and foremost, as intuitively expected, a higher load of UAV UEs will increase all three latency components. However, interestingly, Figure 7.14 shows that the transmission latency increases at a higher rate than the backhaul and computation latencies. Here, we note that the effect that the UAV UE load has on the different latency components relates to two key factors: (a) the relationship between load and latency, and (b) the relationship between the 3D cell partitions and the load as quantified in (7.42). In essence, as the load changes, the 3D cells and the different latency components will vary in a way to minimize the total latency.

168 Resource Management for UAV Networks

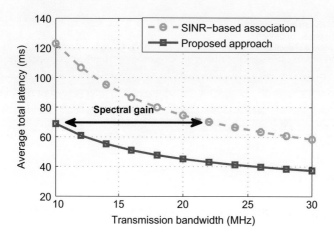

Figure 7.13 Impact of transmission bandwidth on the average total latency.

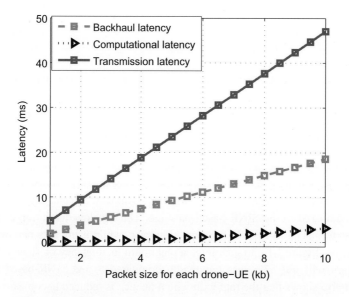

Figure 7.14 Transmission, backhaul, and computation latency resulting from the considered framework vs. load of each UAV UE.

7.2.5 Summary

In this section, we have introduced a novel framework that enables the operation of a fully fledged 3D wireless cellular network. The developed framework allows a network operator to plan the deployment of UAV BSs, in 3D space, so as to meet the target wireless performance metrics of flying UAV UEs. In particular, we have developed a method that allows to jointly deploy a 3D network of UAV BSs, plan the frequencies over this 3D network, and, then, design a latency-minimizing cell association scheme. The key results of this section have shown that the developed solution approach significantly reduces the latency of UAV UEs compared to a conventional, SINR-based cell association baseline. In addition, the developed latency-minimizing cell association techniques can lead to important improvement in the spectral efficiency of 3D cellular networks with UAVs. Naturally, this developed framework can then be extended to account for additional resource management dimensions that include bandwidth allocation and power optimization. Another interesting future research direction is to incorporate a ground, terrestrial network and integrate it into the 3D wireless cellular system that we developed.

7.3 Managing Licensed and Unlicensed Spectrum Resources in Wireless Networks with UAVs

In the previous two sections, we focused primarily on the management of spatial resources, via the development of optimized cell association approaches. However, spectrum is yet another important resource in a UAV-assisted wireless network that must be properly managed. In particular, the spectrum used by flying UAV BSs must be properly shared in a way to meet the needs of the ground users, while also enhancing spectrum efficiency and minimizing interference. Consequently, to shed light on resource management in wireless networks with UAVs, it is imperative to perform spectrum management and analyze how spectral resources can be allocated. Moreover, given the mobility and agility of UAV BSs, it can be potentially beneficial to leverage them to "cache" popular content that can be commonly downloaded by ground users. The use of caching can also help UAV BSs alleviate the need for significant transmissions over the backhaul/fronthaul link between UAV BSs and the core network. Clearly, in a cache-enabled wireless network, with UAV BSs, the network must manage two key resources: (a) spectrum and (b) cached content. To this end, in this section, we focus on this resource management problem within the context of a wireless network assisted by UAV BSs that are also connected wirelessly to a cloud for fronthauling purposes. We particularly focus on a scenario in which the UAVs can alleviate problems of spectrum scarcity by accessing both licensed and unlicensed bands. Indeed, we will study how UAV BSs can leverage LTE over the unlicensed band (LTE-U) capabilities so as to use available WiFi bands to supplement the licensed band resources; whenever such a use of the unlicensed band does not jeopardize the performance of the ground WiFi users.

We here note that studying problems of spectrum and resource management in UAV-assisted wireless networks has been done in [237, 257–263]. However, these prior works often focus on the UAV-to-ground links, without accounting for the presence of fronthaul links between UAV BSs and the core network. Moreover, most of these prior works (except for [257]) do not study the use of LTE-U, jointly with caching, to overcome the spectrum scarcity and fronthaul limitation problems. Meanwhile, the work in [257] does not account for the notion of caching. Hence, there is a need to provide a more comprehensive study of how resources in a wireless network assisted by LTE-U-enabled

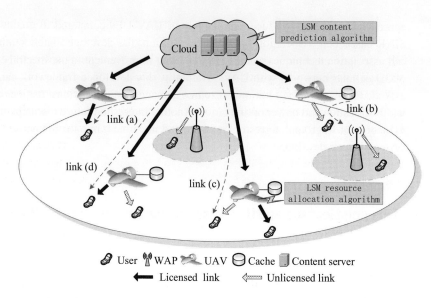

Figure 7.15 Illustration of a system in which multiple, cache-enabled UAV BSs with LTE-U capabilities are deployed for wireless communication purposes.

UAV BSs can be managed properly, while being cognizant of fronthaul limitations and the possibility of caching at the UAV BSs' side.

7.3.1 Model of an LTE-U UAV BS Network

As shown in Figure 7.15, we consider an LTE-U network in which a set $\mathcal{K}$ of K UAV BSs are deployed to serve, over the downlink, a set $\mathcal{U}$ of U LTE-U ground users. In this system, we also consider a ground WiFi network composed of W WiFi access points (WAPs) that are connected to N_w WiFi users. We equip the UAV BSs with storage units that they can use to cache popular content. The UAV BSs will connect to a cloud server via a fronthaul link using (exclusively) the licensed band. The cloud provides connectivity to the core network. Since we focus on an LTE-U scenario, the UAV BSs are assumed to operate in *dual-mode*: They can simultaneously access both licensed and unlicensed spectrum resources. In our model, each UAV BS can allocate at most one type of resource (i.e., licensed or unlicensed) to each ground user. We assume that each unlicensed band can be occupied by either UAV BSs or WAPs. UAV BSs will only be able to use the unlicensed spectrum whenever access to this spectrum does not degrade the data rate of WiFi users below a minimum, guaranteed target rate.

In the considered system, each UAV BS will share the licensed spectral bands among its serviced users. To access the unlicensed spectrum in LTE-U, we adopt a duty cycle method as done in [264]. In such a duty cycle scenario, the UAV BSs will adopt a discontinuous, duty-cycle transmission pattern so as to access the unlicensed spectrum while also maintaining the transmission rate of the WiFi users above a given threshold. Hence, the time slots used to access the unlicensed band will be properly split between

7.3 Managing Licensed and Unlicensed Spectrum Resources

LTE-U and WiFi users. In this regard, LTE-U UAV BSs will transmit for a time fraction ϑ and will be muted for a time fraction $1 - \vartheta$. During the mute-time of the LTE-U UAV BSs, the WiFi transmissions will occur using the standardized carrier sense multiple access with collision avoidance (CSMA/CA) protocol [265].

We assume that all ground users request equally sized contents from a set $\mathcal{N}$ of N contents. Contents are stored at the cloud server of the network, and each content will be of size L. We let $\boldsymbol{p}_i = [p_{i1}, \ldots, p_{iN}]$ be the content request distribution of each user. Here, each element p_{ij} of $\boldsymbol{p}_i$ captures the probability that user i requests content j. To enable caching, we assume that each UAV k has a storage unit that can store a total of C contents drawn from a set $\mathcal{C}_k$ of popular user contents. For the considered LTE-U network with caching, the transmission of a content to a ground user can be done over one of four links: (a) licensed band transmission from the cloud to the UAV BS followed by a licensed band transmission from the UAV BS to the ground user, (b) licensed band cloud-UAV BS transmission followed by an unlicensed band transmission from the UAV BS to the ground user, (c) direct, licensed band transmission from the cache of a UAV BS to the ground user, and (d) direct, unlicensed band transmission from the cache of a UAV BS to the ground user. Clearly, transmission links (c) and (d) showcase the case in which caching is used to offload traffic from the fronthaul and alleviate the congestion on the cloud-UAV BS transmission.

As mentioned earlier, the ground WAPs will use a CSMA/CA scheme with binary slotted exponential backoff. For this case, using standard WiFi models such as in [266], we can find the saturation capacity of the N_w WiFi users that share the unlicensed band:

$$R(N_w) = \frac{P_{\text{tr}}(N_w) P_s(N_w) E[A]}{(1 - P_{\text{tr}}(N_w)) T_\sigma + P_{\text{tr}}(N_w)(T_c + P_s(N_w)(T_s - T_c))}, \quad (7.43)$$

with $P_{\text{tr}}(N_w) = 1 - (1 - \tau)^{N_w}$ and $P_{\text{tr}}(N_w)$ being the probability of having at least one transmission in a time slot and τ being each WiFi user's transmission probability. Here, T_s is the average time that the channel is sensed busy because of a successful transmission, T_c is the average time that the channel is sensed busy by each WAP during a collision, T_σ is the unoccupied slot duration, $P_s(N_w) = N_w \tau (1 - \tau)^{N_w - 1} / P_{\text{tr}}(N_w)$ is the probability of successful transmission, and $E[A]$ is the average size of a packet. We consider standard, distributed coordination function access and RTS/CTS access mechanisms. T_c and T_s can be expressed by [266]:

$$T_s = RTS/C^U + CTS/C^U + (H + E[A])/C^U \\ + ACK/C^U + 3SIFS + DIFS + 4\delta, \quad (7.44)$$

$$T_c = RTS/C^U + DIFS + \delta. \quad (7.45)$$

Here, $H = PHY_{hdr} + MAC_{hdr}$, and C^U represents the bit rate of the WiFi channels. *ACK*, *RTS*, *DIFS*, *SIFS*, *CTS*, and δ represent standard WiFi parameters as given by [266]. To validate the relationship in 7.43, we assume that the following two conditions hold: (a) after completing a successful transmission, any given WiFi user will immediately have a new packet available, and (b) a binary slotted exponential backoff scheme is used.

172 Resource Management for UAV Networks

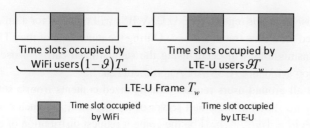

Figure 7.16 Example of how time slot allocations between LTE-U and WiFi users can be done.

Clearly, T_c and T_s will significantly affect the WiFi network saturation capacity of a WiFi network.

Each LTE time slot is composed of T_W WiFi slots. Under the chosen duty cycle approach, on the unlicensed band, UAV BSs can occupy a fraction ϑ of the total T_W time slots. Meanwhile, the remaining fraction $(1-\vartheta)$ of unlicensed time slots will be occupied by WiFi users. Figure 7.16 illustrates this time slot division between WiFi and LTE-U users. Now, we can write the (per user) rate over WiFi:

$$R_w = \frac{R(N_w)(1-\vartheta)}{N_w}. \tag{7.46}$$

In 7.46, N_w represents the total number of WiFi users on the unlicensed band. Given a target data rate requirement γ for each WiFi user, we can use $\vartheta \leq 1 - N_w\gamma/R(N_w)$ to express the fraction of unlicensed band time slots allocated to an LTE-U user.

Having defined the main components of the system model as well as the associated WiFi model, next, we properly model the data rates over the various UAV BS-to-ground user links.

7.3.2 Models for Data Rates and Queuing

Now, we can determine the rates associated with the transmissions of content to the ground users, via the UAV BSs. This includes the cloud-to-UAV BS fronthaul links (ground-to-air links) and the AG links from the UAV BSs to the users. For licensed band transmissions (fronthaul or AG), we assume LOS connections.

The path loss (in dB) experienced by the transmission from UAV BS k to user i over an LOS licensed band link will be [194]:

$$l_{ki}^{\text{LOS}} = 20\log\left(\frac{4\pi d_{ki} f}{c}\right) + \eta^l + \chi_\kappa + \Omega,$$

where $20\log\left(d_{ki}f4\pi/c\right)$ is the free-space path loss with d_{ki} representing the distance between UAV BS k and ground user i, c capturing the speed of light, and f being the carrier frequency, $s\eta^l$ captures additional attenuation factors due to the LOS connections over the licensed band, and χ_κ is a shadow fading Gaussian random variable with zero mean and standard deviation κ. Ω is the small-scale fading power assumed to follow a Rician distribution (see our discussions in Chapter 3 on the use of a Rician channel model for UAVs).

At time t, we can now define the data rate of the downlink licensed-band transmission between UAV BS k and user i

$$R_{lki}(u_{ki}(t)) = u_{ki}(t) F_l \log_2 \left(1 + \frac{P_K 10^{l_{ki}^{\text{LOS}}/10}}{\sum_{j \in \mathcal{K}, j \neq k} P_K 10^{l_{ji}^{\text{LOS}}/10} + P_C h_i + \sigma^2} \right), \quad (7.47)$$

where P_K represents each UAV BS's transmit power, $h_i = g_{Ci} d_{Ci}^{-\alpha}$ where g_{Ci} is a Rayleigh fading channel gain between the cloud and user i with d_{Ci} being the distance between them, F_l is the downlink licensed band bandwidth, and P_C is the cloud's fronthaul transmit power. σ^2 is the Gaussian noise power, and $u_{ki}(t)$ is the fraction of the downlink licensed band allocated from UAV BS k to user i at time t with $\sum_i u_{ki}(t) = 1$. $\sum_{j \in \mathcal{K}, j \neq k} P_K 10^{l_{ji}^{\text{LOS}}/10}$ captures the interference between user i and all UAV BSs other than k. All UAV BSs are assumed to use the same spectrum for content transmission, and they allocate all of their available spectrum resources to their serviced users. As a result, all UAV BSs (except UAV BS k) will interfere with user i.

To compute the data rate of unlicensed band transmissions between UAV BSs and ground users, we use (7.46) to obtain the unlicensed band time slot fraction ϑ that can be occupied by UAV BSs. Given this fraction, the downlink unlicensed band data rate of a given user i associated with UAV BS k will be:

$$R_{uki}(e_{ki}(t)) = e_{ki}(t) \vartheta F_u \log_2 \left(1 + \frac{P_K 10^{l_{ki}^u/10}}{\sum_{j \in \mathcal{K}, j \neq k} P_K 10^{l_{ji}^u/10} + \sigma^2} \right), \quad (7.48)$$

where l_{ki}^u is the LOS path loss over the unlicensed band, F_u is the bandwidth of the unlicensed band, and $e_{ki}(t)$ is the fraction of ϑ with $\sum_i e_{ki}(t) = 1$.

To compute the fronthaul transmission rates, we first equally divide the total UAV BS fronthaul bandwidth F_C among the users that have received contents from the cloud. As a result, the fronthaul rate of a given user that requests a cloud content while being served by UAV BS k will be:

$$R_{Ck}(t) = \frac{F_C}{U_C(t)} \log_2 \left(1 + \frac{P_C L_k}{\sum_{j \in \mathcal{K}, j \neq k} P_K 10^{l_{ki}^{\text{LOS}}/10} + \sigma^2} \right), \quad (7.49)$$

where L_k is the LOS path loss from the cloud to UAV BS k and $U_C(t)$ is the number of the users that receive a content from the cloud at time t. $U_C(t)$ can be computed by the content server when the users request contents.

We can now study the queuing process of the studied system. First, we define variable $V_i(t)$ to capture the random content arrival (number of bits) for user i from the content server at the end of time slot t. Since each user can request at most one content during each time slot t, we have $V_i(t) \in \{0, L\}$. At the beginning of a given time slot t, we can derive the queue length (i.e., number of bits) $Q_i(t)$ of user i as follows [267]:

$$Q_i(t+1) = Q_i(t) - R_{ki}(t) + V_i(t), \qquad (7.50)$$

where $R_{ki}(t)$ is user i's data rate. As shown in Figure 7.15, the content transmission links include: (a) UAV BS-user on the licensed band, (b) UAV BS-user on the unlicensed band, (c) cloud-UAV BS-user on the unlicensed band, and (d) cloud-UAV BS-user on the licensed band. Therefore, we can define the rate of content transmission from UAV BS k to user i as follows:

$$R_{ki}(u_{ki}(t), e_{ki}(t)) = \begin{cases} R_{lki}(u_{ki}(t)), & \text{link (a)}, \\ R_{uki}(e_{ki}(t)), & \text{link (b)}, \\ \frac{R_{uki}(e_{ki}(t))R_{Ck}(t)}{R_{uki}(e_{ki}(t))+R_{Ck}(t)}, & \text{link (c)}, \\ \frac{R_{lki}(u_{ki}(t))R_{Ck}(t)}{R_{lki}(u_{ki}(t))+R_{Ck}(t)}, & \text{link (d)}. \end{cases} \qquad (7.51)$$

where the rate expression of link (c) is obtained from the fact that the time duration of a single data packet transmitted from the cloud to UAV BS k is $1/R_{Ck}(t)$ and a single data packet transmitted from UAV BS k to user i is $1/R_{uki}(t)$. Hence, the transmission data rate from the cloud to user i will be $\frac{1}{1/R_{Ck}(t)+1/R_{uki}(t)}$. In (7.51), links (a) and (b) refer to scenarios in which the content requested by a user i is already cached by UAV BS k. In these cases, the requested content can be directly transmitted from UAV BS k to user i without going through the cloud. In contrast, for links of type (c) and (d), the content of the users is at the cloud and not cached. Hence, for those links, the transmission of content to a user i involves having UAV BS k obtain the data from the cloud (over the backhaul) and then transmitting it to its user. Naturally, we can directly observe that links (a) and (b) can achieve higher data rates compared to links (c) and (d).

To capture the content transmission delay of each user, we adopt the concept of *queue stability*. In essence, a queue $Q_i(t)$ is said to be *rate stable* if [267]:

$$\lim_{t \to \infty} \frac{Q_i(t)}{t} = 0. \qquad (7.52)$$

From [267, Theorem 2.8], we can also see that the queue $Q_i(t)$ is rate stable if $R_{ki}(t) \geq V_i(t)$.

Having define the various performance metrics used in our model, we can now formally pose the resource management problem that will involve spectrum allocation, user association, and content caching.

7.3.3 Resource Management Problem Formulation and Solution

Our goal is to design a resource management framework that can effectively allocate spectrum over the licensed and unlicensed band within the context of a wireless network served by cache-enabled UAV BSs. The objective is to optimize the resource management process in a way to meet the queue stability requirements of all users. This problem can be formulated as an optimization problem whose goal is to maximize the number

7.3 Managing Licensed and Unlicensed Spectrum Resources

of users that have stable queues. This maximization requires finding the optimal association $\mathcal{U}_k$ for each UAV BS k, the licensed band bandwidth allocation as captured by indicators $\boldsymbol{u}_k$, the unlicensed band time slot allocations as captured by variables $\boldsymbol{e}_k$, and the set $\mathcal{C}_k$ of contents that each UAV BS k can potentially cache. Formally, this problem can be posed as follows:

$$\max_{\boldsymbol{u}_k(t), \boldsymbol{e}_k(t), \mathcal{C}_k, \mathcal{U}_k} \sum_{k \in \mathcal{K}} \sum_{i \in \mathcal{U}_k} \mathbb{1}_{\left\{ \lim_{t \to \infty} \frac{Q_i(t)}{t} = 0 \right\}} \\ = \max_{\boldsymbol{u}_k(t), \boldsymbol{e}_k(t), \mathcal{C}_k, \mathcal{U}_k} \sum_{k \in \mathcal{K}} \sum_{i \in \mathcal{U}_j} \mathbb{1}_{\{R_{ki}(u_{ki}(t), e_{ki}(t)) \geq V_i(t)\}}, \quad (7.53)$$

$$\text{s. t.} \quad R_w \geq \gamma, \quad (7.53a)$$

$$\sum_{i \in \mathcal{U}_k} u_{ki}(t) = 1, \quad \forall k \in \mathcal{K}, \quad (7.53b)$$

$$\sum_{i \in \mathcal{U}_k} e_{ki}(t) = 1, \quad \forall k \in \mathcal{K}, \quad (7.53c)$$

where $\mathbb{1}_{\{x\}} = 1$ when x is true and $\mathbb{1}_{\{x\}} = 0$ otherwise, $\mathcal{U}_k$ is the set of users served by UAV BS k, and $\boldsymbol{u}_k(t), \boldsymbol{e}_k(t)$ represent the resource allocation indicators on the downlink licensed and unlicensed bands, respectively.

The first constraint in (7.53) allows the network to maintain the average data rate of each WiFi user above a desired threshold while the second constraint ensures that the licensed band allocation will not exceed each UAV BS's total bandwidth. The last constraint captures the fact that the time slots over the unlicensed band cannot exceed the total number of time slots allocated to the UAV BSs. From (7.51), we can see that $R_{ki}(u_{ki}(t), e_{ki}(t))$ in (7.53) depends on the cached contents, resource allocation, and user association.

The maximization problem in (7.53) is difficult to solve for various reasons. First, content caching and spectrum allocation depend on the UAV BS-user association which, in turn, depends on each user's data rate. These dependencies make the problem challenging to address. Second, problem (7.53) can be easily shown to be non-convex and combinatorial. Third, since caching is involved, UAV BSs must be able to perform some sort of predictions to understand the users' content request distributions. To address these three challenges, as shown in [268], one can adopt machine learning techniques to solve the joint problem of caching and resource allocation. Although many machine learning tools can be used for address predictions and network optimization, here, one can adopt the concept of liquid state machines (LSMs), which is a new type of spiking neural network [269]. LSMs are apropos for the studied model because they are very effective at dealing with time-stamped data, such as the content requests of wireless users. Moreover, LSMs are very effective at dealing with complex problems involving large spaces of continuous variables, as is the case in our resource management problem. Indeed, by using LSM, we will allow the network to properly store and track the users' behavioral information and network state over time. In particular, an LSM-based approach will allow the cloud to exploit user behavior information (stored in LSM) to

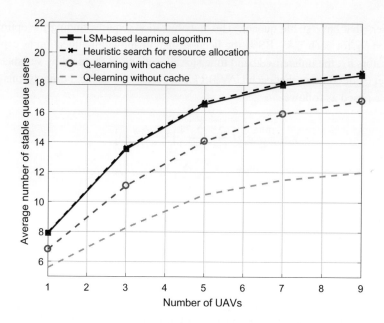

Figure 7.17 Average number of stable queue users as function of the network size.

predict the users' content request distribution and automatically adapt the resource allocation process to any changes in the network environment. Moreover, LSMs are known to have a very effective training process that can be easily run and operated in a wireless network. Using these observations, the work in [268] shows how one can develop an effective prediction and reinforcement learning algorithm for solving the problem in (7.53) in a distributed way. The details of the algorithm and solution are omitted here, and the interested reader is referred to [268]. Next, we will show some representative results on how LSM-based resource management can be effective for deployment in UAV-assisted wireless networks.

7.3.4 Representative Simulation Results

To perform our simulations, we use real data from the *Youku* of *China network video index* to train our LSM on content request distributions. Then, we set up a network simulator in which a circular network area is used with a radius $r = 200$ m in which we deploy $U = 20$ uniformly distributed users and $K = 5$ uniformly distributed UAV BSs. We use the MATLAB LSM toolbox to implement the LSM algorithm [269]. All simulation parameters and training processes are based on [268]. We compare the solution with two schemes: (a) a Q-learning scheme referred to as "Q-learning without cache," and (b) a Q-learning scheme that is complemented by LSM predictions, which we refer to as "Q-learning with cache."

Figure 7.17 presents the average number of users with stable queue resulting from the studied solution and the baselines, for different network sizes. In this figure, in

7.3 Managing Licensed and Unlicensed Spectrum Resources

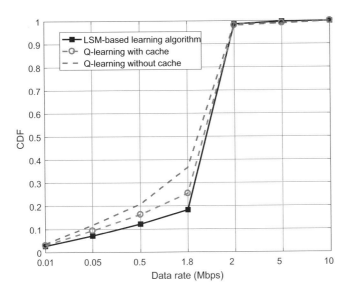

Figure 7.18 Cumulative distribution function of the data rates achieved by the LSM approach and baselines.

addition to the two previously mentioned baselines, we also compared with a heuristic search algorithm that relies on a heuristic for finding the optimal resource allocation while adopting LSM for prediction and user association. From Figure 7.17, we can see that, for 5 UAV BSs, an LSM-based approach can provide up to 17.8% and 57.1% gains in terms of the number of stable queue users compared to Q-learning with cache and Q-learning without cache, respectively. These gains showcase how the ability of LSM to use predictions to solve the caching and resource management problems can lead to improved performance. This advantage of LSM is further corroborated in Figure 7.18, which shows the cumulative distribution function (CDF) of the data rates resulting from all considered solutions. From Figure 7.18, we can first observe that, under all studied solutions, the data rate of all users will be less than 2 Mbps, which is the data rate requirement (per user) used in the simulations. Moreover, Figure 7.18 clearly corroborates the results of Figure 7.17 by showing that an LSM-based approach yields significant improvements in the CDF of the data rates, when compared with the Q-learning baselines.

Next, in Figure 7.19, we assess the convergence properties (in terms of number of iterations) of an LSM-based solution for resource management. Clearly, Figure 7.19 shows that an LSM approach requires 20% fewer iterations to converge compared to Q-learning with cache. This is once again a byproduct of LSM's inherent ability to predict network evolution.

Figure 7.20 shows the impact of the fraction of LTE-U-occupied WiFi time slots on the average number of stable queue users. Figure 7.20 first shows that, as more WiFi time slots are occupied by LTE-U users, more users will meet their stable queue requirements because the UAV BSs will have more slots available on the unlicensed band to allocate

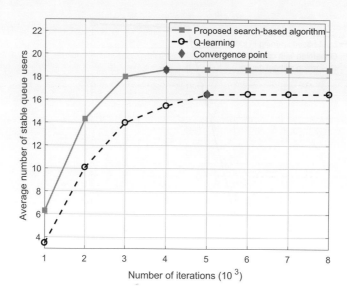

Figure 7.19 Convergence of the learning algorithms.

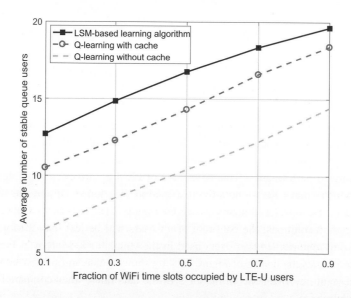

Figure 7.20 Average number of stable queue users as the fraction of WiFi time slots occupied by LTE-U users varies.

to their users. From Figure 7.20, we can also see that LSM yields up to 41.4% and 10% gains in terms of the average number of stable queue users, compared to Q-learning and Q-learning with cache, respectively.

Finally, in Figure 7.21, we study the impact of the fronthaul bandwidth on the transmission rates of the different types of links. Here, we use a randomly chosen user to illustrate the results. Figure 7.21 shows that, the bandwidth of the fronthaul has no

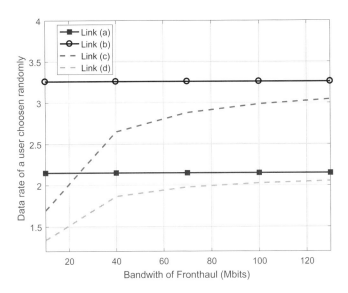

Figure 7.21 Network data rates as the bandwidth of fronthaul varies.

impact on the data rates of links (a) and (b) (because they use cached content) while an increase in the bandwidth leads to better data rates for links (c) and (d). Figure 7.21 also shows that the unlicensed band data rate of links (b) and (c) are higher than the licensed band data rate of links (a) and (d) because using the unlicensed band allows the UAV BSs to avoid the licensed band interference from other UAV BSs and from the fronthaul links.

7.3.5 Summary

In this section, we have shown how the use of UAV BSs with emerging technologies, such as LTE-U and caching, can yield interesting and challenging resource management problems. In particular, we have studied a model in which UAV BSs can use LTE-U and caching to overcome challenges of spectrum scarcity and fronthaul capacity limitations. In the studied model, the UAV BSs must decide which contents to cache, which users to serve, and how to allocate their spectrum across licensed and unlicensed bands, while being cognizant of mutual interference (over the licensed band) and WiFi performance (over the unlicensed band). We have then posed the joint problems of spectrum allocation, caching, and user association, as a non-convex optimization problem. Then, we have provided insights on how to develop predictive, machine learning tools to solve such non-convex resource management problems in large-scale wireless networks with UAV BSs. Our results have shed important light on how complex resource management problems arise in UAV networks and on how one can overcome their complexity and design practical solutions. Naturally, many extensions can be envisioned to this model such as by integrating UAV UEs as well as by accounting for the presence of ground BSs that coexist with the UAV BSs and the WiFi network.

7.4 Chapter Summary

In this chapter, we have shown how the introduction of UAV BSs and UAV UEs in a cellular networking environment will yield a variety of important resource management problems. Such resource management problems include cell association, spectrum sharing, caching, and overall management of various UAV resources that include spatial, spectral, and temporal resources. First, we have shown how unique features of UAV BSs, such as flight time constraints, can impact problems of resource management, in general, and cell association, in particular. Then, we have designed a fully fledged 3D cellular network that integrates both UAV BSs and UAV UEs. For this 3D network architecture, we have shown how various planning and resource management problems become intertwined. We have also shed light on how one can systematically optimize the performance (in terms of latency) of such a 3D cellular system. Then, we have concluded the chapter by studying a joint spectrum allocation, user association, and content caching problem in a network assisted by UAV BSs. We have shown how these three problems are synergistic, and we have designed a learning-based solution to solve them in practical UAV networks. In essence, the models and solutions provided in this chapter can serve as an important basis for designing effective resource management frameworks for 3D cellular networks that integrate a heterogeneous set of UAV BSs, UAV UEs, and ground infrastructure that operate across multiple different radio access technologies and frequency bands.

8 Cooperative Communications in UAV Networks

As we discussed in Chapters 1 and 2, leveraging cooperation among multiple UAVs along with coordinated transmissions is a promising solution for enhancing the performance of wireless networks that incorporate UAV BSs and UAV UEs. In the UAV BS case, multiple UAVs can form a flexible, reconfigurable, and wireless antenna array in the sky [22] within which each UAV acts as an antenna element of the array (as shown in Figure 8.2). A reconfigurable UAV-based antenna array system that acts as a flying BS has a number of key advantages over classical antenna array systems with fixed elements. For instance, the beamforming gain can be maximized by optimizing the position of the UAVs within the array, and beamforming can be done toward any direction in 3D space. Further, with a large space available in the sky, large-gain antenna arrays of UAVs can be created. In the UAV UE scenario, cooperative communication among multiple UAV UEs or BSs (as shown in Figure 8.1) allows boosting coverage and capacity of the network, particularly when UAV UEs coexist with ground users. In this case, coordinated multi-point (CoMP) transmission plays a key role in enabling efficient cooperative communications in cellular-connected UAV UE scenarios.

This chapter will, therefore, study a variety of scenarios that involve cooperative communications for wireless networks with UAVs. We will particularly analyze the role of cooperative communications in improving the connectivity and capacity of UAV UEs leveraging principles of CoMP among BSs, and we will shed light on the use of UAV-based antenna arrays as flying UAV BSs. Specifically, in Section 8.1, we introduce a framework based on CoMP transmission for serving high-altitude cellular connected UAV UEs while mitigating interference. Using tools from stochastic geometry, upper bounds on the content coverage probability are derived providing insights on the overall deployments of UAVs as a function of different system parameters, particularly in the presence of down-tilted ground BS antennas. Then, in Section 8.2, we study how one can effectively use multiple quadrotor UAVs as an aerial antenna array that acts as a single coordinated UAV BS to provide wireless service to ground users. The goal will be to maximize performance while minimizing the airborne service time for communication. We will also characterize the optimal rotor's speed for minimizing the control time using theoretical postulates of bang-bang control theory. The obtained results shed light on some fundamental tradeoffs for leveraging antenna array systems. We conclude the chapter with a summary in Section 8.3.

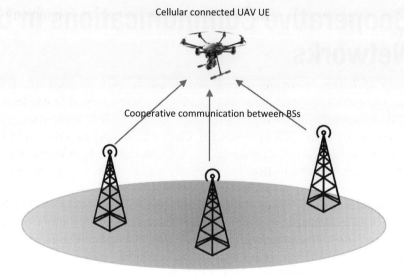

Figure 8.1 Cooperative transmission in cellular connected UAV UE scenario.

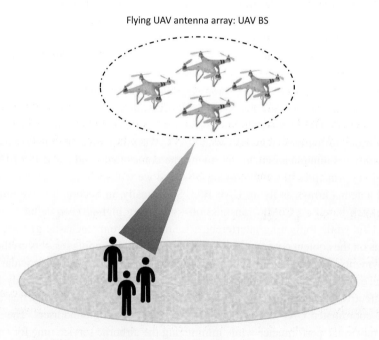

Figure 8.2 Reconfigurable flying antenna array in UAV BS scenario.

8.1 CoMP Transmission in Wireless Systems with Cellular-Connected UAV UEs

In this section, based on our work in [270], we introduce a holistic framework that leverages CoMP transmissions for serving high-altitude cellular-connected UAV UEs while managing cross-cell interference and boosting the received signal-to-interference ratio (SIR). In particular, we consider a network of BSs that are cache-enabled (i.e., can cache content similar to the caching approach of Chapter 5) in which an aerial UAV UE downloads a previously cached content via CoMP transmission from neighboring ground BSs. Using tools from stochastic geometry, we derive a considerably tight upper bound on the content coverage probability as a function of the system parameters. We then demonstrate how the performance that a flying UAV UE can achieve will be impacted by a variety of factors, such as the CoMP collaboration distance, the availability of content, and the target data rate. We then also demonstrate how the down-tilt of the antennas of the ground BS (discussed earlier in Chapter 1) will limit the performance (in terms of coverage probability) that an aerial UAV UE can achieve, even when CoMP is used.

8.1.1 A Model for CoMP in Networks with Aerial UAV UEs

We study a wireless cellular network (e.g., a small cell network) composed of a number of cache-enabled BSs that are distributed based on a homogeneous PPP $\Phi_b = \{b_i \in \mathbb{R}^2, \forall i \in \mathbb{N}^+\}$ whose density is λ_b. In this network, there is a UAV UE that is flying at an altitude h_d and located at $(0, 0, h_d) \in \mathbb{R}^3$. In our model, the BSs are grouped together in multiple (disjoint) clusters in order to serve the UAV UE [271]. In this case, a cluster of BSs is denoted by a circle with radius R_c whose center on the ground is located under the UAV UE (as we can observe in Figure 8.3). The cluster's area is defined as $A = \pi R_c^2$. In this network, BSs located within a cluster can cooperate to transmit cached content to the UAV UE. We note that at high altitudes there can be strong LOS interference from BSs. Thus, we assume that multiple BSs (located within a specific range from the UAV UE) can cooperate to transmit cached contents to the UAV UE.

8.1.2 Probabilistic Caching Placement and Serving Distance Distributions

A BS uses its memory storage capabilities in order to cache content from a file library. Each file in the library includes content catalog that a UAV UE may request. Various files are indexed according to popularity, in a descending order. To determine how content is placed, we adopt a random placement scheme. In this scheme, each content f is independently cached at each BS with a probability c_f, $0 \leq c_f \leq 1$. Moreover, we model BS caching content as a PPP Φ_{bf} whose density is $\lambda_{bf} = c_f \lambda_b$ [151]. Likewise, a BS that does not cache a given content f can be modeled as another PPP $\Phi_{bf}^!$ with density $\lambda_{bf}^\circ = (1 - c_f)\lambda_b$, $\Phi_b = \Phi_{bf} \cup \Phi_{bf}^!$. Now, within a cluster, we can find the probability mass function (PMF) of the number of BSs that will cache content f as follows:

$$\mathbb{P}(n = \kappa) = \frac{(c_f \lambda_b A)^\kappa e^{-c_f \lambda_b A}}{\kappa!}. \tag{8.1}$$

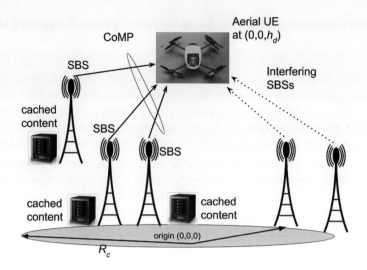

Figure 8.3 Illustration of the proposed system model with ground small BSs (SBSs) and a flying UAV UE.

The model in equation (8.1) is nothing but a Poisson distribution whose mean is $c_f \lambda_b A$.

Considering a cluster with κ BSs with caching capabilities, the distribution of caching-capable BSs is a binomial point process (BPP). In this model, κ BSs are uniformly and independently distributed within the cluster. We use $\Phi_{cf} = \{b_i \in \Phi_{bf} \cap \mathcal{B}(0, R_c)\}$ to represent the set of cooperative BSs that provide content f, with $\mathcal{B}(0, R_c)$ centered WITH radius R_c.

Given a UAV UE located at the origin in $\mathbb{R}^2$, the UAV UE-to-BSs distances will be equal to $\boldsymbol{R}_\kappa = [R_1, \ldots, R_\kappa]$. Now, by conditioning on $\boldsymbol{R}_\kappa = \boldsymbol{r}_\kappa, \boldsymbol{r}_\kappa = [r_1, \ldots, r_\kappa]$, we can find the conditional probability density function (PDF) of the joint distances' distribution, which is given by $f_{\boldsymbol{R}_\kappa}(\boldsymbol{r}_\kappa)$.

Also, the PDF of the horizontal distance r_i between BS i and the UAV UE can be expressed by [151]:

$$f_{R_i}(r_i) = \begin{cases} \frac{2r_i}{R_c^2}, & 0 \leq r_i \leq R_c, \\ 0, & \text{otherwise}, \end{cases}$$

The conditional joint PDF of the serving distances $\boldsymbol{R}_\kappa = [R_1, \ldots, R_\kappa]$ can be determined using the BPP's i.i.d. property:

$$f_{\boldsymbol{R}_\kappa}(\boldsymbol{r}_\kappa) = \prod_{i=0}^{\kappa} \frac{2r_i}{R_c^2}. \tag{8.2}$$

Note that the vertical distance of each BS from the UAV UE is h_{BS}. Here, we use θ_t and θ_B to represent the BS down-tilt angle and the antenna beamwidth in the vertical dimension. In the considered model, the gains of the side lobe and main lobe are represented by G_s and G_m, respectively. The distance between BS i to the UAV UE is $d_i = \sqrt{r_i^2 + (h_d - h_{\text{BS}})^2}$.

8.1.3 Channel Model

The channel gain between BSs and the UAV UE is composed of large-scale fading and small-scale fading, as discussed in Chapter 3. We consider the probabilistic LOS/NLOS model (used in prior chapters and exposed in Chapter 3) for the large-scale fading between BS i and the UAV UE. The channel gain of this model can be given by:

$$\zeta_v(r_i) = A_v G(r_i) d_i^{-\alpha_v} = A_v G(r_i) \bigl(r_i^2 + (h_d - h_{\text{BS}})^2\bigr)^{-\alpha_v/2}, \tag{8.3}$$

where $v \in \{l, n\}$, α_l and α_n represent the path loss exponents for the LOS and NLOS communication links. Also, the path loss constants (considering a 1 m reference distance) are denoted by A_l and A_n. The antenna gain for BS i toward the UAV UE is expressed by:

$$G(r_i) = \begin{cases} G_m, & \text{for } r_i \in \mathcal{S}_{bs}, \\ G_s, & \text{for } r_i \notin \mathcal{S}_{bs}, \end{cases}$$

in which $\mathcal{S}_{bs}$ includes r_is that satisfy $h_{\text{BS}} - r_i \tan(\theta_t + \frac{\theta_B}{2}) < h_d < h_{\text{BS}} - r_i \tan(\theta_t - \frac{\theta_B}{2})$.

To capture the small-scale fading, we use a Nakagami-m fading model with the following PDF:

$$f(\omega) = \frac{2 \frac{m^m}{\eta} \omega^{2m-1}}{\Gamma(m)} \exp\bigl(-\frac{m}{\eta}\omega^2\bigr), \tag{8.4}$$

where m and η denote, respectively, the fading parameter and controlling spread parameter. Considering the fact that there is a dominant LOS link between a BS and the UAV UE, we have $m > 1$.

It can also be shown that the channel power gain distribution has the following PDF:

$$f(\gamma) = \frac{(\frac{m}{\eta})^m \gamma^{m-1}}{\Gamma(m)} \exp\bigl(-\frac{m}{\eta}\gamma\bigr). \tag{8.5}$$

The LOS probability between BS i and the UAV UE located at distance r_i is computed by [272]:

$$\mathbb{P}_l(r_i) = \prod_{n=0}^{\max(p-1,0)} \left[1 - \exp\Bigl(-\frac{(h_{\text{BS}} + \frac{h(n+0.5)}{m+1})^2}{2c^2}\Bigr)\right], \tag{8.6}$$

where a, b, c are the environment-dependent parameters, $h = h_d - h_{\text{BS}}$ and $p = \lfloor \frac{r_i \sqrt{ae}}{1000} \rfloor$.

We will next present a multi-BSs CoMP transmission scheme for mitigating uplink interference as well as enhancing the performance of the UAV UE. In particular, we develop a framework to evaluate the performance of cache-capable CoMP transmissions for cellular-connected UAV UEs.

8.1.4 Analysis of Coverage Probability

We will now characterize the network performance by finding the coverage probability for the UAV UE. In the considered model, the transmit power of each BS is P_t, and a typical UAV UE flies at $(0, 0, h_d) \in \mathbb{R}^3$. The received signal at the UAV UE, assuming κ BSs provide a content f, is given by:

$$P = \underbrace{\sum_{i=1}^{\kappa} P(r_i)\omega_i w_i X_f}_{\text{desired signal}} + \underbrace{\sum_{j \in \Phi_{bf}^! \cap \mathcal{B}(0,R_c)} P(u_j)\omega_j w_j Y_j}_{I_{\text{in}}}$$

$$+ \underbrace{\sum_{k \in \Phi_b \setminus \mathcal{B}(0,R_c)} P(u_k)\omega_k w_k Y_k}_{I_{\text{out}}} + Z, \qquad (8.7)$$

where $P(r_i) = \sqrt{P_t}\zeta_v(r_i)^{0.5}$, $v \in \{l, n\}$, ω_i is the Nakagami-m fading for BS i. Also, w_i is the BS i precoder, and X_f represents the channel input symbol transmitted by multiple BSs. I_{in} is in-cluster interference, I_{out} is the out-of-cluster interference, and Y_j represents the transmit signal of interfering BS j. Moreover, we have:

$$P(u_j) = \begin{cases} P_l(u_j) = \sqrt{P_t}\zeta_l(u_j)^{0.5}, & \text{for LOS,} \\ P_n(u_j) = \sqrt{P_t}\zeta_n(u_j)^{0.5}, & \text{for NLOS,} \end{cases}$$

with u_j is the horizontal distance between the UAV UE and BS j. Z is a circular-symmetric zero-mean complex Gaussian random variable that represents the noise.

We denote the set of interfering BSs by $\Phi_b \setminus \Phi_{cf} = \{b_i \in \{\Phi_b \setminus \mathcal{B}(0, R_c)\} \cup \{\Phi_{bf}^! \cap \mathcal{B}(0, R_c)\}\}$, and we define $\Phi_{cf}^! = \{\Phi_{bf}^! \cap \mathcal{B}(0, R_c)\}$. The set of interfering BSs is shown by $\Phi_b \setminus \Phi_{cf} = \{b_i \in \Phi_b \setminus \mathcal{B}(0, R_c)\}$.

Note that when channel state information is available at the ground BS, the precoder $w_i = \frac{\omega_i^*}{|\omega_i|}$ with ω_i^* being the complex conjugate of ω_i. Considering the independence of X_f, Y_j, and Y_k in (8.7), the SIR for the UAV UE can be derived as follows:

$$\Upsilon_{|r_\kappa} = \sum_{o=0}^{\kappa} \binom{\kappa}{o} \prod_{i=0}^{o} \mathbb{P}_l(r_i) \prod_{j=o+1}^{\kappa} \mathbb{P}_n(r_j) \cdot$$

$$\frac{P_t \left| \sum_{i=1}^{o} \zeta_l^{1/2}(r_i)\omega_i + \sum_{j=o+1}^{\kappa} \zeta_n^{1/2}(r_j)\omega_j \right|^2}{I_{\text{in}} + I_{\text{out}}}, \qquad (8.8)$$

where $\left| \sum_{i=1}^{o} \zeta_l^{1/2}(r_i)\omega_i + \sum_{j=o+1}^{\kappa} \zeta_n^{1/2}(r_j)\omega_j \right|^2$ is the square of weighted sum of κ Nakagami-m random variables. Given the interoperability of a weighted-sum of Nakagami-m random variables, we find its upper bound using Cauchy-Schwarz's inequality:

8.1 CoMP Transmission in Wireless Systems with Cellular-Connected UAV UEs

$$\left| \sum_{i=1}^{o} \zeta_l^{1/2}(r_i)\omega_i + \sum_{j=o+1}^{\kappa} \zeta_n^{1/2}(r_j)\omega_j \right|^2 = \left(\sum_{i=1}^{\kappa} Q_i \right)^2$$

$$\leq \kappa \left(\sum_{i=1}^{\kappa} Q_i^2 \right), \tag{8.9}$$

with $Q_i = \zeta_v^{1/2}(r_i)\omega_i$ being a scaled Nakagami-m random variable (RV), $v \in \{l, n\}$ and $i \in \mathcal{K}_f$. Given $\omega_i \sim$ Nakagami$(m, \eta/m)$, we have $Q_i^2 \sim \Gamma(k_i = m, \theta_i = 2\eta\zeta_v(r_i)/m)$.

In order to find a statistical equivalent PDF of a sum of κ gamma RVs Q_i with different θ_i values, we use the method of sum of gammas second-order moment match [273, Proposition 8].

We can show that the equivalent Gamma distribution ($J \sim \Gamma(k, \theta)$) has the following parameters: $k = \left(\sum_i k_i \theta_i\right)^2 / \sum_i k_i \theta_i^2$ and $\theta = \sum_i k_i \theta_i^2 / \sum_i k_i \theta_i$.

To evaluate the accuracy of our second-order moment approximation, in Figure 8.4 we depict the PDF of the equivalent channel gain. Here, a sum of κ gamma RVs can be approximated by a gamma RV that has the following parameters:

$$k = \frac{m\left(\sum_i \zeta_v(r_i)\right)^2}{\sum_i \left(\zeta_v(r_i)\right)^2} \quad \text{and} \quad \theta = \frac{\eta \sum_i \zeta_v(r_i)}{m \sum_i \zeta_v(r_i)}, \tag{8.10}$$

In this case, the upper bound of the shape parameter k in (8.10) can be given by:

$$k = m\frac{\left(\sum_i \zeta_v(r_i)\right)^2}{\sum_i \left(\zeta_v(r_i)\right)^2} \leq m\frac{\kappa \sum_i \left(\zeta_v(r_i)\right)^2}{\sum_i \left(\zeta_v(r_i)\right)^2} = m\kappa, \tag{8.11}$$

with $m\kappa$ being an integer.

Now, we will derive the coverage probability for the UAV UE. Conditioning on the serving distances r_κ, the coverage probability can be given by:

$$\mathbb{P}_{\text{cov}|r_\kappa} = \mathbb{P}[\Upsilon_{|r_\kappa} > \vartheta] \approx \sum_{o=0}^{\kappa} \binom{\kappa}{o} \prod_{i=0}^{o} \mathbb{P}_l(r_i) \times$$

$$\prod_{j=o+1}^{\kappa} \mathbb{P}_n(r_j) \mathbb{P}\left(\frac{\kappa P_t\left(\sum_{i=1}^{\kappa} Q_i\right)^2}{I_{\text{in}} + I_{\text{out}}} > \vartheta\right), \tag{8.12}$$

$$= \sum_{o=0}^{\kappa} \binom{\kappa}{o} \prod_{i=0}^{o} \mathbb{P}_l(r_i) \prod_{j=o+1}^{\kappa} \mathbb{P}_n(r_j) \mathbb{P}\left(\frac{\kappa P_t J}{I_{\text{in}} + I_{\text{out}}} > \vartheta\right), \tag{8.13}$$

where ϑ represents the SIR threshold. We then have the following result on the unconditional coverage probability of the UAV UE:

THEOREM 8.1 *The UAV UE coverage probability can be expressed by:*

$$\mathbb{P}_{\text{cov}} = \sum_{\kappa=1}^{\infty} \mathbb{P}(n = \kappa) \int_{r_\kappa=0}^{R_c} \mathbb{P}_{\text{cov}|r_\kappa} \prod_{i=0}^{\kappa} \frac{2r_i}{R_c^2} dr_\kappa, \tag{8.14}$$

where $\mathbb{P}_{\text{cov}|r_\kappa}$ shows the conditional coverage probability in (8.16), with $\varpi = \vartheta/\kappa P_t \theta$.

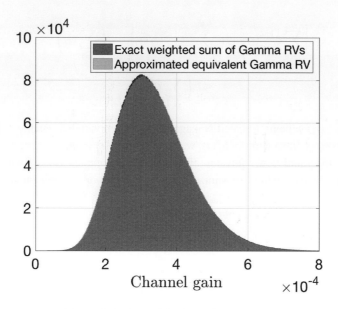

Figure 8.4 Monte Carlo simulation of the PDF of the equivalent gain of channels between cooperating BSs and an aerial UAV UE, including path loss and fading. A PPP realization of density $\lambda_b = 20$ BS/ km^2 is run for a simulated area of 20km^2 with $m = 3$ and $R_c = 200m$.

$$\mathbb{P}_{\text{cov}|r_\kappa} = \sum_{o=0}^{\kappa} \binom{\kappa}{o} \prod_{i=0}^{o} \mathbb{P}_l(r_i) \prod_{i=o+1}^{\kappa} \mathbb{P}_n(r_i) \sum_{k=0}^{k-1} \frac{(-\varpi)^k}{k!} \frac{\partial^k}{\partial \varpi^k}.$$

$$\exp\left(-2\pi \lambda_{bf}^\circ \int_{v=0}^{R_c} (1 - \delta_l \mathbb{P}_l(v) - \delta_n \mathbb{P}_n(v)) v dv\right)$$

$$\times \exp\left(-2\pi \lambda_p \int_{v=R_c}^{\infty} (1 - \delta_l \mathbb{P}_l(v) - \delta_n \mathbb{P}_n(v)) v dv\right). \quad (8.16)$$

Proof This proof is based on our work in [270] and is provided here for guidance

$$\mathbb{P}\left(\frac{\kappa P_t J}{I_{\text{in}} + I_{\text{out}}} > \vartheta\right) = \mathbb{P}\left(\kappa P_t J > \vartheta \left(I_{\text{in}} + I_{\text{out}}\right)\right)$$

$$= \mathbb{E}_{I_{\text{in}}, I_{\text{out}}}\left[\mathbb{P}\left(\kappa P_t J > \vartheta \left(I_{\text{in}} + I_{\text{out}}\right)\right)\right]$$

$$= \mathbb{E}_{I_{\text{in}}, I_{\text{out}}}\left[\mathbb{P}\left(J > \frac{\vartheta}{\kappa P_t} \left(I_{\text{in}} + I_{\text{out}}\right)\right)\right]$$

$$\stackrel{(a)}{\approx} \mathbb{E}_{I_{\text{in}}, I_{\text{out}}}\left[\sum_{k=0}^{k-1} \frac{(\vartheta/\kappa P_t \theta)^k}{k!} \left(I_{\text{in}} + I_{\text{out}}\right)^k e^{-\frac{\vartheta}{\kappa P_t \theta}\left(I_{\text{in}} + I_{\text{out}}\right)}\right]$$

$$\stackrel{(b)}{=} \mathbb{E}_{I_{\text{in}}, I_{\text{out}}}\left[\sum_{k=0}^{k-1} \frac{(-\varpi)^k}{k!} \frac{d^k}{d\vartheta^k} \mathcal{L}_{I_{\text{in}} + I_{\text{out}}|r_\kappa}(\varpi)\right], \quad (8.15)$$

where (a) results from the PDF of a gamma RV and (b) is based on $\varpi = \vartheta/\kappa P_t \theta$ as well as the Laplace transform of the RV $I_{\text{in}} + I_{\text{out}}$.

$$\mathcal{L}_{I_{\text{in}}+I_{\text{out}}|r_\kappa}(\varpi) = \mathbb{E}_{I_{\text{in}},I_{\text{out}}}\left[\exp\bigl(-\varpi(I_{\text{in}}+I_{\text{out}})\bigr)\right]$$

$$= \mathbb{E}\left[e^{-\varpi \sum_{j\in\Phi_{cf}^!} \gamma_j P(u_j)^2}\, e^{-\varpi \sum_{j\in\Phi_b\setminus\mathcal{B}(0,R_c)} \gamma_j P(u_j)^2}\right]$$

$$= \mathbb{E}_{\Phi_b}\left[\prod_{j\in\Phi_{cf}^!} \mathbb{E}_{\gamma_j} e^{-\varpi \gamma_j P(u_j)^2} \prod_{j\in\Phi_b\setminus\mathcal{B}(0,R_c)} \mathbb{E}_{\gamma_j} e^{-\varpi \gamma_j P(u_j)^2}\right]$$

$$\stackrel{(a)}{=} \mathbb{E}_{\Phi_b}\left[\prod_{j\in\Phi_{cf}^!}\left[\left(1+\frac{\varpi P_l(u_j)^2}{m}\right)^{-m}\mathbb{P}_l(u_j)+\right.\right.$$

$$\left(1+\frac{\varpi P_n(u_j)^2}{m}\right)^{-m}\mathbb{P}_n(u_j)\right]\cdot$$

$$\prod_{j\in\Phi_b\setminus\mathcal{B}(0,R_c)}\left[\left(1+\frac{\varpi P_l(u_j)^2}{m}\right)^{-m}\right.$$

$$\left.\left.\cdot\mathbb{P}_l(u_j)+\left(1+\frac{\varpi P_n(u_j)^2}{m}\right)^{-m}\mathbb{P}_n(u_j)\right]\right]$$

$$\stackrel{(b)}{=} \exp\left(-2\pi\lambda_{bf}^\circ \int_{v=0}^{R_c}\bigl(1-\delta_l\mathbb{P}_l(v)-\delta_n\mathbb{P}_n(v)\bigr)v\,dv\right)\cdot$$

$$\exp\left(-2\pi\lambda_p \int_{v=R_c}^{\infty}\bigl(1-\delta_l\mathbb{P}_l(v)-\delta_n\mathbb{P}_n(v)\bigr)v\,dv\right). \quad (8.17)$$

Finally, we can find the UAV UE coverage probability using (8.13), (8.15), (8.17), and (8.1). □

From (8.14), we can see that increasing collaboration distance and the caching probability leads to a higher coverage probability for the UAV UE. Moreover, by increasing the number of caching BSs, the desired signal power will increase.

8.1.5 Representative Simulation Results

Here, we present a number of Monte Carlo simulation results based on our work in [270]. In Table 8.1 we list the simulation parameters.

Figure 8.5 shows the theoretical upper bound on the coverage probability obtained from (8.14) and the simulation results for the exact coverage probability, and the upper bound using Cauchy's inequality. We can clearly observe that the upper bound obtained from the use of the Cauchy-Schwarz inequality is very close to the real, exact coverage probability. While the theoretical upper bound on the coverage probability is not as tight as the Cauchy-based case, it is close to the exact coverage probability. In Figure 8.5, we compare the coverage probability of the CoMP transmission approach with a case in which the UAV UE is served by the nearest ground BS. We can now directly observe that using CoMP for transmitting the same content from multiple BSs substantially improves the UAV UE's coverage probability.

Table 8.1 Simulation parameters.

Description	Parameter	Value
LOS path-loss exponent	α_l	2.09
NLOS path-loss exponent	α_n	3.75
LOS path-loss constant	A_l	-41.1dB
NLOS path-loss constant	A_n	-32.9dB
Antenna main lobe gain	G_m	10dB
Antenna side lobe gain	G_s	-3.01dB
Nakagami fading parameter	m	3
Nakagami spreading factor	η	2
BS antenna height	h_{BS}	30m
Aerial UE altitude	h_d	100m
Area fraction occupied by buildings	a	0.3
Mean number of buildings	e	200 per km^2
Buildings height Rayleigh parameter	c	15
Collaboration distance	R_c	200m
Density of BS	λ_b	20 BS/ km^2
SIR threshold	ϑ	0dB
Down-tilt angle	θ_t	8°
Vertical beamwidth	θ_B	30°
Content caching probability	c_f	1

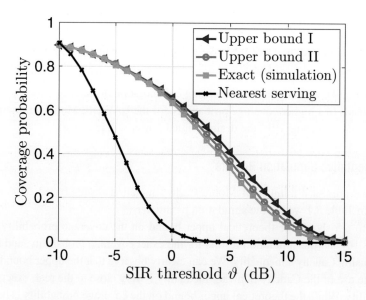

Figure 8.5 The derived upper bound on the coverage probability is plotted versus the SIR threshold ϑ.

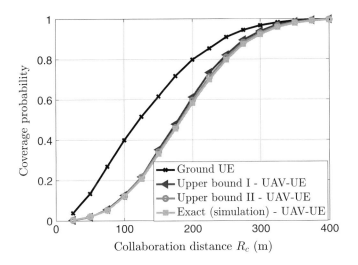

Figure 8.6 Coverage probability versus the collaboration distance R_c for the aerial UAV UE and the ground UE.

In Figure 8.6, we examine the effect of the collaboration distance (R_c) on the coverage probability of the ground users and the UAV UE. This figure shows that increasing the collaboration distance leads to a higher coverage probability. The main reason behind this result is the fact that, with a larger R_c, more BSs will cooperate with each other while serving the UEs. The coverage probability for the ground UE is higher than the UAV UE. However, as the collaboration distance increases, the gap between the coverage probabilities of the UAV UE and of the ground UE become smaller.

Figure 8.7 studies how the coverage probability changes by varying the SIR threshold. We can first see that the coverage probability decreases when the SIR threshold increases. This is because satisfying a higher SIR coverage threshold is less likely due to the fundamental limitations of a communication system. We also evaluate the impact of the caching probability c_f. As we can see from Figure 8.7, by reducing c_f, the coverage probability decreases since the number of caching BSs decreases. In fact, the gain of cooperative communications decreases when the number of caching BSs is smaller.

8.1.6 Summary

In this section, we have discussed a framework for cooperative transmission and probabilistic caching for serving a UAV UE in a cellular-connected UAV system. In particular, we have provided a closed-form expression for upper bound of the content coverage probability. Using Monte Carlo simulations, we have evaluated the theoretical results and demonstrated the tightness of the given approximation for the coverage probability.

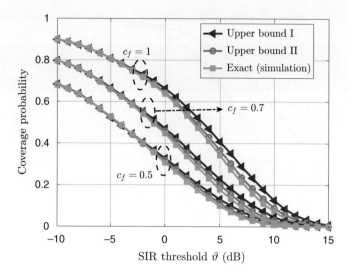

Figure 8.7 Coverage probability versus SIR threshold ϑ for different content caching probability c_f.

Moreover, we have shown how employing CoMP transmission can enhance the coverage probability of the UAV UE. In the next section, we focus on the role of cooperation in creating virtual antenna arrays of UAVs.

8.2 Reconfigurable Antenna Arrays of UAVs: UAV BS Scenario

As we discussed in Chapter 2, one of the promising use cases of UAVs is in creating a flexible, reconfigurable, and wireless antenna array in the sky [22] in which each UAV acts as an antenna element of the array. A reconfigurable UAV-based antenna array system has a number of key advantages over classical antenna array systems in terms of beamforming flexibility and antenna array gain.

In this section, we introduce a framework for deploying a UAV-based antenna array system that acts as a coordinated, fully fledged UAV BS in the sky. This antenna array-based UAV BS can then provide wireless service to ground users. The goal is to design an antenna array of UAVs that serve the ground users within a minimum service time that is composed of transmission time and the control time. We note that minimizing the service time is beneficial from both the users and UAVs points of view. For ground users, lower service time corresponds to less delay as they can be served more quickly. For UAVs service, time is directly related to the flight time and energy consumption. Clearly, by decreasing the service time, the flight time and energy consumption of the flying UAVs decrease accordingly. In order to minimize the service time in the considered UAV-based antenna array system, we consider two main steps. First, we minimize the transmission time by optimizing the positions of UAVs within the array. Then, we

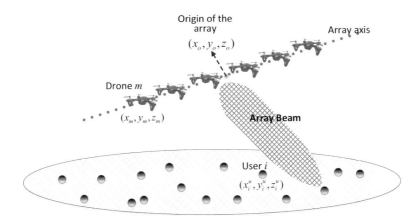

Figure 8.8 Drone-based antenna array.

minimize the control time needed for moving the UAVs while serving different ground users.

8.2.1 UAV-Based Antenna Array in the Sky: A Basic Model

We study a system that encompasses a set $\mathcal{L}$ of L ground users randomly distributed over a certain geographical area. To serve these users, a set $\mathcal{M}$ of M small quadrotor LAP UAVs are deployed and can cooperate to form an aerial UAV BS. In essence, the M UAVs can create a wireless antenna array in the sky that acts as a single UAV BS (with beamforming capabilities), with each element being a UAV, as shown in Figure 8.8. We focus on a symmetric linear antenna array [274], although other generalizations can also be considered. The 3D position of UAV $m \in \mathcal{M}$ and user $i \in \mathcal{L}$ are denoted by $(x_{m,i}, y_{m,i}, z_{m,i})$ and (x_i^u, y_i^u, z_i^u). We consider a minimum separation distance $D_{\min}$ between the two closest UAVs for avoiding a collision between them. The transmitted signal from UAV m has an amplitude a_m and a phase β_m. The distance between UAV m and the origin of the antenna array is $d_{m,i} = \sqrt{(x_{m,i} - x_o)^2 + (y_{m,i} - y_o)^2 + (z_{m,i} - z_o)^2}$ with (x_o, y_o, z_o) being the 3D location of the origin in a Cartesian coordinate. Moreover, for each UAV, the antenna radiation pattern is given by $w(\theta, \phi)$ (in the spherical coordinate), with θ and ϕ being, respectively, the polar angle and the azimuthal angle.

Since ground users are at different locations within the considered area, the UAVs within the antenna array must change their locations in order to serve them. Note that UAVs hover at specific locations to serve a user and move to a new location change the direction of the beam and provide service to a new user. Therefore, in the considered UAV-based antenna array system, we focus on the mechanical beam steering (by moving the UAVs) as opposed to the classical electronic beam steering. The service time is a function of the transmission time, for sending the data, and the control time, for moving and stabilizing the UAVs. Clearly, the transmission time in downlink is inversely related to the data rate of the UAV antenna array. The data rate is a function of the SNR and hence depends on the gain of the antenna array.

Given the importance of the service time for both UAVs and ground users, we aim to minimize it by optimizing and controlling the locations of the UAVs (to minimize the transmission time) and moving them within a minimum control time. For a UAV-to-user link, a dominant LOS model is considered, given the high altitude of the UAV antenna array as well as exploiting beamforming that mitigates the multipath effect (and based on the various arguments provided in previous chapters). The downlink data rate for serving a given ground user i is:

$$R_i(\mathbf{x}_i, \mathbf{y}_i, \mathbf{z}_i) = B\log_2\left(1 + \frac{r_i^{-\alpha} P_t K_o G_i(\mathbf{x}_i, \mathbf{y}_i, \mathbf{z}_i)}{\sigma^2}\right), \tag{8.18}$$

with $\mathbf{x}_i = [x_{m,i}]_{M\times 1}, \mathbf{y}_i = [y_{m,i}]_{M\times 1}, \mathbf{z}_i = [z_{m,i}]_{M\times 1}, m \in \mathcal{M}$ representing the locations of the UAVs servicing user i. Also, B is the transmission bandwidth, and r_i is the distance of the array's origin to user i. The total UAV antenna array's transmit power is given by P_t, the noise power is denoted by σ^2, and K_o is the path loss constant. In 8.18, $G_i(\mathbf{x}_i, \mathbf{y}_i, \mathbf{z}_i)$ represents the array's gain while serving user i.

Now, the total gain of the UAV-based antenna array can be given by:

$$G_i(\mathbf{x}_i, \mathbf{y}_i, \mathbf{z}_i) = \frac{4\pi |F(\theta_i, \phi_i)|^2 w(\theta_i, \phi_i)^2}{\int_0^{2\pi}\int_0^{\pi} |F(\theta, \phi)|^2 w(\theta, \phi)^2 \sin\theta d\theta d\phi}\eta, \tag{8.19}$$

where $0 \leq \eta \leq 1$ is the antenna array's efficiency, and $F(\theta, \phi)$ represents the array factor, which is expressed by [275]:

$$F(\theta, \phi) = \sum_{m=1}^{M} a_m e^{j[k(x_{m,i}\sin\theta\cos\phi + y_{m,i}\sin\theta\sin\phi + z_{m,i}\cos\theta) + \beta_m]}, \tag{8.20}$$

where λ is the wavelength, $k = 2\pi/\lambda$ is the phase constant, and the overall radiation pattern of the antenna array can be computed by $F(\theta, \phi)w(\theta_i, \phi_i)$ [275].

Now, the total service time that the UAV-based antenna array requires in order to connect to the ground users can be formulated as:

$$T_{\text{service}} = \sum_{i=1}^{L} \frac{q_i}{R_i(\mathbf{x}_i, \mathbf{y}_i, \mathbf{z}_i)} + T_i^{\text{crl}}(V, \mathbf{x}_i, \mathbf{y}_i, \mathbf{z}_i), \tag{8.21}$$

where T_{service} represents the total service time and q_i is the load of user i defined as the number of bits that must be transmitted to user i. T_i^{crl} is the control time during which the UAVs adjust their locations according to the location of ground user i. In particular, T_i^{crl} captures the time needed for updating the UAVs' positions from state $i-1$ (i.e., locations of UAVs while serving user $i-1, i > 1$) to state i. The control time is obtained based on the dynamics of the UAVs and is a function of control inputs, external forces, and the movement of UAVs. In fact, each UAV needs a vector of control inputs in order to move from its initial location to a new location while serving different users. For quadrotor UAVs, the rotors' speeds are commonly considered as control inputs.

Therefore, in (8.21), we have $V = [v_{mn}(t)]_{M \times 4}$ with $v_{mn}(t)$ being the speed of rotor n of UAV m at time t. The maximum speed of each rotor is $v_{\max}$. In this case, the control time of the UAVs can be minimized by properly adjusting the rotors' speeds. An important aspect of operating UAV is stability, which needs to be ensured when serving ground users by controlling UAV's rotors' speeds while factoring in wind dynamics.

Our goal is to minimize the total service time of UAVs by finding the optimal locations of the UAVs with respect to the center of the array, as well as the optimal control inputs. More formally, the optimization problem is given by:

$$\underset{X,Y,Z,V}{\text{minimize}} \sum_{i=1}^{L} \frac{q_i}{R_i(x_i, y_i, z_i)} + T_i^{\text{crl}}(V, x_i, y_i, z_i), \qquad (8.22)$$

$$\text{st. } d_{m+1,i} - d_{m,i} \geq D_{\min}, \quad \forall m \in \mathcal{M}\setminus\{M\}, \qquad (8.23)$$

$$0 \leq v_{mw}(t) \leq v_{\max}, \quad \forall m \in \mathcal{M}, w \in \{1, ..., 4\}, \qquad (8.24)$$

where X, Y, and Z are position matrices. In these matrices, row i is vector x_i, y_i, or z_i, $\forall i \in \mathcal{L}$. (8.23) is a constraint for collision avoidance, and (8.24) is related to the maximum rotor's speed constraint.

We can observe that the optimization problem in (8.22) accounts for both transmission time (the first term) and the control time (the second term). Solving this optimization problem is challenging due to its nonlinearity, non-convexity, and mutual dependence between various optimization variables. To solve (8.22), we proceed as follows. In the first step, we minimize the transmission time by optimizing the UAVs' positions within the linear antenna array according to the location of each user. Hence, for L users, we find L sets of locations for the UAV antenna arrays. Then, based on our results in the first step, we minimize the control time by using an optimal control mechanism for moving and stabilizing the UAVs.

8.2.2 Transmission Time Minimization: Optimizing UAV Positions within the Array

Here, for each ground user, we find the optimal locations of the UAVs within the array that ensure a minimum transmission time for serving the user. Considering (8.18), (8.19), and (8.21), the transmission time can be minimized by maximizing the directivity of the UAV antenna array with respect to each user. The array factor for the UAV antenna array with M (assuming M is even) UAVs positioned on the x-axis is written by:

$$F(\theta, \phi) = \sum_{m=1}^{M} a_m e^{j[kx_{m,i} \sin\theta \cos\phi + \beta_m]}$$

$$\stackrel{(a)}{=} \sum_{n=1}^{M/2} a_n \left(e^{j[kd_n \sin\theta \cos\phi + \beta_n]} + e^{-j[kd_n \sin\theta \cos\phi + \beta_n]} \right)$$

$$\stackrel{(b)}{=} 2 \sum_{n=1}^{N} a_n \cos(kd_n \sin\theta \cos\phi + \beta_n), \qquad (8.25)$$

where $N = M/2$ and d_n represents the distance between element $n \in \mathcal{N} = \{1, 2, ..., N\}$ and array's origin. In (a), we used the symmetric property of the array, and in (b), we use Euler's rule. We can now optimize d_n, $\forall n \in \mathcal{N}$ to achieve the maximum array's directivity:

$$\underset{d_n, \forall n \in \mathcal{N}}{\text{maximize}} \frac{4\pi |F(\theta_{\max}, \phi_{\max})|^2 w(\theta_{\max}, \phi_{\max})^2}{\int_0^{2\pi} \int_0^{\pi} |F(\theta, \phi)|^2 w(\theta, \phi)^2 \sin\theta d\theta d\phi}, \qquad (8.26)$$

where $(\theta_{\max}, \phi_{\max})$ are the polar and azimuthal angles for which the array's antenna pattern is maximized.

To solve (8.26), which is a challenging problem due to its highly nonlinear nature, we adopt the perturbation technique [274]. Using this technique, we will be able to provide a suboptimal (with a reasonable accuracy) but tractable solution to (8.26).

UAV Spacing Optimization: A Perturbation Technique

Here, we optimize the distance of UAVs from the origin of the array by using the so-called perturbation technique. To this end, we start with an initial value for distance between adjacent UAVs, and then we determine suitable perturbation values used to update the considered initial value.

The initial distance between UAV n and the array's center is denoted by d_n^0. In this case, the perturbed distance can be given by:

$$d_n = d_n^0 + e_n, \qquad (8.27)$$

where $e_n \ll \lambda$ indicates the perturbation value for UAV n, and λ is the wavelength.

Using (8.27), we can approximate the array factor by:

$$F(\theta, \phi) = 2 \sum_{n=1}^{N} a_n \cos\left(k(d_n^0 + e_n) \sin\theta \cos\phi + \beta_n\right)$$

$$= 2 \sum_{n=1}^{N} a_n \cos\left[\left(kd_n^0 \sin\theta \cos\phi + \beta_n\right) + ke_n \sin\theta \cos\phi\right]$$

$$\overset{(a)}{\approx} \sum_{n=1}^{N} 2a_n \cos\left(kd_n^0 \sin\theta \cos\phi + \beta_n\right)$$

$$- \sum_{n=1}^{N} 2a_n k e_n \sin\theta \cos\phi \sin\left(kd_n^0 \sin\theta \cos\phi + \beta_n\right), \qquad (8.28)$$

where (a) is based on the trigonometric identities when $\sin(x) \approx x$.

Subsequently, we can represent the optimization problem in (8.26) by:

$$\min_{e} \int_0^{2\pi} \int_0^{\pi} F(\theta, \phi)^2 w(\theta, \phi)^2 \sin\theta d\theta d\phi, \qquad (8.29)$$

$$\text{s.t. } d_{n+1}^0 + e_{n+1} - d_n^0 - e_n \geq D_{\min}, \; \forall n \in \mathcal{N} \setminus \{N\}, \qquad (8.30)$$

where $\boldsymbol{e}$ is a vector including all of the perturbation values e_n, $n \in \mathcal{N}$.

Concisely, we consider the following functions that are used in our subsequent analysis:

$$F^0(\theta, \phi) = \sum_{n=1}^{N} a_n \cos\left(kd_n^0 \sin\theta \cos\phi + \beta_n\right), \tag{8.31}$$

$$I_{\text{int}}(x) = \int_0^{2\pi}\int_0^{\pi} x \sin\theta \, d\theta \, d\phi. \tag{8.32}$$

Then, leveraging our results in [22], we can prove the following:

THEOREM 8.2   Our UAV spacing optimization problem in (8.29) is convex. Also, the optimal perturbation vector is determined by solving: [22]:

$$\begin{cases} \boldsymbol{e} = \boldsymbol{G}^{-1}[\boldsymbol{q} + \boldsymbol{\mu}_{\mathcal{L}}], \\ \mu_n\left(e_n - e_{n+1} + D_{\min} + d_n^0 - d_{n+1}^0\right) = 0, \ \forall n \in \mathcal{N} \setminus \{N\}, \\ \mu_n \geq 0, \ \forall n \in \mathcal{N} \setminus \{N\}. \end{cases} \tag{8.33}$$

with $\boldsymbol{G} = [g_{m,n}]_{N \times N}$ being an $N \times N$ matrix, and:

$$g_{m,n} = I_{\text{int}}\bigg(a_m a_n (k \sin\theta \cos\phi w(\theta, \phi))^2$$
$$\times \sin\left(kd_n^0 \sin\theta \cos\phi + \beta_n\right) \sin\left(kd_m^0 \sin\theta \cos\phi + \beta_m\right)\bigg), \tag{8.34}$$

Moreover, $\boldsymbol{q} = [q_n]_{N \times 1}$ with:

$$q_n = I_{\text{int}}\bigg(a_n k \sin\theta \cos\phi w(\theta, \phi) F^0(\theta, \phi)$$
$$\times \sin\left(kd_n^0 \sin\theta \cos\phi + \beta_n\right)\bigg). \tag{8.35}$$

Note that, $\boldsymbol{\mu}_{\mathcal{L}}$ includes Lagrangian multipliers, whose element n is given by $\boldsymbol{\mu}_{\mathcal{L}}(n) = \mu_{n+1} - \mu_n$. μ_n is a Lagrangian multiplier corresponding to constraint n.

Based on the result of Theorem 8.2, the UAV-to-origin distance is updated by:

$$\boldsymbol{d}^1 = \boldsymbol{d}^0 + \boldsymbol{e}^*, \tag{8.36}$$

where $\boldsymbol{d}^1 = [d_n^1]_{N \times 1}$, and $\boldsymbol{d}^0 = [d_n^0]_{N \times 1}$, $n \in \mathcal{N}$.

After $r \in \mathbb{N}$ updates, we have:

$$\boldsymbol{d}^{(r)} = \boldsymbol{d}^{(r-1)} + \boldsymbol{e}^{*(r)}, \tag{8.37}$$

where $\boldsymbol{e}^{*(r)}$ represents the optimal perturbation vector at step r.

In the sequel, using $\boldsymbol{d}^*$, we find the optimal 3D UAVs' positions, which can yield a maximum array directivity for each ground user.

Optimal 3D Locations of UAVs

Now, we seek to maximize the UAV array directivity by optimizing the 3D locations of the UAVs in the array. We introduce (x_i^u, y_i^u, z_i^u) and (x_o, y_o, z_o) to represent, respectively, the 3D locations of user $i \in \mathcal{L}$ and the origin of the UAV antenna array.

Considering the UAV array's center as the origin of the coordinate system, we can present the polar and azimuthal angles of user i as:

$$\theta_i = \cos^{-1}\left[\frac{z_i^u - z_o}{\sqrt{(x_i^u - x_o)^2 + (y_i^u - y_o)^2 + (z_i^u - z_o)^2}}\right], \qquad (8.38)$$

$$\phi_i = \sin^{-1}\left[\frac{y_i^u - y_o}{\sqrt{(x_i^u - x_o)^2 + (y_i^u - y_o)^2}}\right]. \qquad (8.39)$$

In the next theorem (whose proof is found in [22]), we derive the optimal positions of the UAVs.

THEOREM 8.3 The optimal UAVs' positions within the array that maximize the array directivity for serving user i can be determined by [22]:

$$\left(x_m^*, y_m^*, z_m^*\right)^T = \begin{cases} \mathbf{R}_{\text{rot}}\left(d_m^* \sin\alpha_o \cos\gamma_o, d_m^* \sin\alpha_o \sin\beta_o, d_m^* \cos\alpha_o\right)^T, & m \leq M/2, \\ -\mathbf{R}_{\text{rot}}\left(d_m^* \sin\alpha_o \cos\gamma_o, d_m^* \sin\alpha_o \sin\gamma_o, d_m^* \cos\alpha_o\right)^T, & m > M/2, \end{cases} \qquad (8.40)$$

where α_o and γ_o are the initial polar angle and the azimuthal angle of UAV $m \leq M/2$, and $\mathbf{R}_{\text{rot}}$ represents the rotation matrix used for updating UAVs' locations:

$$\mathbf{R}_{\text{rot}} = \begin{pmatrix} a_x^2(1-\delta) + \delta & a_x a_y(1-\delta) - \lambda a_z & a_x a_z(1-\delta) + \lambda a_y \\ a_x a_y(1-\delta) + \lambda a_z & a_y^2(1-\delta) + \delta & a_y a_z(1-\delta) - \lambda a_x \\ a_x a_z(1-\delta) - \lambda a_y & a_y a_z(1-\delta) + \lambda a_x & a_z^2(1-\delta) + \delta \end{pmatrix}, \qquad (8.41)$$

where $\delta = \|\mathbf{q}_i \cdot \mathbf{q}_{\max}\|$, $\lambda = \sqrt{1-\delta^2}$, $\mathbf{q}_i = \begin{pmatrix} \sin\theta_i \cos\phi_i \\ \sin\theta_i \sin\phi_i \\ \cos\theta_i \end{pmatrix}$, $\mathbf{q}_{\max} = \begin{pmatrix} \sin\theta_{\max} \cos\phi_{\max} \\ \sin\theta_{\max} \sin\phi_{\max} \\ \cos\theta_{\max} \end{pmatrix}$. Also, $\mathbf{a} = (a_x, a_y, a_z)^T = \mathbf{q}_i \times \mathbf{q}_{\max}$ is a vector whose elements are a_x, a_y, and a_z.

Using Algorithm 4 (which is based on Theorem 8.3) we can find the optimal UAVs' positions within the array that can yield a minimum transmission time for each user. While serving multiple users one-by-one, the UAVs need to update their locations. Next, we will minimize the control time needed for moving and stabilizing the UAVs.

8.2 Reconfigurable Antenna Arrays of UAVs: UAV BS Scenario

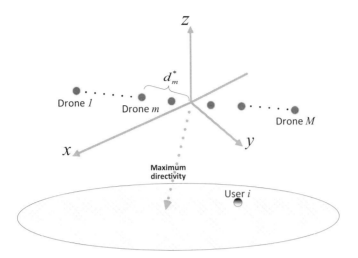

Figure 8.9 Illustrative figure for Theorem 8.3.

Algorithm 4 Optimizing UAVs' locations for maximum array gain toward user i.

1: **Inputs:** Locations of user i, (x_i^u, y_i^u, z_i^u), and origin of array, (x_o, y_o, z_o).
2: **Outputs:** Optimal UAVs' positions, $(x_{m,i}^*, y_{m,i}^*, z_{m,i}^*)$, $\forall m \in \mathcal{M}$.
3: Set initial values for distance between UAVs, $\boldsymbol{d}$.
4: Find $\boldsymbol{e}^*$ by using (8.33)–(8.35).
5: Update $\boldsymbol{d}$ based on (8.36).
6: Repeat steps (4) and (5) to find the optimal spacing vector $\boldsymbol{d}^*$.
7: Use (8.38)–(8.41) to determine (x_m^*, y_m^*, z_m^*), $\forall m \in \mathcal{M}$.

8.2.3 Control Time Minimization: Time-Optimal Control of UAVs

In Section 8.2.2, we determined all the locations where the UAVs need to be while serving different users. Our next step is to develop a way to minimize the control time during which the UAVs move between those predetermined locations. The control time is minimized by optimally adjusting speed of rotors for each quadrotor UAV. In addition, we capture the impact of wind dynamics on the control of the UAV antenna array.

Dynamic Model for Quadrotor UAV

In Figure 8.10, we provide an illustrative figure for a quadrotor drone that uses four rotors for hovering and movement, which is done by controlling the rotors' speeds.

We define (ψ_r, ψ_p, ψ_y) as the roll, pitch, and yaw angles of the UAV located at (x, y, z). The rotors' speeds are denoted by v_i, $i \in \{1, 2, 3, 4\}$. For the considered quadrotor UAV, the total thrust (needed for displacement) and torques (needed for changing orientation) can be computed by: [276]:

$$\begin{pmatrix} T_{\text{tot}} \\ \kappa_1 \\ \kappa_2 \\ \kappa_3 \end{pmatrix} = \begin{pmatrix} \rho_1 & \rho_1 & \rho_1 & \rho_1 \\ 0 & -l\rho_1 & 0 & l\rho_1 \\ -l\rho_1 & 0 & l\rho_1 & 0 \\ -\rho_2 & \rho_2 & -\rho_2 & \rho_2 \end{pmatrix} \begin{pmatrix} v_1^2 \\ v_2^2 \\ v_3^2 \\ v_4^2 \end{pmatrix}, \qquad (8.42)$$

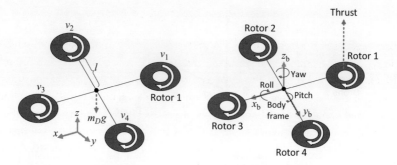

Figure 8.10 A quadrotor UAV.

where T_{tot} is the total upward thrust of the UAV (as shown in Figure 8.10). The torques for the roll, pitch, and yaw movements are, respectively, defined as κ_1, κ_2, and κ_3. In 8.42, ρ_1 and ρ_2 represent the lift and torque coefficients. Finally, l shows the separation distance of each rotor from the UAV's center.

The dynamic equations for a quadrotor UAV can be given by:

$$\ddot{x} = \left(\cos\psi_r \sin\psi_p \cos\psi_y + \sin\psi_r \sin\psi_y\right)\frac{T_{\text{tot}}}{m_D} + \frac{F_x^{\text{W}}}{m_D}, \tag{8.43}$$

$$\ddot{y} = \left(\cos\psi_r \sin\psi_p \sin\psi_y + \sin\psi_r \cos\psi_y\right)\frac{T_{\text{tot}}}{m_D} + \frac{F_y^{\text{W}}}{m_D}, \tag{8.44}$$

$$\ddot{z} = \left(\cos\psi_r \cos\psi_p\right)\frac{T_{\text{tot}}}{m_D} - g + \frac{F_z^{\text{W}}}{m_D}, \tag{8.45}$$

$$\ddot{\psi}_r = \frac{\kappa_2}{I_x}, \tag{8.46}$$

$$\ddot{\psi}_p = \frac{\kappa_1}{I_y}, \tag{8.47}$$

$$\ddot{\psi}_y = \frac{\kappa_3}{I_z}, \tag{8.48}$$

where m_D indicates the UAV's mass, and g represents the gravity, and the wind force has different directions that are given F_x^{W}, F_y^{W}, and F_z^{W}. Moreover, the moments of inertia are shown by I_x, I_y, I_z.

Based on this UAV's dynamic model, we now determine the optimal rotors' speed for which the UAV updates its position from (x_I, y_I, z_I) to (x_D, y_D, z_D) (i.e., from point I to D) within the minimum control time. At time t, the 3D position and the attitude of a UAV are, respectively, denoted by $(x(t), y(t), z(t))$ and $(\psi_r(t), \psi_p(t), \psi_y(t))$. We further define variable $T_{I,D}$ as the total control time that a UAV needs for flying from point I to point D. The time-optimal control problem will now be:

$$\underset{[v_1(t), v_2(t), v_3(t), v_4(t)]}{\text{minimize}} \quad T_{I,D}, \tag{8.49}$$

$$\text{st.} \quad |v_w(t)| \leq v_{\max}, \quad \forall w \in \{1, ..., 4\}, \tag{8.50}$$

$$(x(0), y(0), z(0)) = (x_I, y_I, z_I), \tag{8.51}$$

$$\left(x(T_{I,D}), y(T_{I,D}), z(T_{I,D})\right) = (x_D, y_D, z_D), \tag{8.52}$$

8.2 Reconfigurable Antenna Arrays of UAVs: UAV BS Scenario

$$\left(\dot{x}(T_{I,D}), \dot{y}(T_{I,D}), \dot{z}(T_{I,D})\right) = (0, 0, 0), \quad (8.53)$$

where $[v_1(t), v_2(t), v_3(t), v_4(t)]$ are the speeds of rotors at time instance t, which are less than $v_{\max}$ (i.e., the maximum speed of each rotor). Constraints (8.51)–(8.53) refer to the initial and final positions of the UAV as well as its stability at the final position.

We can now easily see that the optimization problem in (8.49) is nonlinear, non-convex, and contains an infinite number of variables. This, in turn, makes the problem intractable and challenging to solve. One way to reduce the complexity of this optimization problem is to decompose it into two subproblems considering the displacement and orientation changes separately. In order to solve (8.49), we need to use a lemma from the time-optimal control theory [277], as follows.

LEMMA 8.4 (From [277]): Let us consider a moving object during $[0, T]$ with the following state space equations:

$$\dot{x}(t) = Ax(t) + bu(t), \quad u_{\min} \leq u(t) \leq u_{\max}, \quad (8.54)$$

$$x(0) = x_1, \quad (8.55)$$

$$x(T) = x_2, \quad (8.56)$$

with $x(t) \in \mathbb{R}^{N_s}$ and N_s being, respectively, the state vector and the number of elements in the state. $u(t)$ is a control input that is between $u_{\max}$ and $u_{\min}$.

The initial and final states of this moving object are given by x_1 and x_2. The optimal control input for which the state update time is minimized can now be written as follows [277]:

$$u^*(t) = \begin{cases} u_{\max}, & t \leq \tau, \\ u_{\min}, & t > \tau, \end{cases} \quad (8.57)$$

where τ is called the state switching time.

The result given in Lemma 8.4 corresponds to *bang-bang* solution from time-optimal control theory. Based on this solution, the optimal control input adopts either its maximum or minimum value. Using Lemma 8.4 along with a lemma from [22], we provide a solution to (8.49) and find the optimal speeds of rotors at different time instances.

THEOREM 8.5 In order to achieve the minimum control time that a UAV needs to update its location from position $(0, 0, 0)$ to (x_D, y_D, z_D), the speed of its rotors needs to be adjusted according to the following equations [22]:

$$\text{Stage 1:} \begin{cases} v_2 = 0, v_1 = v_3 = \frac{1}{\sqrt{2}}v_{\max}, v_4 = v_{\max}, & \text{if } 0 < t \leq \tau_1, \\ v_4 = 0, v_1 = v_3 = \frac{1}{\sqrt{2}}v_{\max}, v_2 = v_{\max}, & \text{if } \tau_1 < t \leq \tau_2, \\ v_1 = 0, v_2 = v_4 = \frac{1}{\sqrt{2}}v_{\max}, v_3 = v_{\max}, & \text{if } \tau_2 < t \leq \tau_3, \\ v_3 = 0, v_2 = v_4 = \frac{1}{\sqrt{2}}v_{\max}, v_1 = v_{\max}, & \text{if } \tau_3 < t \leq \tau_4. \end{cases} \quad (8.58)$$

$$\text{Stage 2:} \quad v_1 = v_2 = v_3 = v_4 = v_{\max}, \quad \text{if } \tau_4 < t \leq \tau_5. \quad (8.59)$$

$$\text{Stage 3:} \begin{cases} v_2 = 0, v_1 = v_3 = \frac{1}{\sqrt{2}}v_{\max}, v_4 = v_{\max}, & \text{if } \tau_5 < t \le \tau_6, \\ v_4 = 0, v_1 = v_3 = \frac{1}{\sqrt{2}}v_{\max}, v_2 = v_{\max}, & \text{if } \tau_6 < t \le \tau_7, \\ v_1 = 0, v_2 = v_4 = v_{\max}, v_3 = v_{\max}, & \text{if } \tau_7 < t \le \tau_8, \\ v_3 = 0, v_2 = v_4 = \frac{1}{\sqrt{2}}v_{\max}, v_1 = v_{\max}, & \text{if } \tau_8 < t \le \tau_9. \end{cases} \quad (8.60)$$

$$\text{Stage 4:} \quad v_1 = v_2 = v_3 = v_4 = v_{\max}, \quad \text{if } \tau_9 < t \le \tau_{10}. \quad (8.61)$$

$$\text{Stage 5:} \begin{cases} v_2 = 0, v_1 = v_3 = \frac{1}{\sqrt{2}}v_{\max}, v_4 = v_{\max}, & \text{if } \tau_{10} < t \le \tau_{11}, \\ v_4 = 0, v_1 = v_3 = \frac{1}{\sqrt{2}}v_{\max}, v_2 = v_{\max}, & \text{if } \tau_{11} < t \le \tau_{12}, \\ v_1 = 0, v_2 = v_4 = \frac{1}{\sqrt{2}}v_{\max}, v_3 = v_{\max}, & \text{if } \tau_{12} < t \le \tau_{13}, \\ v_3 = 0, v_2 = v_4 = \frac{1}{\sqrt{2}}v_{\max}, v_1 = v_{\max}, & \text{if } \tau_{13} < t \le \tau_{14}. \end{cases} \quad (8.62)$$

$$\text{Stage 6:} \quad v_1 = v_2 = v_3 = v_4 = v_F, \quad \text{if } t > \tau_{14}. \quad (8.63)$$

Moreover, the total control time of the UAV will be equal to:

$$\begin{aligned} T_{I,D} = & \sqrt{2 d_D \left(\frac{m_D}{A_{s2}} - \frac{m_D}{A_{s4}} \right)} \\ & + \frac{2}{v_{\max}} \left[\sqrt{\frac{\Delta \psi_{p,1} I_y}{l \rho_1}} + \sqrt{\frac{\Delta \psi_{r,1} I_x}{l \rho_1}} + \sqrt{\frac{\Delta \psi_{p,3} I_y}{l \rho_1}} \right. \\ & \left. + \sqrt{\frac{\Delta \psi_{r,3} I_x}{l \rho_1}} + \sqrt{\frac{\Delta \psi_{p,5} I_y}{l \rho_1}} + \sqrt{\frac{\Delta \psi_{r,5} I_x}{l \rho_1}} \right], \end{aligned} \quad (8.64)$$

with v_{in} and v_F being the rotor's speeds at initial and final positions of the UAV. $\Delta \psi_{r,i}$ and $\Delta \psi_{p,i}$ represent the roll and pitch changes in Stage i. Also, d_D is the displacement distance of the UAV. $\tau_1, \ldots, \tau_{14}$ are the switching times whose values along with v_F value can be found in [22].

8.2.4 Representative Simulation Results

For simulations, we consider 100 ground users randomly distributed on a geographical area of size 1 m × 1 km. We also consider 10 single-antenna quadrotor UAVs each with an omni-directional antenna. In Table 8.2, we list our simulation parameters. For a benchmark, we consider a UAV antenna array with a fixed separation (half wavelength) between adjacent UAVs.

In Table 8.3, we provide a representative result on the adjacent UAVs separation distances. This table clearly shows that, in the proposed flexible UAV antenna array system, the array is not uniform and the UAV spacing is different for different adjacent UAVs.

In Figure 8.11, we compare the total service time for the flexible UAV antenna array with the fixed array. Clearly, the proposed flexible antenna array system has a better performance compared to the fixed array since it considers optimal UAVs' positions within the array that lead to the maximum array gain. Moreover, we can observe an inherent

8.2 Reconfigurable Antenna Arrays of UAVs: UAV BS Scenario

Table 8.2 Parameters used for simulations.

Parameter	Description	Value
f_c	Carrier frequency	300 MHz
P_i	UAV transmit power	0.1 W
N_o	Total noise power spectral density	−157 dBm/Hz
N	Number of ground users	100
(x_o, y_o, z_o)	Array's center coordinate	(0,0,100) in meters
q_i	Load per user	100 Mb
α	Pathloss exponent	3
I_x, I_y	Moments of inertia	4.9×10^{-3} kg.m^2 [278]
m_D	Mass of each LOS	0.5 kg
l	Distance of a rotor to UAV's center	20 cm
ρ_1	Lift coefficient	2.9×10^{-5} [278]
$\beta_m - \beta_{m-1}$	Phase excitation difference for two adjacent antennas	$\frac{\pi}{5(M-1)}$

Table 8.3 Separation distance of adjacent UAVs in an aerial antenna array with 10 UAVs.

UAVs' separations (cm), f_c = 300 MHz, λ = 1 m	UAVs' separations (cm), f_c = 500 MHz, λ = 0.6 m	Compared to wavelength (λ)
81.9	49.1	81.9 λ
88.7	53.2	88.7 λ
89.8	54.1	89.8 λ
90.7	54.3	90.7 λ
89.8	54.1	89.8 λ
88.7	53.2	88.7 λ
81.9	49.1	81.9 λ

tradeoff between transmission bandwidth and the service time. By increasing the bandwidth, the transmission time (as a major component of the service time) decreases. From Figure 8.11, we can see that, for 10 minutes service time, the bandwidth used for the flexible UAV antenna array is 2/3 of the fixed array.

Figure 8.12 shows how the number of ground users affects the service time in the flexible UAV antenna array system and the fixed array case. As the number of users increases, the transmission time as well as the control time needed for serving the users increases. From Figure 8.12, we can also see that our proposed UAV antenna array system, which acts as a UAV base station, outperforms the fixed-array case for various users. For instance, using our approach, to serve 200 ground users, the flexible array needs to fly for about 28% less compared to the fixed array.

Figure 8.13 illustrates the control, transmission, and service times as a function of the number of UAVs used in the flexible antenna array system. By increasing the number of UAVs (i.e., larger size of the array), the total control time needed for moving the UAVs will increase. However, with more antenna elements within the array, the gain of the UAV-based antenna array system increases. For example, as we can see from Figure 8.13 when the number of UAVs increases from 10 to 30, the time needed for controlling the

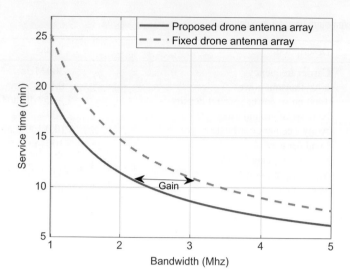

Figure 8.11 Service time vs. bandwidth for the UAV antenna-array and fixed-array cases.

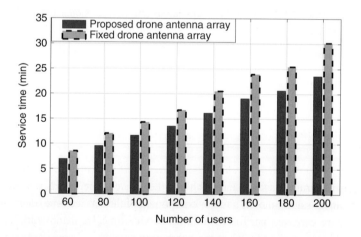

Figure 8.12 Service time vs. number of users for the UAV antenna array and fixed-array (2MHz bandwidth).

UAVs increases by 21%. Nevertheless, in this case, the transmission time for serving the users can be reduced by 37%.

8.2.5 Summary

In this section, we have presented a framework for deploying a UAV-based wireless antenna array system for efficiently and quickly serving ground users. In particular, while providing service to users, we have discussed how to minimize the total service time, which includes transmission time and control time. To minimize the transmission time, we have optimized the positions of the UAVs within the antenna array to achieve

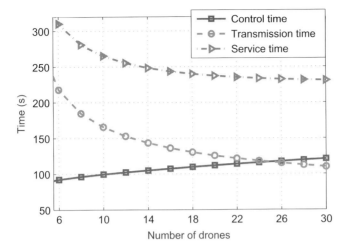

Figure 8.13 Control, transmission, and service times vs. number of UAVs.

a maximum array directivity. To minimize the control time needed for moving and stabilizing the UAVs while serving different users, we have presented an optimal control mechanism based on the time-optimal control theory to dynamically adjust the speed of rotors for each UAV. The results have demonstrated promising advantages of flexible and reconfigurable UAV-based antenna array systems while characterizing inherent design tradeoffs in these systems.

8.3 Chapter Summary

In this chapter, we have described the role of cooperative communication in improving the connectivity and capacity of UAV-enabled wireless networks. In a cellular-connected UAV scenario, we have shown how leveraging principles of CoMP among ground BSs can improve the coverage performance for UAV UEs by reducing the uplink LOS interference. In this case, using tools from stochastic geometry, we have derived the content coverage probability for UAV UEs and provided insights on the overall deployments of UAVs as a function of different system parameters. Additional insights on the benefits of CoMP can also be found in [279]. In the second part of this chapter, we have focused on a UAV BS scenario and presented the promising use of multiple quadrotor UAVs as an aerial antenna array providing wireless service to ground users. In particular, we have described a practical framework for minimizing the airborne service time for wireless communications between the UAV antenna array and ground users. This framework includes optimizing UAVs' positions within the antenna array for maximizing beamforming gain and optimally controlling the motion of the UAVs by exploiting time-optimal control theory. The obtained results have revealed a number of fundamental tradeoffs for leveraging reconfigurable flying antenna array systems. Moreover, our results have shown that there is a very close connection between communications and control when dealing with UAV-based systems. These synergies between communications, control, and cooperation can be further explored in future wireless connected systems (e.g., see the idea of communications and control in networks with swarms of UAV UEs in [280] as well as the role of communications and control in future cellular systems in [26]).

9 From LTE to 5G NR-Enabled UAV Networks

In the previous chapters, we primarily discussed the fundamental challenges and the associated theories and tools for the design of wireless communications and networking solutions with UAV BSs and UAV UEs. In this chapter, we turn to a more practical issue on how to utilize real-world mobile broadband technologies, including LTE and 5G NR, for UAV wireless communications and networking purposes.

Nationwide network coverage is desirable for safely incorporating UAV operations into national airspace. This includes incorporating UAV UEs, UAV BSs, and UAV relays. Developing a clean slate technology and rolling out a new dedicated nationwide network would require extensive and intensive investment in research, product development, testing, field trials, and infrastructure. The extensive efforts required would lead to long time-to-market and may not be economically viable [281]. Mobile networks are already up and running and are providing connectivity for billions of terrestrial devices worldwide. As cost-efficient connectivity solutions with proven track records, they stand ready nationwide to offer UAV connectivity services. The recent few years have seen a surge of activities in utilizing LTE for UAVs, in general, and UAV UEs, in particular [48, 282]. LTE is the dominant 4G mobile technology that is being widely deployed. LTE is being further evolved in 3GPP to meet 5G requirements and will become a 5G wireless access technology in addition to 5G NR. LTE stands ready nationwide in many countries to enable UAV connectivity. 5G NR has the potential of providing more advanced capabilities [283], though its large-scale nationwide commercial deployment may take a few years to complete. As shown in previous chapters, if properly designed, cellular systems, including LTE, 5G, and beyond [26], have a promising potential to integrate UAVs into their operations, both for providing connectivity (UAV BSs) and for communicating with the network (UAV UEs). Indeed, UAVs are seen as an integral component of tomorrow's 6G wireless networks [26].

We begin the chapter with a review of the roles of mobile and cellular technologies for UAV applications, expanding on the discussions of previous chapters. We highlight the use of mobile connectivity and also discuss how mobile technologies can enable the development of new services for UAVs in key areas, such as identification and registration, location-based services, and law enforcement, which complement many of the applications discussed in Chapter 2. Then, throughout Sections 9.2 to 9.4, we discuss LTE-enabled UAVs in more detail, given that LTE is well positioned for initial UAV deployment. We start with a tutorial introduction to LTE in Section 9.2, including basic design principles, system architecture, radio interface protocols, and physical layer time-frequency structure. In Section 9.3, we focus on UAV UE use cases in which UAVs act as LTE UE and discuss the key connectivity issues associated with using terrestrial LTE networks to connect UAV UE. These first few section discussions go more in depth into some of the issues discussed in Chapter 1. In particular, we elaborate more on the role of real-world LTE technology in integrating UAV UEs. We also touch upon some performance-enhancing solutions that can optimize LTE connectivity for providing improved performance for UAV UEs while protecting the performance of terrestrial mobile devices. As discussed in Chapters 1 and 2, BSs mounted on UAVs can provide on-demand connectivity in hotspots and emergency communication in natural disaster zones. As a result, in Section 9.4, we turn our attention to UAV BSs, particularly LTE-enabled UAV BSs, and we discuss the key practical connectivity issues associated with deploying a UAV-based LTE network. In Section 9.5, we discuss 3GPP standardization efforts on connected UAVs that aim to address the anticipated usage of mobile technologies by UAVs and regulatory requirements. Next, we discuss 5G NR-enabled UAVs in Section 9.6. The discussion includes a primer on 5G NR essentials, how 5G NR can provide superior UAV connectivity performance, and the roles of network slicing and network intelligence for identifying, monitoring, and controlling UAVs in the 5G era. Finally, Section 9.7 concludes the chapter with a short summary and future outlook for mobile technologies-enabled UAVs. This chapter, in essence, is a practical complement to the more fundamental discussions on cellular-connected UAVs done in the previous chapters.

9.1 Mobile Technologies-Enabled UAVs

9.1.1 Connectivity Aspects

Cellular networks can provide wide-area, secure, reliable, low-latency, high data rate mobile connectivity to enable a full continuum of consumer and enterprise UAV use cases, as elaborated in Chapters 1 and 2. For example, broadband connectivity allows cellular-connected UAV UEs to live stream the data or video it captures. Example applications include movie and documentary filming, broadcasting of news events, surveillance, cargo delivery, and infrastructure inspection and surveys. In particular, the wide-area secure connectivity provided by cellular networks is a key enabler for many beyond visual LOS UAV applications. In addition, temporary cellular coverage during special public events can be provided by BS mounted on UAV, as discussed in Chapters 1 and 2.

During natural disasters such as floods, earthquakes, and storms, mobile technologies-enabled UAVs can fly beyond visual LOS, collect real-time data about the disaster zones, and transmit the information back to the first responder agencies via the cellular network if the cellular infrastructure is functional. If the cellular infrastructure is damaged by the disasters, a group of UAVs may serve as BSs to provide temporary cellular connectivity or act as relays between the devices in the disaster zones and nearby operational terrestrial BSs.

Cellular technologies developed by 3GPP are based on standards from an industry-wide consensus. 3GPP has developed 4G LTE and the first release of 5G NR standards and is working on further evolution of LTE and NR standards. The 3GPP standards provide a global, inter-operable, and scalable platform for the UAV ecosystem. The licensed spectrum further empowers cellular networks to provide reliable, quality UAV connectivity. In addition, the mobile connections are encrypted and secure, which can help meet high standards of data protection and privacy in UAV applications.

The latest mobile technologies, including LTE and 5G NR, have been designed, and are being further evolved, to connect a wide range of things, including massive IoT and connected vehicles. These have laid a solid foundation for the initial deployment of UAVs. The main types of communications that can be supported are summarized as follows:

- *Communication between UAVs and ground control system:* With cellular networks being the backbone of communications systems for the UAV operations, the UAV operators can maintain connections with their UAVs for command and control as well as payload communication to ensure safe and proper operations.
- *Communication among UAVs:* The direct device-to-device communication feature (also known as sidelink or vehicle-to-vehicle communication) can be adopted for UAV identification and collision avoidance [284].
- *Communication between UAV and air traffic management systems:* Cellular networks can provide secure, reliable communication channels for UAVs to transmit, for example, tracking data to the air traffic management system and receive the latest information, such as airspace constraints, geo-fencing, and alerts from the air traffic management system.

To address the anticipated need of cellular connected UAV, 3GPP conducted a study on enhanced LTE support for aerial vehicles in 2017 [21] and introduced enhancements in its Release 15 to improve the support of LTE technologies for UAVs [285].

9.1.2 Services beyond Connectivity

As elaborated in previous chapters, mobile technologies can play a very prominent role in the development of new UAV services in addition to providing wireless connectivity. In this section, beyond the applications we discussed in Chapter 2, we further describe a few exemplary services, including identification and registration, location-based services, and law enforcement, which will particularly benefit from mobile technologies.

Mobile technologies can assist with UAV identification and registration. For the identification of handset devices, the international mobile equipment identity (IMEI) is used in cellular networks. For the identification of the access service subscriptions, the international mobile subscriber identity (IMSI) stored on a subscriber identification module (SIM) card is used in cellular networks. Consumers need to provide proof of identification in some countries to register for the cellular services. Likewise, it is necessary to identify the UAV device and the service subscription. One possible approach is to use IMEI to identify the UAV device and the IMSI to identify the service subscription. It is also necessary to associate a UAV with its owner or pilot to comply with UAV regulations in many countries. Mobile technologies may be applied to UAV registration as well. UAV identification and registration may be necessary for both UAV BSs and UAV UEs.

Mobile technologies can assist with UAV positioning and localization, including independent verification of the UAV location for use by authorized users (e.g., air traffic control, public safety agencies). A basic requirement for any UAV air traffic management system is the ability to obtain the location information of the UAV. Most LTE chipsets today contain an integrated global navigation satellite system (GNSS) receiver. A UAV may locate its position via GNSS and report the location information to the ground control center or air traffic management system. The GNSS solution alone may not be reliable due to potential spoofing and jamming. 3GPP has introduced a rich set of positioning methods to LTE, such as enhanced cell identity (E-CID), observed time difference of arrival (OTDOA), UL time difference of arrival (UTDOA), and assisted GNSS (A-GNSS) supported within a common location service architecture [286]. Mobile network-based positioning can be used to locate a UAV and is well positioned to provide independent verification of the location information reported by a UAV.

By assisting with UAV identification, registration, and tracking, mobile networks can help UAV operations comply with law enforcement requirements. Regulatory bodies are looking into UAV identification and tracking programs that would allow authorized users to query the identity and metadata of a UAV and its owner or operator. One such effort has been carried out by the FAA unmanned aerial system (UAS) identification and tracking aviation rulemaking committee (ARC) [287]. To meet the needs of business, security, and public safety, 3GPP is also studying the use cases and potential requirements for the remote identification and tracking of UAS linked to a 3GPP subscription [288]. Another inherent advantage of cellular networks is their support of lawful interception of communications, which is important for law enforcement agencies to ensure safe UAV operations and protect public safety.

9.2 Introduction to LTE

LTE, as its name suggests, is a wireless access technology that has undergone long-term evolution, starting from the first version in 3GPP Release 8 approved back in 2007 and going through a continuing evolution to the latest Release 16. LTE, also known as the evolved universal terrestrial radio access (E-UTRA), is built on an industry-wide consensus and represents a huge collaborative effort in the wireless industry. In this section, we provide a tutorial introduction to LTE to help understand how the wide range of LTE capabilities can be exploited to support UAV applications. The introduction in this section is by no means exhaustive. We refer interested readers to the excellent 4G LTE book [289] and the corresponding 3GPP technical specifications for a more in-depth understanding of the sophisticated LTE technology.

9.2 Introduction to LTE

9.2.1 Design Principles

LTE was designed by 3GPP from a clean slate to meet the performance requirements of new services for mobile devices. It was enabled by the advancement of mobile technologies and went hand-in-hand with advancement in other technologies such as processor, memory, color displays, and cameras. The main design targets of LTE included close to Gbps data rate, latency reduction, increased spectral efficiency, and high spectrum flexibility. These design considerations heavily influenced the main design principles and choices behind LTE standards, which are described in the following list.

- *OFDM transmission*: The use of OFDM as the fundamental modulation waveform is a distinct feature of LTE compared to the CDMA-based wireless access technologies in the 3G era. OFDM is an attractive transmission technology for broadband communications and can flexibly support different multi-antenna techniques. While the LTE DL is based on OFDM, the LTE UL is based on single-carrier frequency division multiple access (SC-FDMA), in which the OFDM modulator is preceded by a DFT precoder. SC-FDMA is a technique for reducing the cubic metric of UL signals to achieve lower power amplifier cost and higher power efficiency.
- *Channel-dependent scheduling and rate adaption*: The multiple access scheme for LTE is OFDMA, which can be used to assign different subsets of subcarriers to different individual users. The overall time-frequency resource in LTE is dynamically shared among the users. At each time instant, the scheduler can decide which part of the shared resource should be assigned to a user. By taking into account the channel conditions of the users, the allocation of the shared resource can be performed to favor users with good channel conditions while maintaining fairness. This type of channel-dependent scheduling harnesses multiuser diversity from a system perspective. The modulation and coding scheme for a scheduled user can be chosen to adapt to the corresponding channel condition. The channel-dependent scheduling and rate adaption in LTE help enhance overall system capacity.
- *Multi-antenna techniques*: LTE provides extensive support for different multi-antenna transmission techniques. Multiple antennas can be used at the transmitter for transmit diversity and transmit beamforming. Multiple antennas can be used at the receiver for receive diversity and receive beamforming. Spatial multiplexing and multiuser MIMO are also supported in LTE. In the LTE DL, ten transmission modes are available: Transmission mode 1 is used for single antenna transmission, while the other nine are associated with different multi-antenna transmission schemes. How the multi-antenna schemes are used is under network control. The use of the multi-antenna techniques helps improve link robustness, coverage, spectral efficiency, and system capacity.
- *Flexible spectrum and deployment*: LTE supports flexible spectrum and deployment scenarios. Both frequency division duplex (FDD) and time division duplex (TDD) are supported to enable operation in paired and unpaired frequency bands. FDD supports both full duplex and half-duplex operation at the terminal. The half-duplex FDD operation does not require a duplex filter at the terminal and thus can help reduce terminal

cost. A wide range of carrier bandwidths ranging from 1.4 MHz to 20 MHz are supported in LTE. To limit implementation complexity, the radio frequency requirements are defined only for six channel bandwidths: 1.4 MHz, 3 MHz, 5 MHz, 10 MHz, 15 MHz, and 20 MHz, which were chosen based on the known spectrum migration and deployment scenarios. Nonetheless, they offer enough flexibility for LTE deployment in different bands and can meet the requirements of different operators that may have different spectrum resources. Note that the channel bandwidths can go beyond 20 MHz by exploiting carrier aggregation techniques that enable multiple component carriers to be aggregated and jointly used for transmission.

- *Flat system architecture*: In addition to the air interface, the system architecture in LTE is also simplified with fewer nodes in a less hierarchical structure, leading to a flat radio and core network architecture. The architecture in GSM relies on circuit-switching, and later packet-switching is added to the circuit-switching in general packet radio services (GPRS). The 3G universal mobile telecommunications system (UMTS) keeps the dual-domain concept (circuit and packet) on the core network side. The voice service is traditionally supported via circuit-switched core in mobile systems before 4G LTE. In contrast, LTE has a single packet-switched core, evolved packet core, to support all services, including voice-based on Internet Protocol (IP). The LTE system architecture is described in more detail in the next section.

9.2.2 System Architecture

The LTE system architecture, known as the evolved packet system (EPS), consists of two parts: the evolved packet core (EPC) and the radio access network (RAN). Figure 9.1 gives an illustration of the basic architecture of the EPS, where the UE is connected to the EPC with E-UTRA access. The EPC handles non-radio-related functionality, such as access control, packet routing and transfer, and mobility management. The RAN handles

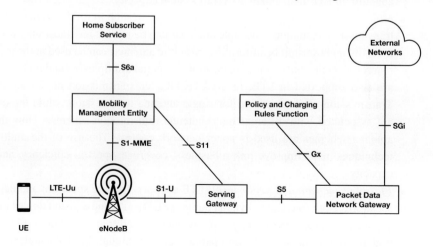

Figure 9.1 An illustration of the basic architecture of the EPS with E-UTRA access.

radio-related functionality, such as scheduling, link adaptation, and hybrid automatic repeat request (ARQ).

The EPC has a "flat" architecture to achieve efficient handling of the data traffic. The user data (i.e., user plane) is separated from the signaling (i.e., control plane) in the EPC. This functional split facilitates network dimensioning. The main nodes in the EPC are briefly described as follows.

- *Serving gateway (S-GW)*: S-GW is the user-plane node connecting the LTE RAN to the EPC. It acts as the mobility anchor when terminals move across different cells and across different 3GPP radio networks. S-GW transports IP data packets between the LTE RAN and the EPC.
- *Packet data network gateway (P-GW)*: P-GW connects the EPC to external IP data networks, such as the internet. P-GW transports packets to and from the external IP data networks. It also performs various other functions, such as IP address allocation, quality-of-service enforcement, and packet filtering. It is logically connected to S-GW.
- *Mobility management entity (MME)*: MME is the control-plane node of the EPC. It mainly manages terminal access and mobility management functions (tracking, paging, roaming, and handover). It is the termination point of the non-access stratum (NAS).
- *Policy and charging rules function (PCRF)*: PCRF interfaces with P-GW to support quality-of-service enforcement and is also responsible for charging.
- *Home subscriber service (HSS)*: HSS is a database storing subscriber information. It interfaces with MME to support mobility management, call and session setup, service authorization, and user authentication.

The LTE RAN is also flat with one type of node known as the eNodeB that terminates the air interface protocol. The eNodeB is connected to the EPC via S1 interface: the user-plane part, S1-U, carries data traffic between the eNodeB and the S-GW, and the control-plane part, S1-MME, carries signaling between the eNodeB and the MME. The eNodeBs are connected via X2 interface among themselves. The X2 interface is mainly used to support handover, intercell interference management, and multicell radio resource management.

9.2.3 Radio Interface Protocols

The LTE radio interface is based on a layered architecture. The user-plane RAN protocols consist of packet data convergence protocol (PDCP), radio link control (RLC), medium access control (MAC), and physical layer (PHY). The MAC, RLC, and PDCP together can be referred to as "layer 2" in the protocol stack, while the PHY layer is often referred to as "layer 1." The control-plane RAN protocols include an additional radio resource control (RRC) layer known as "layer 3."

Figure 9.2 presents the overall RAN protocol stack. Note that the NAS layer in the control plane and the IP and application layers in the user plane are not part of RAN

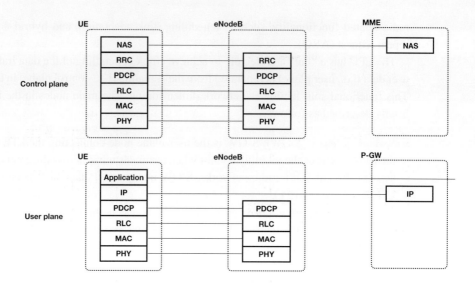

Figure 9.2 An illustration of the overall LTE RAN protocol stack.

protocol stack but are included in the figure for completeness. The RAN protocol entities are summarized in the following list.

- **RRC**: The RRC layer performs access stratum control-plane functions, including the broadcast of system information, the transmission of paging messages, connection management, mobility management, measurement configuration and reporting, among others. RRC messages are transmitted using signaling radio bearers.
- **PDCP**: The PDCP layer performs functions mainly including IP header compression, data ciphering, integrity protection for control-plane signaling, in-sequence delivery, and duplicate removal. There is one PDCP entity for each radio bearer.
- **RLC**: The RLC layer performs functions mainly including segmentation and concatenation, ARQ, duplicate detection, and in-sequence delivery. There is one RLC entity per radio bearer.
- **MAC**: The MAC layer performs functions mainly including priority handling of logical channels, mapping of logical channels to transport channels, hybrid ARQ, and scheduling.
- **PHY**: The PHY layer is responsible for the actual transmission and reception of transport blocks. Its main functions include modulation, coding, multi-antenna mapping, and layer 1 control functionality.

The MAC layer uses logical channels to provide services to the RLC layer. A logical channel is associated with a type of information it carries. The PHY layer uses transport channels to provide services to the MAC layer. A transport channel is defined by how and with what characteristics the data are transmitted. Each transport channel is mapped to a corresponding physical channel. A physical channel is defined by the set of time-frequency resources used for the transmission.

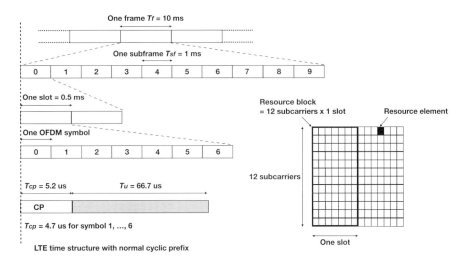

Figure 9.3 An illustration of the LTE time and frequency structure.

9.2.4 Physical Layer Time-Frequency Structure

OFDM is the fundamental transmission scheme in LTE. Figure 9.3 shows the basic LTE time and frequency structure. The normal subcarrier spacing in LTE OFDM equals 15 kHz for both downlink (DL) and uplink (UL). In LTE specifications, the basic time unit is $T_s = \frac{1}{15000 \times 2048}$ seconds, which can be considered the sampling interval of an FFT-based implementation with an FFT size $N_{FFT} = 2048$ for the 15 kHz subcarrier spacing. In other words, the nominal sampling rate is $f_s = \frac{1}{T_s} = 30.72$ MHz.

In the time domain, LTE transmissions are organized into radio frames, and the length of a radio frame is $T_f = 307200 \cdot T_s = 10$ ms. A radio frame is equally divided into ten subframes, and the length of a subframe is 1 ms. A normal subframe is equally divided into two slots. Each slot consists of a number of OFDM symbols with cyclic prefixes. The duration of an OFDM symbol with 15 kHz subcarrier spacing is $T_u = 2048 \cdot T_s \approx 66.7$ μs. Two different cyclic prefix lengths are supported: normal and extended cyclic prefixes. With extended cyclic prefixes, each slot consists of six OFDM symbols. The length of an extended cyclic prefix is $T_{ecp} = 512 \cdot T_s \approx 16.7$ μs. With normal cyclic prefixes, each slot consists of seven OFDM symbols. The length of a normal cyclic prefix equals $T_{cp} = 160 \cdot T_s = 5.2$ μs in the first OFDM symbol of a slot, and $T_{cp} = 144 \cdot T_s = 4.7$ μs in the subsequent OFDM symbols of the slot.

The physical resource can be described by a time-frequency resource grid. Each column and each row of the resource grid correspond to one OFDM symbol and one subcarrier, respectively. The smallest unit in the time-frequency resource grid is a resource element composed of one subcarrier over one OFDM symbol. A resource block consists of 12 consecutive subcarriers in the frequency domain and one slot in the time domain. Scheduling decisions in LTE can be made in every subframe. The normal basic scheduling unit is a resource-block pair consisting of two time-consecutive resource blocks in the same subframe.

9.3 UAV as LTE UE

LTE networks can offer wide-area, secure, and quality wireless connectivity to enhance the safety of UAV operations beyond visual LOS range. With increasing height above the ground, the radio environment changes. Using terrestrial cellular networks to provide connectivity to UAV UEs leads to new challenges, as briefly discussed in Chapter 1 and developed in some of the other previous chapters. In this section, we further elaborate on the key connectivity issues associated with using terrestrial LTE networks to connect UAV UE. We also discuss performance-enhancing solutions that can optimize LTE connectivity for providing improved performance for UAV UEs while protecting the performance of terrestrial mobile devices.

9.3.1 Coverage

Cellular networks have been traditionally designed and optimized for terrestrial communication. Cell sites have been planned and selected to provide terrestrial coverage and serve the increasing terrestrial traffic demand. A common objective is to minimize the cost of the network while meeting the terrestrial coverage and capacity demand. The configuration of a cellular network is set to optimize the terrestrial coverage. One particularly important configuration is BS antenna tilt. As shown through our early analysis in Chapter 8, the antenna tilt of the BS antennas will have important implications on the performance of cellular-connected UAV UEs. In essence, the tilt of a BS antenna represents the antenna inclination relative to a reference pointing direction. BS antennas in a cellular network are usually tilted down by a few degrees to concentrate the transmit powers toward the ground to reduce intercell interference. There are two types of antenna tilt: mechanical tilt and electrical tilt. The former is achieved by physically tilting down the antenna, while the latter is achieved by changing the phases of the antenna elements in an array antenna.

With down-tilted BS antennas, UAV UE may be served by the side lobes of BS antennas. Figure 9.4 presents a synthesized BS antenna pattern. The antenna array consists of one column of eight pairs of cross-polarized antenna elements, where the vertical antenna element space normalized by the wavelength is 0.8. The side lobes are illustrated in the synthesized BS antenna pattern in Figure 9.4. We can see that even the antenna gain of the strongest side lobe is about 14 dB lower than the antenna gain of the main lobe. Due to the presence of antenna nulls in the side lobes, the strongest received signal at UAV UE may come from a faraway BS instead of the geographically closest BS.

The radio channel between a UAV UE flying in the sky and a ground BS usually enjoys high likelihood of LOS. The more benign propagation condition can make up for antenna gain reductions, even though the UAV UE may be served by the side lobes. Consider a rural scenario where sites are placed on a hexagonal grid with 37 sites and 3 cells per site. The LTE system bandwidth is 10 MHz at 700 MHz carrier frequency. Each BS has two cross-polarized antennas with 6 degrees of downtilt at the height of 35 m. The BS antenna pattern is modeled according to Figure 9.4. Figure 9.5 shows

9.3 UAV as LTE UE

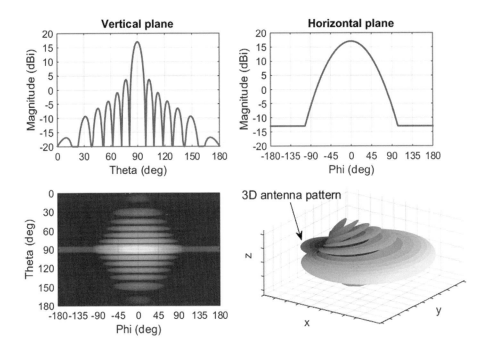

Figure 9.4 An illustration of a synthesized BS antenna pattern: theta denotes zenith angle, and phi denotes azimuth angle. © IEEE. Reprinted, with permission, from [48].

the DL coupling gain (antenna gain plus path gain) at three different altitudes: 1.5 m (ground level), 40 m (5 m above the BS antenna height), and 120 m (close to the FAA altitude limit of 400 ft for small UAV [290]). From the DL coupling gain distributions, we can see that that the free-space propagation can make up for antenna gain reductions for UAV UE. The fifth percentile DL coupling gains at the altitudes of both 40 m and 120 m are in fact higher than the fifth percentile DL path gain at the ground level of 1.5 m. We can also see that the variance of the DL coupling gains are smaller above the BS antenna height, while the DL coupling gains on the ground level are more spread out.

The coverage of using existing terrestrial cellular networks to provide connectivity to UAV UEs may become insufficient above certain heights. To provide coverage higher in the sky, enhancements to the existing terrestrial cellular networks may be needed. For example, additional antennas pointing toward the sky may be installed at selected cells to provide more seamless coverage at higher heights. These observations generally align with our Chapter 8 analysis.

9.3.2 Interference

The more favorable propagation conditions in the sky lead to strong received signal powers – not only the desired signal powers but also the co-channel interference signal powers, as discussed in Chapter 1 and studied in Chapter 6. In particular, when a UAV

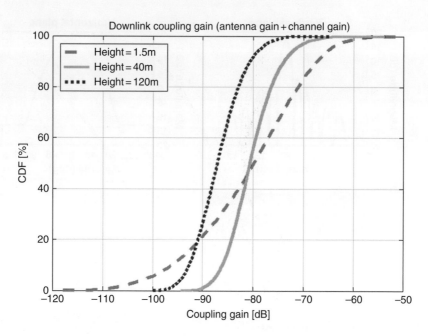

Figure 9.5 An illustration of DL coupling gain distributions versus UAV UE heights. © IEEE. Reprinted, with permission, from [48].

UE is flying well above BS antennas, it may have more LOS propagation conditions to multiple neighboring BSs. In such a scenario, an UL signal transmitted from the UAV UE may cause interference to multiple neighboring BSs. The increased UL interference, if not properly controlled and managed, may cause performance degradation to the devices on the ground. Similarly, due to the more LOS propagation conditions, DL signals transmitted from multiple neighboring BSs may cause strong DL interference to the UAV UE.

Under the same setup as in Figure 9.5, Figure 9.6 shows the DL SINR distributions at three different altitudes: 1.5 m, 40 m, and 120 m. We can see that the SINR values at the altitudes of both 40 m and 120 m are statistically lower than the SINR values at the ground level of 1.5 m. Specifically, at the operating point of 20% resource utilization, the median SINR values at the altitudes of 40 m and 120 m are 10.9 dB and 11.3 dB lower than the median SINR at the ground level, respectively. These results show that more LOS propagation conditions also lead to stronger interfering signals from non-serving cells to the UAV UE.

Intercell interference is not a new issue. A rich set of tools in terms of both standards and implementation have been studied and developed for LTE to deal with interference. One possible interference mitigation tool is coordinated multi-point transmission and reception (and its variants). The new challenge here is that a UAV UE may receive interfering signals from more ground BSs in the DL, and its UL signals are visible to more cells due to more LOS propagation conditions. As a result, CoMP techniques may have to be performed across a larger set of cells to mitigate the interference issues

9.3 UAV as LTE UE

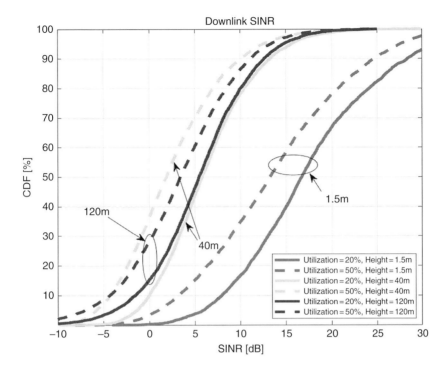

Figure 9.6 An illustration of DL SINR distributions versus UAV UE heights. From right to left (at 50%), the first two curves correspond to the height of 1.5 m; the third and the sixth curves correspond to the height of 40 m; and the fourth and the fifth curves correspond to the height of 120 m. © IEEE. Reprinted, with permission, from [48].

at the cost of increased coordination complexity. In Chapter 8, we provided an initial study of CoMP for UAV UEs, which can be used as a basis for future research in this area.

Interference can also be handled by receiver techniques, such as interference rejection combining and network-assisted interference cancellation and suppression. UAV may be equipped with multiple antennas, which can be used to cancel or suppress the interfering signals from more ground BSs. With multiple antennas, beamforming that enables directional signal transmission or reception to achieve spatial selectivity is also an effective interference mitigation technique.

A simpler interference mitigation solution would be to partition radio resources so that aerial traffic and terrestrial traffic are served with orthogonal radio resources. The static radio resource partition may not be efficient since the reserved radio resources for aerial traffic may be underutilized. If UAV operators can provide supplemental data, such as flight routes and UAV positions to the network operators, such data can be utilized for more dynamic and efficient radio resource management. It would also be possible to reduce interference by designing wireless-aware UAV UE trajectories, as studied in Chapter 6.

Uplink power control is yet another powerful interference mitigation technique. An optimized setting of UL power control parameters may be applied to limit the excessive UL interference generated by UAV UE. Optimized UL power control can reduce interference, increase spectral efficiency, and benefit UAV UE as well as terrestrial devices. An initial view on the impact of power control on interference was provided in Chapter 6 (jointly with trajectory design).

9.3.3 Mobility Support

Mobility support is a distinct feature of cellular networks and why mobile operators can command higher cellular subscription fees than other forms of telephony and data access [291, 292]. Seamless and robust mobility support for UAVs moving in the sky is imperative for maintaining communication service continuity, which is important not only for good user experience but also for safe control and operation of a UAV. With increasing height above the ground, the radio propagation environment changes. The change in the radio propagation environment and other factors, such as down-tilted BS antennas, result in different signal and interference characteristics for cellular connected UAV, as described in the previous two sections and in Chapter 1. These may pose new mobility management challenges for UAVs. Several key questions need to be properly answered:

- The down-tilted BS antennas may result in fragmented cell association patterns in the sky. In particular, the strongest signal may come from a faraway BS that may be chosen by the UAV UE as its serving BS, while the strongest site is usually the closest one at the ground level. Would the fragmented cell association pattern in the sky result in more handovers and potentially more handover failures?
- The overall SINR level in the sky is significantly worse than on the ground. The reduced SINR might lead to a higher probability of handover commands and measurement reports being lost. This may result in a higher risk of radio link failure (RLF) and failed handover.

Mobility support for cellular devices and UAVs is a complex issue that involves many detailed aspects. During an early 3GPP study on mobility enhancements for heterogeneous networks [293], 3GPP developed a simplified mobility modeling methodology, which is yet sophisticated enough and can serve as a good reference model for mobility performance evaluation in cellular networks. This model was also used by 3GPP during the study on LTE-connected UAV UEs. In the sequel, we introduce this 3GPP mobility modeling methodology for LTE networks.

Handover failure modeling employs the previously mentioned RLF criteria and procedures. An RLF occurs when the UE cannot establish or maintain a stable connection to the serving cell. RLF is detected in LTE upon expiry of the timer T310; or upon indication from RLC that the maximum number of retransmissions has been reached; or upon random access problem indication from MAC while none of the timers, including T300, T301, T304, and T311, are running [294]. We focus on the timers T310 and T311 and refer to [294] for a detailed description of the other timers that are less relevant for our discussion herein.

When monitoring the radio link, a UE periodically computes wideband channel quality indicator (CQI). If the CQI drops below a threshold Q_{out}, lower layers indicate *out-of-sync* to higher layers that count subsequent out-of-sync indications. A maximum number of consecutive out-of-sync indications denoted by N310 can be configured. If N310 consecutive out-of-sync events are indicated, the UE starts timer T310, whose expiry would trigger RLF. If the CQI is above another threshold Q_{in}, lower layers indicate *in-sync* to higher layers that count subsequent in-sync indications. A maximum number of consecutive in-sync indications denoted by N311 can be configured. If N311 in-sync events are indicated, the UE stops timer T310. The timer T310 may also be stopped upon triggering the handover procedure or upon initiating the connection reestablishment procedure. At the expiry of T310, if security is activated, the UE initiates the connection reestablishment procedure; otherwise, the UE goes to the RRC IDLE state. The timer T311 starts upon the RRC connection reestablishment procedure. It stops when selecting a suitable E-UTRA cell or a cell using another radio access network. At the expiry of T311, the UE enters RRC IDLE state.

In the basic PHY processing configuration for the radio link monitoring in non to - discontinuous reception (DRX) mode, the PHY sample rate for evaluating out-of-sync and in-sync is typically once every 20 ms. The PHY samples are filtered linearly over a sliding window. The sliding window lengths are typically 200 ms for evaluating out-of-sync and 100 ms for evaluating in-sync, respectively.

For the purpose of modeling, the handover procedure is divided into three states, as shown in Figure 9.7.

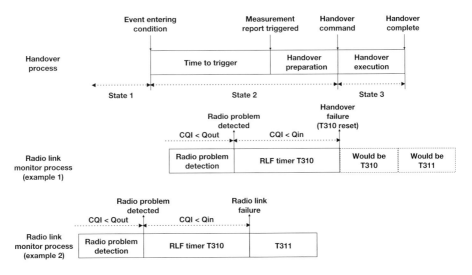

Figure 9.7 An illustration of handover modeling in LTE networks.

- State 1 is the state before the event A3 entering condition is satisfied. The event A3 is triggered when a neighbor cell becomes better than the serving cell by an offset.
- State 2 is the state after the entering condition of event A3 is satisfied, but before the UE successfully receives a handover command.
- State 3 is the state after the UE receives the handover command, but before the UE successfully sends a handover complete message.

An RLF may occur either in State 1 or State 2. The RLF performance may be measured by the *RLF rate*, which is defined as the average number of RLF occurrences per UE per second.

A handover failure may occur either in State 2 or State 3. In State 2, the UE is attached to the source cell. In State 3, the UE is attached to the target cell. The handover failure causes may be divided into three categories.

- *RLF in State 2*: In this case, the channel quality of the serving cell has become bad such that an RLF is triggered before the handover process to the target cell can be executed. This is considered handover failure in the modeling.
- *Physical DL control channel (PDCCH) failure in State 2*: In this case, timer T310 is running when the UE expects to receive a handover command in Sate 2. Though RLF has not been declared, the channel quality of the serving cell has become bad such that the timer T310 is running. Meanwhile, UE's measurement report is triggered and the UE is expecting to receive a handover command. Due to the poor link quality, the source cell may not receive the UE's measurement report or the UE may not receive the handover command from the source cell, resulting in handover failure. PDCCH failure modeling is used to model this type of handover failure. The PHY sample rate for evaluating PDCCH failure is typically once every 10 ms. The PHY samples are typically filtered linearly over a sliding window of 200 ms.
- *PDCCH failure in State 3*: In this case, the target cell's DL filtered average wideband CQI is less than Q_{out} at the end of the handover execution time in State 3. As a result, the target cell signal quality is not good enough to complete the handover procedure. PDCCH failure modeling is used to capture this type of handover failure. The PHY sample rate for evaluating PDCCH failure should be at least two samples during the typical handover execution time of 40 ms. The PHY samples are typically averaged to evaluate whether or not the PDCCH failure occurs.

The handover performance may be measured by *handover failure rate*, which is defined as the ratio of the number of handover failures to the number of handover attempts (including both successful handovers and failed handovers).

If a UE switches connection from a cell A to another cell B via a handover and later switches connection from cell B back to cell A via another handover, and if the time the UE connects to cell B is less than a minimum time of stay, such an event is known as ping-pong. The handovers involved in a ping-pong event may be considered unnecessary handovers, since the UE does not stay connected in the target cell long enough before it is handed back to the original source cell.

Next, we describe UE mobility models. UE is initially dropped at a random location in the simulation area. An initial random moving direction is generated for the UE. The UE then moves in a straight line in the selected direction at a constant speed. When the UE hits the simulation boundary, either wrap-around or bouncing-circle model can be applied. In the wrap-around model, when the UE hits the wrap-around contour, it will enter the simulation area from a different point on the wrap-around contour and continue moving in a straight line at a constant speed with the moving direction determined based on the wrap-around methodology. In the bouncing-circle model, a bouncing circle is defined within the simulation area. When the UE hits the bouncing circle, it will bounce back and continue moving in a straight line at a constant speed with a new randomly generated moving direction.

Based on the 3GPP mobility modeling methodology, mobility performance of LTE-connected UAV has been evaluated. Example evaluation results may be found in [21, 295].

9.3.4 Latency and Reliability

As discussed throughout this book, reliable and low-latency communication is imperative for many UAV applications. For example, command and control links need to be robust and reliable, and the packets should be successfully delivered within some latency bound (depending on the use case) with high probability. In this section, we focus on the latency and reliability performance of UAV command and control links. Latency and reliability may be measured in different ways and at different layers. Concretely, in this section, we follow the latency and reliability definitions used by 3GPP during the study on scenarios and requirements for 5G access technologies [296].

Latency metrics include control plane latency and user plane latency. Control plane latency refers to the time it takes to move from a battery-efficient state, such as idle mode, to the start of a continuous data transfer state, such as the active mode. User plane latency refers to the time it takes to successfully deliver an application layer packet from the radio protocol layer 2/3 service data unit (SDU) ingress point to the radio protocol layer 2/3 SDU egress point with the radio interface. When measuring the user plane latency, it is assumed that neither the transmitter nor the receiver is restricted by DRX. Reliability is defined as the success probability of delivering an application layer packet of X bits from the radio protocol layer 2/3 SDU ingress point to the radio protocol layer 2/3 SDU egress point with the radio interface within L seconds.

Traffic characteristics of command and control are expected to be different from those of payload communication. 3GPP usually adopts FTP-based traffic models for performance evaluation. The FTP models may be suitable for payload communication but are less suitable for command and control. During the 3GPP study item on enhancing LTE support for aerial vehicles [21], it was assumed that the command and control traffic had similar characteristics as those of voice over IP (VoIP) traffic. In the used traffic model for command and control, packets arrive periodically with a period of 100 ms and the packet size is fixed to be 1250 bytes.

Table 9.1 Reliability simulation results for command and control traffic.

Metric	Number of used PRBs	1.5 m	30 m	50 m	100 m	300 m
Reliability (%)	6	86.81	76.66	16.85	8.49	4.22
	15	98.86	99.79	99.64	99.15	91.91
	25	99.35	99.91	99.98	99.89	99.9
	50	99.62	99.95	99.98	99.99	99.99
Resource utilization ratio (%)	6	40.91	56.71	89.92	94.97	96.23
	15	11.05	11.26	22.54	29.77	47.27
	25	6.21	5.36	7.51	8.98	11.43
	50	2.74	2.41	2.65	2.78	2.92

Representative Simulation Results

Table 9.1 collects some reliability simulation results contributed to the 3GPP study item [21]. In Table 9.1, reliability is defined with packet size $X = 1250$ bytes and latency bound $L = 50$ ms. Table 9.1 also presents the corresponding resource utilization ratio statistics. Resource utilization ratio is defined as the fraction of utilized radio resources averaged over time, frequency, and cells. It is a key performance indicator that can reflect the interference level in the network.

The reliability performance was simulated in the DL of an urban macro cellular network, where sites are placed on a hexagonal grid with 19 sites and 3 cells per site. The inter-site distance is 500 m. Each BS has two cross-polarized antennas with 10 degrees of downtilt at the height of 25 m. The LTE system bandwidth is 10 MHz at 2 GHz carrier frequency. In each cell, there are 5 aerial UEs. The aggregate aerial traffic demand per cell is $1250 \times 8 \times 10$ bps per UE $\times$ 5 UEs per cell = 500 kbps per cell. In the evaluation, it was assumed that the scheduler partitioned the radio resources so that aerial traffic and terrestrial traffic were scheduled in orthogonal frequency resources. With this partition, the signals to terrestrial UEs and the signals to aerial UAV UEs did not interfere. However, aerial UEs in a cell still experienced interference from neighbor cells since the neighbor cells may use the same radio resource to serve other aerial UEs connected to neighboring cells.

The reliability performance was evaluated at different fixed heights: 1.5 m, 30 m, 50 m, 100 m, and 300 m. Note that the height of 30 m is close to the BS antenna height (25 m). In the evaluation, different numbers of physical resource blocks (PRBs), 6 PRBs, 15 PRBs, 25 PRBs, and 50 PRBs, were used to serve the aerial traffic. These numbers of PRBs are the same as the supported system bandwidths of LTE. So the evaluation results are also relevant in the case where a dedicated LTE carrier is deployed to provide connectivity for low-altitude UAV UEs.

From Table 9.1, we can see that when 6 PRBs are used to serve the aerial traffic, it is not possible to meet the 50 ms latency bound with a high confidence level (e.g. 90%) even at the ground level of 1.5 m. The reliability numbers are not high for all the heights evaluated. The corresponding resource utilization ratios summarized in Table 9.1 help explain the results. At the ground level of 1.5 m, the resource utilization ratio is already

40.91%. As the height increases to 50 m, the resource utilization ratio becomes close to 90% and increases further to ~95% as the height further increases. These results indicate that serving the aerial traffic demand of 500 kbps per cell with 6 PRBs is challenging.

When the number of PRBs used to serve the aerial traffic is increased to 15, it is possible to meet the 50 ms latency bound with a high confidence level (~99%) at the heights of 1.5 m, 30 m, 50 m, and 100 m. At the height of 300 m, the reliability is reduced to 91.91%. By examining the corresponding resource utilization ratios, we can see that the resource utilization ratios are lower than 30% for the heights of 100 m and below, and thus the DL interference experienced at aerial UE is moderate. In contrast, the resource utilization ratio is increased to 47.27% at the height of 300 m, suggesting that the DL interference experienced at aerial UE is stronger.

When the number of PRBs used to serve the aerial traffic is further increased to 25, we can see that the resource utilization ratios are lower than 12% at all the heights, which implies that the DL interference experienced at aerial UE is minor. In this case, it is possible to meet the 50 ms latency bound with an even higher confidence level (~99.9%). When the number of PRBs used to serve the aerial traffic is further increased to 50, the resource utilization ratios are further decreased to below 3% and the reliability performance is further improved.

Remarks on the Reliability Results

The results in Table 9.1 indicate that it is possible to achieve a high reliability (such as 99.9%) if the network uses enough dedicated frequency resources to serve aerial traffic. There is a tradeoff between reliability performance and the number of PRBs used for aerial command and control. We find that using 15 PRBs to serve the aerial traffic can provide ~99% reliability at the height of 1.5 m, 30 m, 50 m, or 100 m, and ~90% reliability at the height of 300 m. Note that the reliability performance was evaluated under a relatively high traffic demand: there are 5 aerial UAV UEs per cell and each aerial UE has periodic packet arrivals with a fixed packet size of 1250 bytes and a period of 100 ms. In the initial deployment of a low-altitude UAV, it is likely that the demand of aerial command and control traffic is much lower. As a result, fewer PRBs would be needed when the traffic demand is lower.

A general trend we observe from the resource utilization ratios in Table 9.1 is that as the height increases from 30 m to 300 m, the resource utilization ratio increases for the same offered command and control traffic. Take the case with 15 PRBs, for example. To achieve similar reliability performance (~99%), the resource utilization ratio is increased about two times when the height increases from 30 m to 50 m, and about 3 times when the height increases from 30 m to 100 m.

A key lesson from the reliability evaluation results is that when the resource utilization ratio is low, the DL interference experienced at aerial UE is not severe, which makes it possible to deliver a small data packet within the 50 ms latency bound with a high reliability. Though this lesson is drawn from a specific interference mitigation technique, i.e., using dedicated frequency resources to serve aerial traffic, we expect that this lesson is true in a more general sense. In particular, it is expected that any interference mitigation technique that can lead to satisfactorily received signal quality would facilitate delivering low-latency high-reliability connectivity services to UAV UEs.

In this section, we have focused on a simple interference mitigation solution in which orthogonal frequency resources are used to serve aerial traffic and terrestrial traffic. The static frequency resource partition may not be efficient since the allocated frequency resources for aerial traffic may be underutilized. If supplemental data such as flight routes and positions of aerial UEs are known to the network, such information can be utilized to achieve more dynamic and efficient radio resource management. Other resource management solutions, such as those discussed in Chapters 6, 7, and 8, can also be adopted to further enhance these results.

9.4 UAV as LTE BS

As discussed previously, one can use UAV BSs to provide temporary connectivity to congested areas (e.g., hotspots), hard-to-reach areas, as well as areas affected by an emergency or disaster. Indeed, UAV BSs are becoming a crucial part of public safety networks. Multiple UAV BSs together can provide temporary connectivity to a designated area on the ground. As shown in our studies of Chapters 4–8, UAV BSs can indeed provide effective wireless connectivity solutions. In this section, we elaborate further on the key connectivity issues associated with deploying a network of UAV BSs that adopt cellular technologies such as LTE.

As in the case of a UAV as LTE UE, interference management is a critical issue for using UAVs as LTE BSs. In [297], the authors considered deploying UAV BS as part of an LTE heterogeneous network for public safety communications. LTE Release-10 enhanced intercell interference coordination (eICIC) and LTE Release-11 further enhanced intercell interference coordination (FeICIC) were applied to mitigate intercell interference. The results of [297] showed that with optimized UAV BS locations, the reduced power subframes in FeICIC can provide considerably higher fifth-percentile spectral efficiency than the almost blank subframes in eICIC. In [298], the authors investigated interference management in using UAVs as BSs for emergency communication in natural disaster zones, where part of the communication infrastructure is damaged. Simulations were used in [298] to analyze how the throughput performance can be improved by exploiting the inherent mobility nature of UAVs. In the previous chapters, we have also discussed many other use cases for UAV BSs.

LTE-U and licensed-assisted access (LAA) are the technology choices for unlicensed spectrum. Fair coexistence is the main consideration for unlicensed spectrum. As discussed in Chapter 7, in [257], the authors considered the load balancing issue between LTE-U UAV BSs and WiFi access points on the ground using a game theoretic approach. In the design, UAV BSs equipped with LTE-U used regret-based learning for dynamic duty cycle selection. The proposed approach targeted ensuring satisfactory throughput performance for all users. Our analysis in Chapter 7 demonstrated how one can use cache-enabled UAV BSs to serve ground users over both licensed and unlicensed bands.

Although in Chapter 7 we studied analytically how one can design a 3D cellular network with UAV BSs, it is important to also shed light on practical considerations for such a network. In this regard, the deployment of EPC to support an on-demand LTE network formed out of UAV BSs is a major design issue. The EPC and RAN are usually connected via wires in terrestrial LTE networks. A similar approach can be used for an LTE network composed of UAV BSs whereby the EPC is deployed on the ground, the RAN is mounted on the UAV BS, and the EPC and UAV are tethered by wires. This approach, however, limits the deployment flexibility and may not scale well to support a network with multiple UAV BSs. Alternatively, the EPC on the ground and the RAN mounted on the UAV BS can be connected by wireless communication. The backhaul wireless connectivity should be carefully designed to achieve requirements of reliability, communication range, and capacity. Another alternative is to deploy both EPC and RAN at the UAV BS, as proposed in [299]. In this alternative, the entire EPC is implemented in one single entity and located on each UAV BS. Realizing this design, however, needs to address various challenges, including limited compute resource of UAVs and mobility management.

As discussed in Chapter 1, it is worth stressing again that a third use case of UAVs, namely UAV as LTE relay can be considered and has also been studied in the literature, though strictly speaking, it does not belong to the category of UAV as LTE BS. In [300], the authors proposed to use a swarm of UAVs as relays to compensate temporary overload or site outage in LTE networks. The aerial relay placement, number of aerial relays, and transmit powers of aerial relays were discussed and analyzed. The results showed that interference aware positioning of aerial relays could increase spectral efficiency in overload and outage scenarios. In [301], the authors studied UAV as LTE relay to provide enhanced LTE connectivity to a ground user from a terrestrial BS. The communication layer of the customized integrated UAV relay was based on *OpenAirInterface*. To maximize the throughput performance, the authors also proposed a placement algorithm that updated the position of the UAV relay in real time based on user location and wireless channel condition. Many of the designs that we performed in previous chapters of UAV BSs can, in general, be extended to the UAV relay use case.

9.5 3GPP Standardization on Connected UAV

The 3GPP ecosystem is well positioned to support UAV operation, as detailed in Section 9.1. Meanwhile, regulators are investigating safe UAV operation programs so that UAV can harmoniously coexist with commercial air traffic and the general public. Regulatory efforts have been taken in the areas of registration and licensing programs and safety and performance standards. To address the anticipated usage of mobile technologies by UAV and regulatory requirements, 3GPP has been conducting a series of works to enhance cellular standards to better support UAV operation. UAV is often referred to as aerial UE or aerial vehicle in the 3GPP works. In the sequel we follow these terminologies to discuss the 3GPP works on connected UAV.

9.5.1 3GPP Release-15 Study Item on LTE-Connected UAV

To understand the potential of LTE for aerial vehicles, 3GPP conducted a study on enhanced LTE support for aerial vehicles in 2017. The study assessed the performance of Release-14 LTE networks used for serving aerial vehicles and identified potential enhancements for better handling of aerial traffic and not impacting the performance of terrestrial devices. This study focused on LTE with the intention that the lessons learned will be applied to 5G NR. The outcomes of the study can be found in the 3GPP TR 36.777 [21].

Evaluation Scenarios

Compared to a terrestrial UE, an aerial UAV UE exhibits different behaviors and experiences different radio conditions. To accommodate the new unique class of aerial devices, 3GPP revisited its performance evaluation framework, which was predominantly constructed for terrestrial devices prior to this study item.

When an aerial UAV UE is introduced to an LTE network, the system performance for both the aerial UAV UE and legacy terrestrial UE needs to be evaluated. For indoor terrestrial UEs and outdoor terrestrial UEs, 3D urban macro (UMa) and 3D urban micro (UMi) evaluation scenarios were developed in LTE Release 12 (TR 36.873 [105]) to support full-dimensional MIMO (FD-MIMO) with 2D antenna arrays. Similar models were also introduced in NR and extended to support carrier frequencies from 0.5 GHz to 100 GHz in TR 38.901 [105]. In addition to UMa and UMi, an evaluation model for rural macro (RMa) deployments was also developed in NR. The UMa scenario is intended to emulate network deployment scenarios in urban areas with eNodeB (eNB) antennas mounted above the rooftop levels of surrounding buildings. The UMi scenario targets urban deployments where the eNB antennas are mounted below the rooftop levels of surrounding buildings. The RMa scenario is used to model larger cell sizes in rural areas where the eNB antennas are mounted on the top of towers. Since aerial UE is studied in LTE networks, the UMa and UMi scenarios in LTE are relevant for the study. The rural deployment scenario is regarded as an important scenario for aerial UAV UEs, and thus the RMa scenario defined in NR is also considered in the study.

An aerial UE is essentially considered as a different type of an outdoor UE. Since aerial UAV UEs do not exist in the traditional UMa, UMi, and RMa models, these models must be extended to include aerial UEs with heights generally well above ground level. The new models with aerial vehicles are referred to as UMa-AV, UMi-AV, and RMa-AV. In the traditional UMa and UMi scenarios, two types of terrestrial UE are considered: indoor terrestrial UE and outdoor terrestrial UE. The former accounts for 80% of the terrestrial devices, and the remaining 20% are outdoor terrestrial UEs. The heights of outdoor terrestrial UEs are fixed at 1.5 m, while the heights of indoor terrestrial UEs vary up to a maximum height of 22.5 m. The study considers aerial UEs with a height up to 300 m. Performance statistics are collected for aerial UE with heights uniformly distributed between 1.5 m and 300 m. The study also considers performance at fixed aerial UE heights of 50 m, 100 m, 200 m, and 300 m. To study the impact of supporting aerial UEs with different densities in a cell, aerial UE ratios of 0%, 0.67%, 7.1%, 25%, and 50% are considered, assuming that the total number of UEs per cell is 15.

In UMa-AV, UMi-AV, and RMa-AV, the channel modeling for terrestrial UE follows the existing 3GPP channel models. The channel models are extended for aerial UE to capture the different propagation conditions in the sky. The general principle is to adopt an aerial UE height-dependent channel modeling approach. When the height of an aerial UE is within the applicability height range of the terrestrial 3GPP channel models, existing terrestrial 3GPP channel models are used for the aerial UE. New channel models are developed and used for aerial UEs with heights outside the applicability height range of the terrestrial 3GPP channel models. The details of the channel models can be found in TR 36.777 [21].

Identified Problems and Solutions

Under the evaluation scenarios and channel models described in the previous section, extensive simulations were performed during the study. The evaluation results, supplemented by field trial data, indicate that aerial UEs may cause UL interference to more cells and observe DL interference from more cells than terrestrial UEs. This is because an airborne aerial UAV UE experiences LOS propagation conditions to more cells with higher probability than terrestrial UE does. The extra interference is generally manageable for low aerial UAV UE density, for example, when the number of aerial UAV UEs is no more than one per cell. Also, it is observed that the performance is generally better in rural environments than in urban environments.

Due to the distinct features of aerial UEs, it is important that mobile networks can identify if a UE is an aerial UE or a regular ground UE to provide the right service optimization for the aerial UE while protecting the performance of the ground UE from the potential interfering signals from aerial UEs. For a legitimate aerial UE, standard mechanisms can be enforced so that it can be recognized by the networks. For example, the aerial UE can be required to have a SIM card that is registered for aerial usage if it would like to use cellular connection. It is also necessary to identify a "rogue" aerial UE that is not properly registered with the network. This may occur when a normal ground UE is attached to a UAV and being flown in the network. This phenomenon is being observed in the field and has drawn much attention from mobile operators, since flying a UAV with regular terrestrial UE may generate excessive interference to the network and may not be allowed by regulations in some regions.

To address the identified problems and needs, 3GPP studied interference detection and mitigation techniques, mobility enhancements, and aerial UE identification.

Interference detection: Interference detection is related to flying mode recognition because both UL and DL interference increase when an aerial UAV UE is above a certain height. Interference detection can also be used as a trigger for applying interference mitigation. Potential solutions are broadly categorized into either UE-based or network-based solutions. UE-based solutions may utilize UE measurements of reference signals received power (RSRP), reference signals received quality (RSRQ), reference signals SINR (RS-SINR), mobility history reports, speed estimation, timing advance adjustment values, and location information. Take UE measurements, for example. The triggering of measurement reports can be linked to the changing interference condition, e.g., a measurement report is triggered when RSRP or RSRQ values of multiple cells are above a threshold. Network-based solutions rely on the inter-eNB information exchange. Such information may include UE measurement reports and UL reference signal configuration. These solutions have requirements on backhaul over a large number of eNBs.

Downlink interference mitigation: The objective of DL interference mitigation is to reduce the interference level that an aerial UE experiences. The various DL interference mitigation techniques evaluated during the study include FD-MIMO, directional antennas at UE, receive beamforming at UE, intra-site joint transmission CoMP, and coverage extension (for enhancing synchronization and initial access performance of aerial UE). These solutions do not require additional specification work. Another scheme of coordinated transmission of control and data from multiple cells was briefly discussed. It, however, requires more study to evaluate its performance as well as specification impacts.

Uplink interference mitigation: The objective of UL interference mitigation is to reduce the interference level that a terrestrial UE experiences due to aerial UAV UEs transmitting in the air. The various UL interference mitigation techniques evaluated during the study include FD-MIMO, directional antennas at UE, and power control-based mechanisms. The first two do not require additional specification work since FD-MIMO is supported in LTE since Release 13 and the use of directional antennas at a UE is an implementation choice. Some of the power control-based mechanisms require minor specification change.

Mobility enhancements: In the simulated baseline networks without interference mitigation techniques, the mobility performance of an aerial UAV UE is shown to be worse than that of a terrestrial UE, especially when the aerial UE density is large. The mobility simulation results further show that aerial UE in the RMa-AV scenario experiences better mobility performance than in the UMa-AV scenario. Applying interference mitigation techniques to reduce the interference levels in the network can improve the aerial UE's mobility performance. The mobility algorithms can be improved to provide better mobility support for an aerial UAV UE. For example, handover procedures may be improved by considering conditional handover and optimization of handover-related parameters considering aerial UE's location information, airborne status, and flight path plans. The mobility algorithmic improvements may also include enhancements to the existing measurement reporting mechanisms, such as introducing new events and enhancements of event-triggering conditions.

Aerial UAV UE identification: A combination of subscription information and radio capability indication from the UE can be used for aerial UE identification. The subscription information can be signaled from MME to eNB to indicate whether the user is authorized to operate for aerial usage. Aerial UE can indicate its support of aerial-related functions introduced in Release 15 via radio capability signaling to the eNB. Another issue studied is aerial UE flight mode detection. Potential mechanisms include explicit indication of flight mode by UE, UE-based reporting using altitude information on altitude information, and interference detection related techniques.

Overall Conclusion of the Study Item

Drawing on comprehensive analysis, extensive simulations, and field trial data, the 3GPP study concludes that LTE networks are capable of serving aerial UAV UEs, but there may be challenges related to UL and DL interference as well as mobility support. The challenges become more visible when the density of the aerial UE is high. Both implementation solutions and specification enhancements are identified to address these issues. To serve aerial UE more efficiently and limit the impact on terrestrial UE, solutions based on specification enhancements are beneficial. To this end, 3GPP conducted a follow-up work item in Release 15, which is described in the next section.

9.5.2 3GPP Release-15 Work Item on LTE-Connected UAV

During the study item on enhanced LTE support for aerial vehicles and UAVs, various performance-enhancing solutions were identified and evaluated. The follow-up Release-15 work item aimed to specify the features that could improve the efficiency and robustness of terrestrial LTE networks for providing aerial connectivity services, particularly for low-altitude UAVs. The key features introduced in the work item include:

- subscription-based aerial UE identification and authorization;
- flying mode detection based on height and location reporting;
- interference detection based on measurement reporting;
- flight path reporting; and
- open loop power control enhancements.

We will now describe each feature in more detail.

Subscription-based aerial UE identification and authorization: Support of aerial UE function is stored in the user's subscription information in the home subscriber server. The HSS transfers this information to the MME from where it can be provided to the eNB via the S1 application protocol. In addition, for X2-based handover, the source eNB can include the subscription information in the X2 application protocol handover request message to the target eNB. For the intra and inter MME S1-based handover, the MME provides the subscription information to the target eNB after the handover procedure. The eNB may then combine this information with radio capability indication from the aerial UAV UE in order to identify whether the aerial UE has been authorized to be connected to the E-UTRAN network while flying.

Flying mode detection based on height and location reporting: The flying mode detection is a separate issue. Input to flying mode detection is event-triggered height and location reporting. A new configurable event within radio resource management with height threshold is introduced for Release-15 aerial UE. When a UE is configured with this event, a report is triggered when the UE's altitude crosses a configured altitude threshold. In addition to flying mode detection, the exact height information is considered useful as E-UTRAN may choose to reconfigure, for example, measurements for the UE when it crosses a height threshold.

Interference detection based on measurement reporting: The flying mode detection is also related to interference detection as the interference conditions for flying aerial UAV UEs are different from aerial UEs in terrestrial mode. For interference detection, which may also serve as input to flying mode detection, an enhancement to existing events triggered RSRP/RSRQ/RS-SINR reports was introduced. The UE may be configured to trigger an event such as A3, A4, A5, which all consider neighbor cell measurements, so that measurement report is triggered when the measured RSRPs/RSRQs/RS-SINRs of multiple cells are above the configured threshold. For example, event A3 is triggered when the measured RSRP of a neighbor cell becomes better than the measured RSRP of the serving cell by a certain amount. The enhanced triggering would require, for example, that three neighbor cells' RSRP values become higher than the serving cell's RSRP value by a certain amount.

Flight path reporting: The support for E-UTRAN to request flight path information from UE using RRC signaling was introduced in Release 15. In the request, eNB may configure the maximum number of waypoints that the UE can include in the report. Further, the configuration indicates whether the time stamp per waypoint can be reported. UE reports the flight path if the information is available. In the *RRCReconfigurationComplete* message, the UE may indicate the availability of flight path report. However, support for indicating updates or changes of the flight path plan that UE may receive via application layer is not supported in Release 15.

Open loop power control enhancements: Open loop power control is one of the techniques that can be used to mitigate UL interference from aerial UE. The nominal received power P_0 and fractional path loss compensation factor α are two open loop power control parameters that were studied during the study item in Release 15.

- In LTE up to Release 14, the parameter α can only be configured in a cell-specific manner. Given that the degree of the UL interference caused by aerial UE may differ from one aerial UE to another, it is desirable to introduce a UE-specific alpha parameter. In Release 15, the parameter α can be configured in a UE-specific manner for the physical UL shared channel.
- In LTE up to Release 14, the parameter P_0 consists of both cell-specific and UE-specific components. The UE-specific component of the parameter P_0 had a value range from -8 dB to $+7$ dB up until LTE Release 14. In Release 15, the value range of the parameter P_0 has been extended to the range from -16 dB to $+15$ dB to provide better flexibility in setting open loop power control parameters on a UE-specific basis.

9.5.3 3GPP Release-16 Study Item on Remote UAV Identification

Remote UAV identification and tracking are imperative for authorized parties such as law enforcement, public safety, and air traffic control agencies to query the identity and metadata of a UAV and its controller via UAS traffic management (UTM). As described in the previous Section 9.5.2, 3GPP introduced subscription-based aerial UE identification and authorization. In Release 16, 3GPP continues further study into device identification and auxiliary information as part of a further study on service requirements [288].

The objective of the study item is to identify the use cases and potential requirements for the remote identification and tracking of UAV linked to a 3GPP subscription. The studied aspects include content of identification data, availability of identification data, and use of identification data. Content of identification data may include UAV identifier, route data, location, and controller information. Availability of identification data may include access authorization, privacy, latency, and reliability. Use of identification data may include tracking, data retention, and authorization to operate.

The outcome of the study is documented in 3GPP TR 22.825 [302]. There are ten use cases with potential service requirements identified in TR 22.825. These include:

- use case for initial authorization to operate;
- use case for live data acquisition by UTM;
- use case for data acquisition by law enforcement;
- use case for enforcement of no-fly zones;
- use case for distributed closed-field separation service;
- use case for local broadcast of UAS identity;
- use case for differentiation between UAV-specific UE and regular UE attached to UAV;
- use case for cloud-based NLOS UAV operation;
- use case for UAV fly range restriction; and
- use case for the UAS-based remote inspection.

As an example, we introduce the use case for initial authorization to operate. Consider a switched-on UAS made up of a UAV and a UAV controller. The onboard UE authenticates with the mobile network and sends UAV data and identifiers to the UTM to request permission. The request may include flight authorization, access to mobile data services while flying, and using certain services provided by the UTM. The UAV controller needs to go through similar authentication and authorization process so that the mobile network and UTM can associate the UAV with its controller. Depending on the flight mission and required services, different levels of authentication and authorization may be required before the UAS becomes fully operable. For example, in order to use the UTM services, such as flight tracking and collision avoidance, the UAS may need to carry out an additional application-level authentication and authorization process.

As another example, we look at the use case for data acquisition by law enforcement. An authorized official may want to query a UTM for information of an active UAS. For example, police may receive a nuisance complaint about a UAV, and thus the police may query the UTM by providing the information about the geographic area where the reported UAV is flying. The UTM may then return data of all the active UASs in the queried area. For a UAS, such data may include the identity of the UAV, identity of the UAV controller, identity of the UAV operator, flight data, and live location. More data may be further provided upon request of the authorized official.

Local broadcast of a UAV identity can be a backup means for remote UAS identification and tracking when the network coverage is not available. This would enable an authorized official equipped with appropriate equipment to discover an active UAS within proximity. The authorized official may then query the UTM by providing the received identity to obtain more information about the UAS from the UTM.

For the detailed description, scenario, and potential service requirements of the other use cases, we refer interested readers to [302]. The study also includes additional considerations, such as lawful interception and security. It is concluded in the study that 3GPP should create normative service requirements based on the identified potential requirements to better serve UAS ecosystems with cellular connectivity.

9.6 Towards 5G NR-Enabled UAVs

LTE stands ready nationwide in many countries to enable cellular connected UAVs and help create innovative services for the UAV industry. Many of the lessons learned from LTE-enabled UAVs will be adapted for 5G NR-enabled UAV. Although the next-generation mobile technology, 5G NR, is still in its infancy, it promises much more capabilities, including delivering enhanced 3D coverage, higher transmission rate, lower latency, customized end-to-end QoS guarantee, and network intelligence [283]. The improved capabilities of 5G networks will enable large-scale UAV deployments with more diverse UAV uses, including both UAV as 5G NR UE and UAV as 5G NR BS.

9.6.1 A Primer on 5G NR

In this section, we provide a short primer of 5G NR essentials. We refer interested readers to [283, 303] and the corresponding 3GPP technical specifications for a more in-depth understanding of 5G NR.

5G NR aims to address a variety of usage scenarios from enhanced mobile broadband (eMBB) to ultra-reliable low-latency communications (URLLC) to massive machine type communications (mMTC). 5G NR can meet the performance requirements set by the ITU for international mobile telecommunications for the year 2020 (IMT-2020) [304]:

- 20 Gbps DL peak date rate and 10 Gbps UL peak date rate in the eMBB usage scenario;
- 30 bps/Hz DL peak spectral efficiency and 15 bps/Hz UL peak spectral efficiency in the eMBB usage scenario;
- 100 Mbps DL 5%ile user experienced data rate and 50 Mbps UL 5%ile user experienced data rate in the dense urban eMBB test environment;
- 0.3 bps/Hz, 0.225 bps/Hz, and 0.12 bps/Hz DL 5%ile user spectral efficiency in the indoor hotspot, dense urban, rural eMBB usage scenario, respectively; 0.21 bps/Hz,

0.15 bps/Hz, and 0.045 bps/Hz UL 5%ile user spectral efficiency in the indoor hotspot, dense urban, rural eMBB usage scenario, respectively;
- 9 bps/Hz, 7.8 bps/Hz, and 3.3 bps/Hz per transmission reception point average spectral efficiency in the indoor hotspot, dense urban, rural eMBB usage scenario, respectively; 6.75 bps/Hz, 5.4 bps/Hz, and 1.6 bps/Hz per transmission reception point average spectral efficiency in the indoor hotspot, dense urban, rural eMBB usage scenario, respectively;
- 10 Mbps/m^2 DL area traffic capacity in the indoor hotspot eMBB test environment;
- 4 ms user plane latency in the eMBB usage scenario; 1 ms user plane latency in the URLLC usage scenario; 20 ms (10 ms encouraged) control plane latency in the eMBB and URLLC usage scenarios;
- 1,000,000 devices per km^2 in the mMTC usage scenario;
- high network energy efficiency (qualitative measure) with support of a high sleep ratio and long sleep duration;
- 10^{-5} success probability of transmitting a layer 2 protocol data unit of 32 bytes within 1 ms in channel quality of coverage edge for the urban macro URLLC test environment;
- 1.5 bps/Hz normalized UL traffic channel link data rate for mobility speed up to 10 km/h in the indoor hotspot eMBB usage scenario; 1.12 bps/Hz normalized UL traffic channel link data rate for mobility speed up to 30 km/h in the dense urban eMBB usage scenario; 0.8 bps/Hz and 0.45 bps/Hz normalized UL traffic channel link data rate for mobility speed up to 120 km/h and 500 km/h, respectively, in the rural eMBB usage scenario;
- 0 ms mobility interruption time in the eMBB and URLLC usage scenarios;
- 100 MHz minimum bandwidth; up to 1 GHz for operation in higher frequency bands (e.g., above 6 GHz).

The design of 5G NR is forward compatible, which will allow 3GPP to smoothly introduce new technology components in the future for currently unknown use cases. Key 5G NR technology components include ultra-lean transmission, support for low latency, advanced antenna technologies, and spectrum flexibility.

The two main types of architecture in 5G NR are non-standalone (NSA) and standalone (SA). In the NSA architecture, a UE is connected to both LTE eNB and 5G NR NodeB (gNB) via E-UTRA – NR dual connectivity (EN-DC), where the eNB acts as a master node and the gNB acts as a secondary node [305]. In an NSA operation, LTE is used for initial access and mobility handling while the SA version can be deployed independently from LTE.

OFDM with cyclic prefix is used for both DL and UL transmissions. The use of DFT-spread OFDM (DFT-S-OFDM) is also supported for single-layer UL transmission. To flexibly support different deployment scenarios and a wide range of carrier frequencies, NR adopts flexible subcarrier spacing of $2^\mu \cdot 15$ kHz ($\mu = 0, 1, 2, 3, 4$): 15 kHz, 30 kHz, and optionally 60 kHz subcarrier spacing for data channels in sub-6 GHz frequency bands referred to as frequency range 1 (FR1); 60 kHz and 120 kHz subcarrier spacing for data channels in above 24 GHz frequency bands referred to as frequency range 2

(FR2). The cyclic prefix is approximately 4.7 us for the 15 kHz subcarrier spacing, and it inversely scales with the subcarrier spacing. In the time domain, a subframe consists of 2^μ slots and each slot consists of 14 OFDM symbols.

In NR, rate compatible quasi-cyclic low-density parity-check (LDPC) coding is used for data channels, while Reed-Muller block coding and cyclic redundancy check (CRC) assisted polar coding are used for control channels. The modulation schemes include binary and quadrature phase shift keying (B/QPSK) and quadrature amplitude modulation (QAM) of orders 16, 64, and 256 with binary reflected Gray mapping.

NR supports FDD, TDD with semi-statically configured UL/DL configuration, and dynamic TDD. Transmissions can be scheduled to start at any OFDM symbol in a slot and last only a fraction of a slot needed for the communication. This type of "mini-slot" transmission can facilitate low-latency use cases and minimize interference.

9.6.2 Superior Connectivity Performance

5G NR, capable of meeting the ambitious IMT-2020 performance requirements, can provide superior connectivity performance for 5G NR-enabled UAVs, including enhanced coverage, faster transmission, and lower latency. From the lessons learned from LTE-enabled UAVs, as well as our studies in the previous chapters, we know that a UAV UE served by terrestrial networks tends to experience more interference, cell coverage irregularities, and complex neighbor cell relationships. These lead to higher coverage complexity and more challenging mobility management issues for UAV UE. In a 5G NR network, some selected cells can be equipped with gNB antennas pointing toward the sky to improve aerial coverage. Neighbor cell coverage and mobility management strategies can be customized to achieve coverage optimization in the sky. Information, such as UAV flight path, if available, can be exploited for connectivity management and optimization.

5G NR supports advanced antenna technologies, including massive MIMO. For single-user MIMO, it supports a maximum of eight DL transmission layers and four UL transmission layers, which can increase data rate for the UAV. 5G NR is particularly designed to support the scheduling of many users on the same time-frequency resource with multi-user MIMO. This can improve the overall system capacity, facilitating the support of UAV UE in a 5G NR network. Beam management is introduced in 5G NR. The beam directions can be improved by utilizing beam-sweeping spatial filters at the transmitter and receiver. Beamforming can increase the SINR and further improve the system capacity and coverage.

5G NR supports large bandwidth transmission and reception. The maximum bandwidth of an NR carrier is 100 MHz in sub-6 GHz frequency bands (FR1) and 400 MHz in above 24 GHz frequency bands (FR2), respectively. Carrier aggregation (CA) of up to 16 NR carriers can be further utilized to achieve wider bandwidth. In 5G NR, CA can be flexibly configured with both intra-band CA and inter-band CA, self-carrier scheduling and cross-carrier scheduling, and different subcarrier spacing choices for the aggregated carriers in inter-band CA.

With advanced antenna technologies and wide bandwidths, 5G NR can provide various levels of DL and UL high data rates in different frequency bands with different configurations. In particular, 5G NR is well positioned to serve UAV UE transmitting high-definition images or videos for augmented reality (AR) and VR immersive experiences that require multi-Gbps data rate. Some recent studies on using UAV UEs for such applications have been done in [33].

9.6.3 Service Differentiation with Network Slicing

As described in Section 9.6.1, 5G needs to support a wide variety of applications with different service requirements, while being more flexible, cost- and energy-efficient for service delivery. Network slicing is one of the key technology components to enable flexible and scalable 5G mobile networks [306]. Network slicing allows multiple logical networks to be created and run on top of a common shared physical infrastructure. The slices of the network are isolated from each other in the control and user planes. Each slice is a complete end-to-end logical network consisting of network capabilities and the associated resources for serving a particular service category or even individual customers [307].

A network slice consists of a radio access network part and a core network part. 3GPP has defined system architecture and functional aspects to support network slicing in a next-generation radio access network (NG-RAN) and a 5G core network (5GC) [308, 309]. Network slicing is supported with the basic idea that different protocol data unit sessions can be constructed for the traffic on different slices. By scheduling and providing different configurations in layers 1 and 2, the network can realize the different slices. Under network slicing, customers can be regarded as belonging to different tenant types. Each tenant has its associated service requirements that determine the corresponding eligible slice types for the tenant.

With network slicing, it is possible to logically distinguish the service and radio resource management for a UAV UE from those for the terrestrial terminals. Service differentiation for the different UAV use cases discussed in Chapter 2 can be enabled by, for example, using different slices to support command and control signaling and different application data services. The physical network can be partitioned at an end-to-end level to allow optimized grouping of UAV traffic, such as low-latency UAV traffic and high data rate UAV traffic, and to isolate them from terrestrial traffic of different characteristics.

Take UAV command and control, for example. Network slicing is well positioned to address the latency, reliability, and security requirements of UAV command and control. A network slice can be constructed to use network elements at appropriate locations to minimize the length of the communication path to reduce propagation delay. By using prioritized scheduling and optimized configurations, air-interface latency, core network latency, and processing delay can be reduced. These collectively reduce the end-to-end latency for UAV command and control. The slicing reserves resources that may include hardware, software, and radio resources, which can help improve reliability. Security and data privacy may benefit from network slicing due to, for example, the isolation of distinct slices.

9.6.4 Network Intelligence

5G networks are evolving into highly complex systems that are beyond the capability of humans to fully comprehend and control. Traditional networking and data analysis approaches are becoming incompetent to keep pace with the growing complexity in the 5G systems. Networks already generate huge amounts of data today and will generate much more with the growing scale and multitude of interactions of the 5G systems. Network intelligence with machine learning, big data analytics, and artificial intelligence [221] are being developed for better extracting information out of the data and realizing automated network control and management.

3GPP is conducting a series of works toward network intelligence. 3GPP has introduced the stage 3 definition of the network data analytics function (NWDAF) services in Release 15 [310]. In Release 16, 3GPP continues to study enablers for network automation [311] and radio access network-centric data collection utilization [312]. The ITU telecommunication standardization sector also established a focus group in November 2017 to study machine learning for future networks including 5G [313]. Though machine learning, big data analytics, and artificial intelligence in communications networking are still in their infancy, they are becoming essential to achieve network intelligence and automation to ultimately realize proactive, self-aware, self-adaptive, and predictive networking [221].

5G networks with network intelligence are well positioned to efficiently and effectively identify, monitor, and control UAV. Network intelligence at different layers in the 5G architecture can enable data processing for various purposes. Local learning and decision making, combined with centralized data consolidation, can achieve efficient UAV identification and management. For example, each local BS can collect radio measurement data, the time series of user mobility events, and associated contextual information. The data and knowledge can be shared across sites to construct airborne radio environment distribution models that can be used to identify UAV UEs with machine learning methods. To assist with UAV regulations and air traffic management, centralized intelligence is imperative to obtain a comprehensive global understanding of networks. This can help, for example UAV flight path planning to avoid coverage holes, congestion, collisions, and no-fly zones. In some of the previous chapters, we have also shown how machine learning can be used to perform effective resource management for various UAV use cases.

9.7 Chapter Summary

Cellular networks have connected tens of billions of devices on the ground in the past decades and are now ready to support the flying robots – UAVs – in the sky. Mobile technologies can underpin the UAV ecosystem by providing a wide range of capabilities and features to identify, track, and control the growing fleet of UAVs from takeoff to landing. The wide-area, quality, secure connectivity offered by cellular networks is essential for extending the UAV operation range beyond visual LOS. Mobile technologies can play a role in the development of new UAV services in addition to providing wireless connectivity.

9.7 Chapter Summary

The existing mobile LTE networks targeting terrestrial usage can support the initial UAV deployment, but there may be challenges such as interference and mobility. Enhancements and the next generation 5G networks will provide more efficient connectivity for wide-scale UAV deployments. Mobile technologies, based on evolving global standards, will be the essential foundation for the vibrant global growth of the UAV ecosystem. We envision that 5G networks and beyond (e.g., 6G) will seamlessly integrate UAVs, in all their three roles: UAV BSs, UAV UEs, and UAV relays.

10 Security of UAV Networks

In the previous chapters, we focused on showcasing how enabling wireless communications with UAVs, for both UAV BSs and UAV UEs use cases, can lead to a suite of important research problems pertaining to communications and networking. However, equipping UAVs with communications capabilities will also expose them to a broad range of security threats. Indeed, the advantages of UAVs, which include their agility and ability to communicate over LOS links, render them vulnerable to a plethora of security attacks that include both cyber threats, such as jamming spoofing, as well as physical threats in which an adversary can direct the control system of the UAVs or simply attempt to physically destroy the UAV. Hence, it is imperative to study and analyze the security of UAV-equipped networks and to introduce new defense solutions that can help secure UAVs against the aforementioned cyber-physical security threats.

In consequence, the goal of this chapter is to provide a succinct overview on the security challenges of UAV-based networks. To this end, in Section 10.1, we start by providing a general overview on the various security threats facing UAV systems, ranging from communication channel attacks to information attacks and Global Positioning System (GPS) spoofing attacks. Then, in Section 10.2, we develop, using game theory, a generic framework that can provide cyber-physical security for UAV applications, such as delivery systems. We conclude the chapter in Section 10.3 with key remarks on the security of UAV systems.

10.1 Overview on UAV Security Problems

In this first section, we discuss a number of important security threats that can jeopardize the operation of a network with UAVs. Cyber-physical attacks on UAVs can vary widely as they can target different components within the UAV system. For example, some attacks can target the connection between a UAV and its GSs [314]. Other attacks can target a UAV's information either by intercepting the sent information or injecting false data [315]. Denial of service is yet another form of attack that aims at preventing a UAV from performing its designated service [316]. Finally, some attacks can target specific components in the UAV such as its GPS receiver either by jamming or spoofing the authentic GPS signals [317]. Note that there are also attacks that can target a group of UAVs to disrupt the connectivity of their UAV networks (UAVNs) [318].

Communication channel attacks: Attacks to the communication channel of a UAV system can take many forms such as disrupting the connection between the UAV and the GS [315]. In this type of attack, the attacker can prevent the GS from communicating and controlling the UAV in order to steal the UAV or to make the drone operator lose the UAV. If the attacker wants to steal the UAV, it will first hijack the UAV-GS connection, and then it will send its own control signals to the UAV in order to lead the UAV to a place where it can be captured, this is known as a *fly-away attack* [319]. A similar communication attack can happen if the UAV is being controlled by a mobile application, in this case, the attacker will de-authenticate the UAV from its legitimate mobile device and then establish a new connection with a malicious mobile device.

Information attacks: Attacks to the information of UAVs can have a wider range of effects on the targeted UAV. Eavesdropping on the transmitted information between a UAV and its communicating receiver (e.g., BS or another UAV) is one of the most basic attacks that an attacker can launch to access the UAV's private information. Due to the lack of strong encryption of the transmitted data (as limited by the UAVs' computational capabilities), eavesdropping can have serious effects based on the sensitivity of the transmitted data. Naturally, here one can envision several solutions to eavesdropping ranging from developing lightweight cryptographic algorithms to exploring physical layer security solutions [320, 321]. While eavesdropping is a passive attack on the UAV's information, false data injection represents another important and *active* attack against a UAV's information. In false data injection, the attacker can transmit manipulated information to a UAV by masquerading the identity of the real control center. This form of attack is particularly effective when no authentication is used between the UAV and the control center. A more powerful data injection attack is known as a man-in-the-middle attack [314] in which the attacker intercepts the message sent from the control center, alters it, and resends it to the target UAV. The goal of data injection attacks can be to mislead the target UAV to perform harmful tasks or to prevent it from performing its intended task.

Denial-of-service (DoS) attacks: DoS attacks are well-known security threats in computer networks and wireless networks [322] in which the attacker floods the network with requests in order to exhaust the network resources and, thus, prevents the legitimate users from obtaining the service. In UAV networks, DoS attacks can prevent a UAV from performing its mission, particularly for scenarios in which UAVs receive requests from users. For instance, many UAVs are used as aerial base stations to provide the necessary cellular connectivity to users in emergency situations or in time-sensitive applications, such as real-time video streaming in big events [323]. Such UAVs are prone to DoS attacks where the attacker can send malicious requests to affect the service with the legitimate users. Another variant of DoS is known as a distributed denial of service (DDoS) attack in which the attacker uses multiple devices to send the requests so it is harder for the network to identify the malicious users. The notion of a DoS attack can also be used by an adversary to compromise the wireless link between a UAV and its controller [324] either by transmitting a large control request to cause buffer-overflow at the UAV or sending multiple control signals in parallel to prevent the UAV from receiving its authentic control signal.

GPS attacks: The next type of attack targets a UAV's GPS receiver. GPS jamming is a common attack against UAVs in which the attacker transmits high power signals to prevent a UAV from receiving the GPS signals that are used to determine the location. While there might not be a universal, effective defense against GPS jamming, the authors in [325] proposed a technique to determine the jammer's location in order to stop the jamming source. GPS spoofing is yet another powerful attack that can target UAVs' GPS receivers. In GPS spoofing attacks, an attacker transmits fake GPS signals to the UAV's GPS receivers. These signals are transmitted with slightly higher power than the authentic GPS signals so the UAV will lock on to these fake signals and determine its location incorrectly. The attacker can benefit from this by sending the UAV to another predetermined location where it can be captured [317], which is known as a capture via a GPS spoofing attack.

The effect of a GPS spoofing attack is determined by the type of the attack. If the attacker does not seek to maintain a covert attack, it can theoretically impose any location on the UAV with the risk of being detected if the UAV is using a spoofing detection techniques. On the other hand, if the attacker is launching a covert attack, it will be limited by the changes it can impose on a UAV's location in order not to be immediately captured by the spoofing detection techniques. This limit is determined by the instance drifted distance [326], which depends on the GPS spoofing technique adopted by the UAV. To illustrate the effects of GPS spoofing attacks, we conduct a few simulations that rely on the UAV GPS spoofing model in [327]. For instance, in Figure 10.1, we show the capture possibility of a group of 5 UAVs subject to a GPS spoofing attack and managed by a common drone operator. In this scenario, the attacker can spoof the GPS signals of one UAV at each time step. Similarly, one UAV can update its location, using its neighboring UAVs' locations at each time step. In Figure 10.1, the drone operator chooses which UAV must update its location, at each time step, in

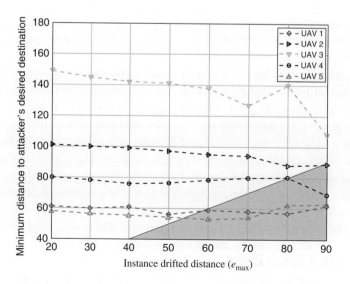

Figure 10.1 The effect of changing the instance drifted distance on the UAVs capture possibility under deterministic strategies.

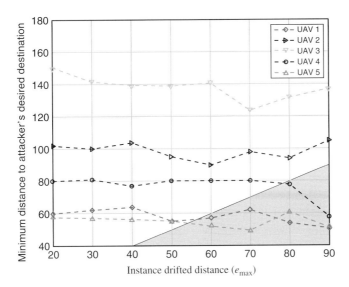

Figure 10.2 The effect of changing the instance drifted distance on the UAVs capture possibility under random strategies.

a predetermined order. In particular, the drone operator will choose all the UAVs in sequential order, starting from UAV 1 at the first time step. We can see that the attacker is able to start capturing UAVs 1 and 5 when the instance drifted distance equals 60 m, it will be also able to capture UAV 2 when the instance drifted distance equals 90 m. Similarly, Figure 10.2 shows a case in which the drone operator chooses a random UAV to update its location, at each time step. In Figure 10.2, we can see that the attacker is able to capture UAVs 1 and 5 when the instance drifted distance is 60 m. However, it can capture UAV 4 starting from the instance drifted distance of 80 m, which is worse than using the deterministic strategies. These simulation results clearly showcase how one can study GPS spoofing attacks on UAVs and, then, design corresponding defense strategies. For more insights on such designs, we refer the reader to the work in [327].

Other attacks: Finally, there are several other attack types that can target a group of UAVs to disrupt their network connectivity [318]. These attacks have similar effects to the attacks on wireless sensor networks (WSNs) [328], mobile ad hoc networks (MANETs) [329], and vehicular ad hoc networks (VANETs) [330]. Although there are some differences between these types of networks in terms of the available resources, the amount of transmitted information, and the number of nodes, their similarities can enable some defense mechanisms to be ported from a system to another, after modifying it to suit the nature of the new system.

10.2 Security of UAV UEs in Delivery Systems

Following our broad overview of UAV security problems, our next step is to develop a general framework to study and analyze UAV security for a very specific scenario pertaining to UAV delivery systems. For instance, as mentioned in Chapter 2, UAVs will admit a plethora of real-world applications. In particular, UAV UEs will be central to many foreseen smart city applications. Such applications particularly include UAV delivery systems [331, 332], such as Amazon's Prime Air and Google's Project Wing, as well as the use of UAV UEs for search and rescue missions. In such applications, UAV UEs are primarily tasked with achieving a time-sensitive mission that requires them to move from a given origin to a destination. Along their travel, UAV UEs will have to communicate with ground infrastructure (e.g., BSs and GSs) as well as with other UAVs. This ability to communicate, coupled with their mobility and agility, renders the UAV UEs of a delivery system highly susceptible to cyber-physical attacks. On the cyber level, as discussed in Section 10.1, adversaries can attempt to jeopardize the delivery mission by taking control of the UAV through false data injections or by compromising the communication system of the UAVs, through jamming or DoS attacks [333–335]. Meanwhile, given that the FAA limited the flight of UAVs to about 400 ft, UAV UEs will then be within the range of civilian rifles that can be used to physically attack them [336]. Such physical attacks can seriously jeopardize the mission of the UAVs and, thus, lead to catastrophic consequences for the UAV operator.

Due to these cyber and physical vulnerabilities, in addition to the works discussed in Section 10.1, the authors in [333–335] have also attempted to identify the various vulnerabilities of UAV systems (particularly on the cyber side) and, then, provide security solutions to overcome those vulnerabilities. However, the majority of these prior studies still focus on the cyber vulnerability of generic UAV systems and do not take into account some of the unique features of UAV delivery systems [337, 338], such as their need for timely delivery and their vulnerability to physical threats. To overcome this gap in the literature, in this section, we will introduce a framework, built on game theory and prospect theory, to analyze and understand the cyber-physical vulnerabilities of UAV delivery systems. This framework, based on our work in [339], will shed light on how UAV operators can properly manage the security of their UAV delivery systems.

10.2.1 Modeling the Security of a UAV Delivery System

We study the security of a UAV delivery system similar to Amazon Prime Air in which the delivery system operator will dispatch a UAV UE to deliver online goods to a target destination. In this studied system, once a delivery is requested, the operator will dispatch a UAV UE to deliver the purchased good from a given origin (e.g., a warehouse) O to a destination D (e.g., the customer premises). The UAV delivery system operator will seek to minimize the delivery time and associated costs for sending its UAV UE from O to D. Therefore, it will often seek to choose the shortest path between origin and destination. However, in our system, we consider the presence of an attacker that can be located at one of multiple "danger points" (such as locations i and j in Figure 10.3) in order to attack the UAV UE and compromise its mission. Here, a danger point is a geographical location along a given path from O to D in which the UAV's cyber or physical capabilities are exposed to the attacker. For example, high-rise buildings or high hills can be potential danger points. These danger points represent a threat source because

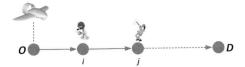

Figure 10.3 Threat points from warehouse (O) to customer location (D).

they can lead to physical proximity and possibly a direct LOS between the UAV and the adversary. Thus, they enable targeting a traversing UAV with physical (such as shooting the UAV) and cyber (such as jamming) attacks.

In this section, we assume that the UAV UE belongs to a legitimate operator and that the attacker is a malicious entity. However, as discussed in [340], the developed framework can also accommodate the case in which the UAV is being used for a nefarious mission (e.g., to compromise the security of an airport) and the attacker (e.g., a government agency) is trying to stop this malicious UAV. In our considered security model, whenever the attacker is successful, it will be able to completely compromise the UAV UE (e.g., by destroying it or rendering it out of order). Hence, once an attack is successful, the operator will have to resend its product from O to D (using a new UAV), which, in turn, leads to substantial delivery delays. As a result, the probability of a successful attack will directly impact the expected delivery time of the product. Consequently, to guarantee a timely delivery in the presence of potential adversaries, the UAV operator can no longer rely on the shortest physical path, which can be potentially risky. Instead, it must choose alternative paths that can be less risky and can lead to better delivery times. As shown in Figure 10.4, we define a directed graph $\mathcal{G}(\mathcal{N}, \mathcal{E})$ to model the UAV delivery paths between the origin and the destination. This graph has N nodes in the set $\mathcal{N}$, which represent the danger point locations connected via E edges in set $\mathcal{E}$. We assume that UAVs can fly in nearly unconstrained locations[1] and, hence, the number of paths between O and D can be infinitely large. Each such path will cross a subset of danger points (which can also be shared among various paths). Given that we focus on the security aspect, we can capture the large set of all possible paths via the set of danger points traversed by each path. Since UAV delivery operators seek to minimize their delivery time, we assume that the movement of any UAV UE between two danger points m and n will follow the shortest edge between those two vertices. Hence, in our graph $\mathcal{G}$, to define an edge between any two nodes, we will only use the shortest path between those nodes.

In consequence, the graph $\mathcal{G}(\mathcal{N}, \mathcal{E})$ can be viewed as a compact model for the security of a UAV delivery system. In particular, this graph is used to represent the continuous space between O and D using danger points and the shortest edges that link them. For each edge $e_k \in \mathcal{E}$ that links 2 danger points m and n, we define t_k as the time that the UAV needs to fly from m to n over edge e_k. For each danger point $n \in \mathcal{N}$, we define p_n as the probability that the attack performed from n is successful.

---

[1] Note that, in the future, the FAA may regulate the UAVs and require them to fly on predefined airways.

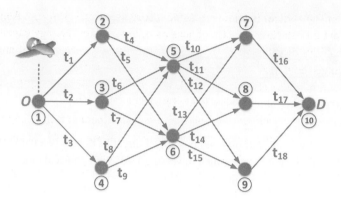

Figure 10.4 A graph representation for a UAV delivery system.

Let $\mathcal{H}$ be the set of H paths (with no repeated vertices) from O to D in $\mathcal{G}$. In essence, each element $h \in \mathcal{H}$ represents a sequence of danger points that constitute a path from O to D. Since each h has a unique set of nodes and associated edges, we can represent h by its sequence of traversed nodes, which is nothing but a subset of $\mathcal{N}$. For example, from Figure 10.4, $h_1 \triangleq (1, 2, 5, 7, 10)$ represents a sample path. We can now define an $(H \times N)$ path-node incidence matrix $\boldsymbol{L}$. Every element $l_{hn}, \forall h \in \mathcal{H}, \forall n \in \mathcal{N}$ of this metric will be equal to 1 if $n \in h$ and 0 otherwise. We also define a distance function $f^h(.): h \to \mathbb{R}$ (over path $h \in \mathcal{H}$). This function outputs the time needed for a node $n \in h$ to reach the destination from the origin, by following a given path h. For example, in Figure 10.4, $f^{h_1}(5) = t_1 + t_4$ where $h_1 \triangleq (1, 2, 5, 7, 10)$.

In this model, we will refer to the UAV operator, U, as an *evader* whose goal is to choose an optimal, preferred path (from O to D) for its UAV that minimizes the expected delivery time T while potentially evading a successful attack. Meanwhile, we refer to the attacker A as an *interdictor* whose goal is to select a danger point from which to attack the UAV and interdict its path. By doing so, the interdictor can potentially compromise the UAV and, hence, maximize the delivery time T. Given that the decisions of the operator and attacker are largely intertwined, it is appropriate to model their interactions using concepts from game theory [341]. In particular, as discussed next, we formulate a *zero-sum network interdiction game* [342] to study the security of this considered UAV delivery system.

10.2.2 UAV Security as a Network Interdiction Game

In our considered system, the UAV operator's goal is to find a randomized path selection strategy, captured via an optimal probability distribution $\boldsymbol{y} \triangleq [y_1, y_2, ..., y_H]^T \in \mathcal{Y}$ chosen over the set $\mathcal{H}$ of all possible flight paths from O to D where $\mathcal{Y} = \{\boldsymbol{y} \in \mathbb{R}^H : \boldsymbol{y} \geq 0, \sum_{h=1}^{H} y_h = 1\}$. The operator has an incentive to randomize its path selection because, otherwise, if the path selection is done deterministically, it will make it trivial for the attacker to attack the chosen path and compromise the UAV. Here, we also note that the choice of a probability distribution over the set of actions (as opposed to a pure, deterministic choice) is known in game theory as a *mixed strategy*. Analogously, we assume that the attacker will also choose an optimal mixed strategy captured by an optimal probability distribution $\boldsymbol{x} \triangleq [x_1, x_2, ..., x_N]^T \in \mathcal{X}$ over the set of attack locations (i.e., danger points) $\mathcal{N}$. Here, we define $\mathcal{X} = \{\boldsymbol{x} \in \mathbb{R}^N : \boldsymbol{x} \geq 0, \sum_{n=1}^N x_n = 1\}$. Recall that, at a given location n, an attack will not always be successful, instead, the probability of a successful attack is captured by p_n.

For any successful attack, we assume that the drone operator will need to resend a new UAV UE to replace the compromised one. Hence, in our model, whenever a UAV UE arrives at a given node $n \in h$ of a path h, this UAV may continue its flight along path h normally with a probability $1 - p_n$ or, alternatively, it will be successfully interdicted by the adversary with probability p_n. The case in which the UAV UE is successfully attacked will be viewed to be equivalent to having the UAV virtually sent back to the origin O (because in practice a new UAV UE will be sent to replace the compromised one). For simplicity, we assume that for any product that is being redispatched after a successful attack, law enforcement agencies will have already secured path h (that was taken during the first attempt) and, hence, the operator can safely send its substitute UAV and item over path h without any security threats. We can now define the expected delivery time T using the previously defined mixed strategies, as follows:

$$T = \sum_{h \in \mathcal{H}} \sum_{n \in \mathcal{N}} y_h x_n [l_{hn} p_n (f^h(n) + f^h(D)) + (1 - l_{hn} p_n) f^h(D)]$$
$$= \sum_{h \in \mathcal{H}} \sum_{n \in \mathcal{N}} y_h x_n [l_{hn} p_n f^h(n) + f^h(D)]. \tag{10.1}$$

We can now also define an $(H \times N)$ matrix $\boldsymbol{M}$. Each element m_{hn} of this matrix will be given by:

$$m_{hn} = l_{hn} p_n f^h(n) + f^h(D) \ \forall h \in \mathcal{H} \text{ and } n \in \mathcal{N}. \tag{10.2}$$

Consequently, we can formally write the expected delivery time as follows:

$$T = \boldsymbol{y}^T \boldsymbol{M} \boldsymbol{x}. \tag{10.3}$$

Remark 10.1 In the considered system, we consider two losses that stem from the compromise of a given UAV: (a) economic losses pertaining to the monetary value of the UAV and its item, and (b) delivery time delays. Given that UAV delivery system operators will have very stringent delivery times, any delays will be highly detrimental to the reputation of the operator and its drone delivery program. Hence, minimizing delivery delays will be one of the primary objectives of UAV delivery system operators. As a result, hereinafter, we do not account for the economics losses, and we restrict our attention to the delays incurred by prospective cyber-physical attacks.

In the studied security problem, the operator primarily seeks to minimize its expected delivery time T while the adversary seeks to maximize T. Clearly, this is a zero-sum game that can be formally posed as a min-max problem (P_1):

$$(P_1): \quad T^* = \min_{\boldsymbol{y}} \max_{\boldsymbol{x}} \boldsymbol{y}^T \boldsymbol{M} \boldsymbol{x}, \tag{10.4}$$

$$\text{s.t. } \boldsymbol{1}_N \boldsymbol{x} = 1, \ \boldsymbol{1}_H \boldsymbol{y} = 1, \ \boldsymbol{x} \geq 0, \ \boldsymbol{y} \geq 0, \tag{10.5}$$

where $\mathbf{1}_N \triangleq [1, ..., 1]^T \in \mathbb{R}^N$ and $\mathbf{1}_H \triangleq [1, ..., 1]^T \in \mathbb{R}^H$. In (P_1), the constraints primarily pertain to restraining x and y to $x \in \mathcal{X}$ and $y \in \mathcal{Y}$. Meanwhile, the adversary's delivery time-maximization problem is nothing but the max-min counterpart of (P_1) introduced later in (10.14).

The distributions y and x chosen by the operator and attacker to solve their zero-sum game are known as *security strategies* [341]. Security strategies primarily consider scenarios in which an opponent seeks to inflict worst-case scenarios. For example, in (10.4), the operator will consider that the response of the adversary to any path selection strategy y will encompass the choice of an attack strategy $x \in \mathcal{X}$ that yields the worst-case scenario, i.e., the highest possible expected delivery time.

To find a solution for this zero-sum network interdiction game, we follow well-established approaches for solving zero-sum matrix games [341]. For instance, from (10.4), we observe that the maximization is done as a function of a given y. In other words, choosing an optimal $x \in \mathcal{X}$ is directly dependent on y. We can, therefore, rewrite (10.4) as follows:

$$\min_{y \in \mathcal{Y}} u_1(y), \tag{10.6}$$

where $u_1(y) = \max_{x \in \mathcal{X}} y^T M x \geq y^T M x \ \forall x \in \mathcal{X}$.

Since $\mathcal{X}$ is an N-dimensional simplex, then, we can rewrite the last inequality as follows:

$$M^T y \leq \mathbf{1}_N u_1(y). \tag{10.7}$$

We now make the change of variables $\hat{y} = y/u_1(y)$ and, then, we reformulate the min-max problem, (P_1), as a linear programming (LP) problem (P_2):

(P_2):
$$\min_{y \in \mathbb{R}^H} u_1(y) \tag{10.8}$$

s.t. $\quad M^T \hat{y} \leq \mathbf{1}_N,$ (10.9)

$\quad \hat{y}^T \mathbf{1}_H = 1/u_1(y),$ (10.10)

$\quad y = \hat{y} u_1(y), \hat{y} \geq 0.$ (10.11)

As proven in [343, Chapter 2], (10.8)–(10.11) can be reduced to a standard maximization problem (P_3):

(P_3): $\quad \max_{\hat{y}} \hat{y}^T \mathbf{1}_H$ (10.12)

s.t. $\quad M^T \hat{y} \leq \mathbf{1}_N, \hat{y} \geq 0.$ (10.13)

The optimal $\hat{y}$ can then be found by solving (P_3). Then, as per (10.10), this solution can be used to find $u_1(y)$. Thus, given $u_1(y)$ and $\hat{y}$, we can derive the optimal y as per (10.11).

We can now perform a similar analysis and transformation for the attacker's max-min problem. To do so, we first define the attacker's objective function as follows:

$$\max_{x \in \mathcal{X}} \min_{y \in \mathcal{Y}} y^T M x. \tag{10.14}$$

From (10.14), we can observe that the minimization is here performed for a given vector x. Hence, we define:

$$u_2(x) = \min_{y \in \mathcal{Y}} y^T M x \text{ and } \hat{x} = x/u_2(x). \tag{10.15}$$

Analogous to what we did for the operator's min-max problem, we can reduce the max-min in (10.14) to a standard minimization:

$$(P_4): \quad \min_{\hat{x}} \hat{x}^T \mathbf{1}_N, \tag{10.16}$$

$$\text{s.t.} \quad M\hat{x} \geq \mathbf{1}_H, \hat{x} \geq 0. \tag{10.17}$$

By solving (P_4), as done in (10.10), we find the optimal $\hat{x}$ and use it to derive $u_2(x)$:

$$\hat{x}^T \mathbf{1}_N = 1/u_2(x). \tag{10.18}$$

Consequently, given the optimal $\hat{x}$ and $u_2(x)$, we can now derive the optimal x as per (10.15).

From a game-theoretic perspective, solving LP problems (P_3) and (P_4) leads to a so-called *mixed-strategy Nash equilibrium* (NE) of the game, which is formally defined next:

DEFINITION 10.1 The strategy profile (y^*, x^*), is an NE (also known as a saddle-point equilibrium [SPE]) if and only if:

$$(y^*)^T M x^* \leq (y)^T M x^* \quad \forall y \in \mathcal{Y}, \tag{10.19}$$

$$(y^*)^T M x^* \geq (y^*)^T M x \quad \forall x \in \mathcal{X}. \tag{10.20}$$

We can now use the solutions of (P_3) and (P_4) to find the expected delivery time T^* at the NE as shown in the following Proposition 10.2:

PROPOSITION 10.2 *The solution strategies (y^*, x^*) constitute an NE of the network interdiction game, and the solutions of LP problems (P_3) and (P_4) result in value functions $\mu_1(\hat{y}^*) = (\hat{y}^*)^T \mathbf{1}_H$ and $\mu_2(\hat{x}^*) = (\hat{x}^*)^T \mathbf{1}_N$ satisfying $\mu_1(\hat{y}^*) = \mu_2(\hat{x}^*) = 1/T$.*

In summary, the studied problem is formally a finite zero-sum network interdiction game (defined over matrix M). In this game, for a mixed strategy pair (y, x), the expected payoffs of the operator U and the attacker A will be, respectively, given by $\Pi_A(y,x) = -\Pi_U(y,x) = y^T M x = T$. It is well-known that, for any finite zero-sum game, if y' is a mixed security strategy for the first player and x' is a mixed security strategy for the second player, then the NE of the game is given by (y', x') [343]. As a result, given that y^* and x^* are mixed security strategies for our finite, zero-sum UAV network interdiction game, then, $(y^*$ and $x^*)$ will be an NE of the game.

Proceeding from (10.10) and given the equivalence between (P_2) and (P_3), we can now write:

$$u_1(y^*) = [(\hat{y}^*)^T \mathbf{1}_H]^{-1} \Rightarrow u_1(y^*) = 1/\mu_1(\hat{y}^*). \tag{10.21}$$

However, by definition of $u_1(y)$ and T^*, we have:

$$u_1(y^*) = \min_{y \in \mathcal{Y}} u_1(y) = \min_{y \in \mathcal{Y}} \max_{x \in \mathcal{X}} y^T M x = T^*. \tag{10.22}$$

Therefore, using (10.21) and (10.22), we can find:

$$\mu_1(\hat{y}^*) = (\hat{y}^*)^T \mathbf{1}_H = 1/u_1(y^*) = 1/T.$$

Similarly, we can show the following:

$$\mu_2(\hat{x}^*) = (\hat{x}^*)^T \mathbf{1}_N = 1/u_2(x^*) = 1/T.$$

From this analysis, we can see that our game can admit multiple NEs (i.e., multiple security strategies for every player). Nonetheless, because the game is zero-sum, then all the equilibria will yield an equal expected delivery time, as shown in [341]. In addition, these NEs will be interchangeable [341], i.e., if (y^*, x^*) and (y', x') are two NEs, then, (y^*, x') and (y', x^*) are also NEs.

Note that, in this subsection, we analyzed the UAV security problem using conventional game theory (CGT) in which players are assumed to be objective and rational. In practice, given that humans will be involved in two ways – (a) the operator of the UAV system will involve humans in the planning of its UAV delivery system, and (b) the adversary will likely be a human decision maker – we can revisit the solutions by taking into account the bounded rationality of humans. Such an analysis under bounded rationality is done next.

10.2.3 Security of UAV Delivery Systems in Presence of Human Decision Makers

As discussed, in CGT, it is assumed that each player (operator or adversary) will evaluate the likelihood of achieving a certain delivery time objectively and rationally by using *expected values* to quantify the benefit of a pair of strategies (y, x). The ability to compute expectations is a tenet of *expected utility theory* (EUT) in which it is assumed that players in a game are rational and can objectively compute probabilistic outcomes, such as those in (10.1) and equivalently in (10.3).

Nonetheless, many empirical studies, including the Nobel-prize winning prospect theory experiments done by Kahneman in [344], have shown that, in the real world, human decision makers will not act in a fully rational manner. In particular, when faced with risk and uncertainty (as is the case of our UAV security game), the decision-making processes of human individuals can significantly deviate from the fully rational case of EUT and CGT. For instance, as demonstrated in the field of prospect theory (PT) [344], human players often evaluate probabilistic outcomes in a subjective manner. Such subjectivity will naturally appear in our UAV security setting due to various factors. On the one hand, both the UAV operator and the adversary can have their own, subjective perceptions on the probability of success of an attack at any given danger point. In this regard, the risk level of any given UAV path or the potential damage that a given attack at a given location can cause will be subjectively assessed by the operator and the adversary and, hence, the objective expectations on the delivery times that were used in the previous section may no longer hold. On the other hand, the way in which a given value for the delivery time is evaluated (by the operator or the attacker) will be subjective and done differently from EUT. For instance, one of the main performance metrics of a UAV delivery system is to achieve a very low delivery time. In particular, it is critical for a

10.2 Security of UAV UEs in Delivery Systems

UAV operator to meet the target delivery time, T^o, that it has promised to achieve. For example, Amazon Prime Air promises a delivery time of *less than 30 minutes* [332]. Therefore, in a real-world UAV delivery system, the delivery time is not an absolute quantity. Instead, it is measured with respect to a reference point T^o since an increase in the expected delivery time above T^o will have significant consequences on the reputation and effectiveness of the UAV delivery system. For example, if Amazon Prime Air cannot consistently meet its promised delivery time of $T^o = 30$ minutes, then its UAV delivery program may fail and will be significantly affected. Meanwhile, when UAVs are used in search and rescue or emergency medicine delivery [337, 338] missions, the slightest delay can have catastrophic outcomes. Clearly, one of the shortcomings of using CGT for analyzing UAV delivery system security is that it perceives the calculated expected delivery time as an absolute and objective quantity that the operator and attacker objectively use to choose their mixed strategies. As discussed, in practice, the delivery time is a relative quantity (with respect to T^o) that the players will subjectively assess.

We adopt the framework of prospect theory [344] in order to explicitly factor in these subjective perceptions into our UAV security game. In this regard, we leverage the so-called weighting and framing effects from prospect theory. The weighting effects allows us to capture the fact that the players will subjectively assess probabilistic outcomes. In particular, PT studies showed that human players tend to underweigh high probability outcomes and overweigh low probability outcomes. Meanwhile, the concept of PT framing will take into account the fact that the players will analyze their delivery time, not as an absolute, raw quantity, but rather as a relative quantity with respect to a reference point.

To integrate these PT effects into our game, instead of merely deriving the delivery time T in expectation as was done in the previous section, we will define a valuation function $V_z(T)$ for $z \in \{U, A\}$ that the operator, U, or the attacker, A, will subjectively define for any given T. By using (10.1), we can define this valuation function as follows (for $z \in \{U, A\}$):

$$V_z(T) = \sum_{h \in \mathcal{H}} \sum_{n \in \mathcal{N}} y_h x_n \left[v_z \left(l_{hn} \omega_z(p_n) f^h(n) + f^h(D) - R_z \right) \right]. \quad (10.23)$$

In equation (10.23), $\omega_z(.): [0, 1] \to \mathbb{R}$ is a nonlinear weighting function that captures the PT weighting effect and $v_z(.): \mathbb{R} \to \mathbb{R}$ is a nonlinear value function that will incorporate the PT framing effect. The weighting function in (10.23) captures the subjective perception that each player has of the likelihood of occurrence of probabilistic outcomes. The outcomes in our game correspond to the achieved delivery time whenever U selects a path $h \in \mathcal{H}$ and the adversary targets node $n \in h$ for its attack. The achieved delivery time is clearly a probabilistic outcome since it depends on the underlying probability of a successful attack. In essence, whenever the operator chooses path h and the attacker selects node $n \in h$, they will achieve a delivery time of $(f^h(n) + f^h(D))$ with probability p_n and a delivery time of $f^h(D)$ with probability $(1 - p_n)$. Hence, instead of objectively measuring the probabilities of these two outcomes, in the presence of human decision makers, our two players will perceive a weighted value of these probabilities.

For instance, under the PT weighting effect, any player $z \in \{U,A\}$ will view the probability with which the delivery time would be equal to $f^h(n) + f^h(D)$ (when U selects h and A selects $n \in h$) as $w_z(p_n)$, which is a nonlinear mapping that transforms the objective probability p_n into a subjective weight $w_z(p_n)$. This PT effect captures the fact that, in the real world, human players tend to underweight high probability outcomes and overweight low probability outcomes. To accurately model the subjective probability perceptions of each player $z \in \{U,A\}$, we define a weight $w_z(p_n)$ based on the Prelec function that is very popular in PT:

$$w_z(\sigma) = \exp(-(-\ln(\sigma)_z^\gamma)), \ 0 < \gamma_z < 1, \tag{10.24}$$

where γ_z is a *rationality parameter* that quantifies the distortion between player z's subjective and objective probability perceptions. In this context, a small value for γ_z implies lower rationality and larger subjectivity. When $\gamma_z = 1$, $\omega_z(p_n)$ reduces to the rational probability p_n. Meanwhile, when γ_z is close to 0, we get the fully irrational case.

Moreover, the value function in (10.23) also incorporates the PT concept of framing to capture how the operator and attacker value outcomes as gains and losses with respect to a delivery time reference point R_z (which can, for example, correspond to T^o) rather than as absolute quantities. Once the framing effect is included, the value function of the UAV operator will be defined as follows:

$$v_U(a_U) = \begin{cases} \lambda_U(a_U)^{\beta_U}, & \text{if } a_U \geq 0, \\ -(-a_U)^{\alpha_U}, & \text{if } a_U < 0, \end{cases} \tag{10.25}$$

$$\text{where} \quad a_U = l_{hn}\omega_U(p_n)f^h(n) + f^h(D) - R_U. \tag{10.26}$$

In 10.25, the parameters λ_U, β_U, and α_U are positive constants (with $\lambda_U > 1$) and $\omega_U(.)$ is also a Prelec weighting function. In our game, since the operator is a minimizing player, then $a_U \geq 0$ correspond to losses for the operator and $a_U < 0$ corresponds to gains for operator. The structure of this function allows us to capture key PT properties: (a) the operator will consider the value of a certain delivery time to be a gain or a loss depending on how it compares to a subjective reference point R_U (e.g. T^o), and (b) under PT, losses loom larger than gains, as measured by the loss multiplier λ_U in (10.25). This captures the fact that the operator will exaggerate the impact of not meeting its promised target delivery time. This is due to the potentially detrimental consequences that excessive delays will have on the reputation and operability of the UAV delivery system.

Next, we also define a value function for the attacker A that incorporates the PT framing effect while factoring in the fact that the attacker is a maximizer:

$$v_A(a_A) = \begin{cases} -\lambda_A(-a_A)^{\beta_A}, & \text{if } a_A < 0, \\ (a_A)^{\alpha_A}, & \text{if } a_A \geq 0, \end{cases} \tag{10.27}$$

$$\text{where} \quad a_A = l_{hn}\omega_A(p_n)f^h(n) + f^h(D) - R_A. \tag{10.28}$$

Moreover, to account for PT effects, we introduce the $(H \times N)$ matrices $\boldsymbol{M}^{U,\text{PT}}$ and $\boldsymbol{M}^{A,\text{PT}}$ whose elements are, respectively, given by ($\forall h \in \mathcal{H}, \forall n \in \mathcal{N}$):

$$m^{U,\text{PT}} = v_U \left(l_{hn} \omega_U(p_n) f^h(n) + f^h(D) - R_U \right), \quad (10.29)$$

$$m^{A,\text{PT}} = v_A \left(l_{hn} \omega_A(p_n) f^h(n) + f^h(D) - R_A \right). \quad (10.30)$$

Consequently, the vendor will find its optimal mixed strategy by solving the following problem, (P_5):

$$\min_{y \in \mathcal{Y}} \max_{x \in \mathcal{X}} y^T \boldsymbol{M}^{U,\text{PT}} x, \quad (10.31)$$

while the defender solves the following problem, (P_6):

$$\max_{x \in \mathcal{X}} \min_{y \in \mathcal{Y}} y^T \boldsymbol{M}^{A,\text{PT}} x. \quad (10.32)$$

In a real-world UAV delivery system, neither the UAV operator nor its adversary will have complete information on each others' subjectivity levels. Therefore, it is reasonable to assume that, in such a security setting, each player will assume that its opponent will be choosing the most damaging (i.e., worst case) strategy. We have captured this fact, respectively, by the min-max and max-min formulations of (P_5) and (P_6). Thereby, we can reduce (P_5) and (P_6), respectively, into standard maximization and minimization problems as done in the EUT case. However, in contrast to the CGT case, our analysis here will not lead to an equilibrium because $\boldsymbol{M}^{U,\text{PT}}$ and $\boldsymbol{M}^{A,\text{PT}}$ are different [343]. An analysis of the PT equilibrium can only be done if certain aspects of the models are modified, as discussed in [340].

10.2.4 Representative Simulation Results

In our simulations, we focus on the directed graph shown in Figure 10.4 with $N = 10$ vertices and $E = 18$ edges. Moreover, we define $[t_1, t_2, ..., t_{18}] \triangleq [3, 3, 3, 6, 6, 3, 6, 6, 6, 8, 6, 8, 10, 10, 10, 14, 12, 14]$ and $[p_1, p_2, ..., p_{10}] \triangleq [0, 0.2, 0.4, 0.2, 0.4, 0.4, 0.5, 0.8, 0.5, 0]$. We then number the different paths as follows: $[1, 2, ..., 18] \triangleq [(2, 5, 7), (2, 5, 8), (2, 5, 9), (2, 6, 7), (2, 6, 8), (2, 6, 9), (3, 5, 7), (3, 5, 8), (3, 5, 9), (3, 6, 7), (3, 6, 8), (3, 6, 9), (4, 5, 7), (4, 5, 8), (4, 5, 9), (4, 6, 7), (4, 6, 8), (4, 6, 9)]$ where, since node 1 (O) and node 10 (D) are part of all paths, a path (i, j, k) corresponds to $(1, i, j, k, 10)$. Unless stated otherwise, we choose the PT parameters as follows: $\lambda_A = \lambda_U = 5$, $\beta_U = \beta_A = 0.8$, and $\alpha_U = \alpha_A = 0.2$.

We first study the path lengths and path selection strategies (under CGT and PT) of the operator for various rationality parameters. These results are shown in Figure 10.5. First, Figure 10.5a presents the length of each possible path (from origin to destination) in $\mathcal{H}$. From Figure 10.5a, we can observe that path 8 is the shortest path by paths 2 and 14. Meanwhile, in Figure 10.5b, we present the operator's optimal path strategy. Figure 10.5b demonstrates that, under CGT, the shortest path (path 8) is not chosen with the highest probability. For instance, we can clearly see that paths 7 and 9 are more likely to be selected by the operator because of the high risk ($p_8 = 0.8$) associated with the shortest path, path 8. In contrast, as seen from Figure 10.6a (in this figure, we let

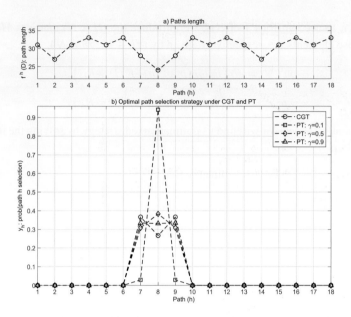

Figure 10.5 a) Length of each path in set $\mathcal{H}$. b) Optimal path selection in the CGT and PT cases, for various values of the rationality parameter.

$\gamma = \gamma_U = \gamma_A$), the weighting effect will flatten the perceived probabilities. Indeed, from this figure, we can clearly see the impact of PT weighting on the probabilities whereby high probabilities ($p_n > 0.4$) are underweighted and low probabilities are overweighted. Moreover, we can see that a very irrational UAV operator (for $\gamma = 0.1$) will see the probability of a successful attack as almost equally likely at all nodes in the graph. As a result, from Figure 10.5b, we can observe that, in the PT scenario, the operator will be more likely to travel through the shortest path. Indeed, for a very low rationality level (e.g., $\gamma = 0.1$), the operator will view all paths to be of equal risk and, thus, it will use the shortest with a likelihood of 94%. In addition, Figure 10.6b presents the optimal attack strategy of the adversary when γ changes. We can see that, for the CGT case, the optimal strategy of the attacker is to randomize between nodes 7, 8, and 9 while assigning the highest probability to node 8, which is part of the shortest path (that is very risk, i.e., $p_8 = 0.8$. In contrast, for the PT cases, the attacker will primarily target nodes 5 and 8, which are both part of the shortest path.

Figures 10.5 and 10.6 demonstrate that the PT weighting effect as well as the rationality parameter will have a significant impact on the achieved delivery time because they will affect the strategy choices of both players. For instance, Figure 10.7 presents the variation in the achieved expected delivery time for $\gamma \in \{0.1, 0.5, 0.9\}$. In Figure 10.7, lower rationality levels lead to higher delivery times. Indeed, when γ decreases from 0.9 to 0.1, the resulting delivery time will increase by about 11%. Note that Figure 10.7 assumes $T^o = R_U = R_A = 30$. Hence, the distorted perception of probability leads to a selection of risky path strategies whose expected delivery times will exceed the desired target. Such delays, as discussed earlier, can have major consequences on

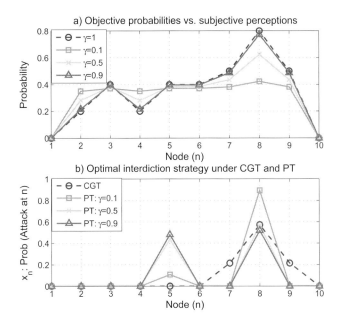

Figure 10.6 a) Objective and subjective perceptions of p_n. b) Optimal attacker strategy (for CGT and PT) when the rationality parameter varies.

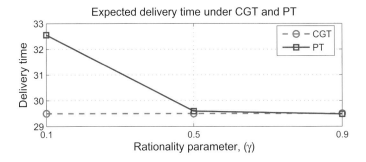

Figure 10.7 Expected delivery time as a function of the PT rationality parameter, γ.

the UAV operator's system, particularly for search and rescue and critical applications [337, 338]. We recall that the delivery time used in our model is, in fact, the expected *flight time* of the UAV when under attack. The actual achieved delivery time will also include additional delays pertaining to processing and re-handling (these added delays can mathematically seen as additive constants).

In Figure 10.8, for $R_U = 30$, we study how the loss parameter λ_U can impact the probability of choosing the shortest path and the achieved expected delivery time. In essence, a higher value for λ_U implies that the associated player will amplify its losses further, i.e., it is more averse to loss. In our UAV scenario, an increase in the value of λ_U implies that the consequences of not meeting a target delivery time will be amplified

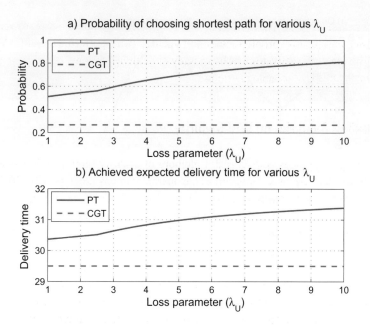

Figure 10.8 Variation with respect to λ_U of a) probability of choosing the shortest path, and b) achieved expected delivery time.

for the operator. As a result, higher values of λ_U will lead the operator to be more risk-seeking, and, hence, it will tend to select more risky path strategies when those strategies have shorter path lengths. Figure 10.8a clearly shows that, as λ_U, the operator will be more likely to select the shortest path. For example, increasing λ_U from 1 to 10 leads to an increase in the probability of choosing the shortest path from 0.51 to 0.81. This risky path selection strategy will, in turn, lead to an increase in the achieved delivery time, as clearly seen from Figure 10.8b. An important observation here is that, under the subjective behavior observed by PT, the expected delivery time exceeds that under CGT as well as the target delivery time. Hence, this shows that the subjective perception of probabilities and outcomes by the vendor can impair its chosen path strategies, incurring delays to the delivery time.

10.2.5 Summary

In this section, we have provided a preliminary study on the security of a UAV-based delivery system. First, we have discussed how the cyber and physical functions of a UAV will render it susceptible to cyber-physical attacks. We have then devised a generic framework that can be used to study the impact of security threats on the performance of a UAV delivery system. We have used a game-theoretic framework to study this performance, and we have shown how the behavior of the operator and the adversary can impact the overall delivery time of the system. In particular, using prospect theory, we have particularly studied cases in which the players have bounded rationality and can act in a risk-seeking or risk-averse manner. From the study done in this section, we can clearly see that security problems are intertwined with the performance of UAV applications. This, in turn, motivates a holistic study of communications, security, and network performance for UAV-enabled systems.

10.3 Concluding Remarks on UAV Security

In this chapter, we have shown that a plethora of security threats on UAVs exist, and those threats can seriously compromise the operation of any network with UAVs. Therefore, in order to effectively deploy UAVs in the context of the various applications discussed in Chapter 2, it is imperative to: (a) identify the types of threats that target the UAV system, and (b) devise defense mechanisms to thwart those threats. Hence, in this chapter, we have first identified several types of security threats within a UAV system, and we have discussed their potential effects. Examples include eavesdropping attacks, jamming, and GPS spoofing attacks. For the GPS spoofing case, we have provided preliminary simulations that show how various parameters impact the attack effectiveness and how a drone operator can leverage those parameters to improve its defense system. Then, we have provided a comprehensive framework for instilling security in UAV systems, in general, and delivery systems with UAV UEs, in particular. We have used notions from game theory and prospect theory to design our framework, and we have discussed the effects of various network parameters, particularly the effect of human factors. In a nutshell, maintaining the security of UAV systems is a necessity for delivering their promised applications. Indeed, without proper security measures, adversaries can compromise the UAV system and leverage it for nefarious purposes. As a result, designing effective UAV systems requires addressing challenges at the intersection of communications, networking, and security, as demonstrated in this book.

References

[1] S. O'Donnell, "A short history of unmanned aerial vehicles," *Consortiq Blog*, June 2017. [Online]. Available: https://consortiq.com/media-centre/blog/short-history-unmanned-aerial-vehicles-uavs

[2] R. P. Hallion, *Taking Flight: Inventing the Aerial Age, from Antiquity through the First World War*. Oxford, UK: Oxford University Press, 2003.

[3] R. D. Layman, *Naval Aviation in the First World War: Its Impact and Influence*. Annapolis, MD, USA: Naval Institute Press, 1996.

[4] The Drone Enthusiast, "The history of drones (drone history timeline from 1849 to 2019)," 2019. [Online]. Available: www.dronethusiast.com/history-of-drones/

[5] D. Hunt, "World War 1 history: The Kettering bugworld's first drone," October 2017. [Online]. Available: https://owlcation.com/humanities/World-War-1-History-The-Kettering-Bug-Worlds-First-Flying-Bomb

[6] D. Gettinger and A. H. Michel, "Drone sightings and close encounters: An analysis," Center for the Study of the Drone, Bard College, Annandale-on-Hudson, NY, USA, 2015.

[7] Airbus, "Zephyr," 2003. [Online]. Available: www.airbus.com/defence/uav/zephyr.html

[8] Google, "Project Loon," 2018. [Online]. Available: https://x.company/projects/loon/

[9] A. Fotouhi, H. Qiang, M. Ding, M. Hassan, L. G. Giordano, A. Garcia-Rodriguez, and J. Yuan, "Survey on UAV cellular communications: Practical aspects, standardization advancements, regulation, and security challenges," 2018. Available: arxiv.org/abs/1809.01752, 2018.

[10] C. Stöcker, R. Bennett, F. Nex, M. Gerke, and J. Zevenbergen, "Review of the current state of UAV regulations," *Remote Sensing*, vol. 9, no. 5, p. 459, 2017.

[11] U. Challita, A. Ferdowsi, M. Chen, and W. Saad, "Machine learning for wireless connectivity and security of cellular-connected UAVs," *IEEE Wireless Communications*, vol. 26, no. 1, pp. 28–35, February 2019.

[12] Y. Zeng, R. Zhang, and T. J. Lim, "Wireless communications with unmanned aerial vehicles: Opportunities and challenges," *IEEE Communications Magazine*, vol. 54, no. 5, pp. 36–42, May 2016.

[13] A. Orsino, A. Ometov, G. Fodor, D. Moltchanov, L. Militano, S. Andreev, O. N. Yilmaz, T. Tirronen, J. Torsner, G. Araniti et al., "Effects of heterogeneous mobility on D2D-and drone-assisted mission-critical MTC in 5G," *IEEE Communications Magazine*, vol. 55, no. 2, pp. 79–87, 2017.

[14] Y. H. Nam, B. L. Ng, K. Sayana, Y. Li, J. Zhang, Y. Kim, and J. Lee, "Full-dimension mimo (fd-mimo) for next generation cellular technology," *IEEE Communications Magazine*, vol. 51, no. 6, pp. 172–179, June 2013.

[15] 3GPP, "Study on elevation beamforming/full-dimension (FD) MIMO for LTE," *TR 36.897*, May 2017.

[16] W. Lee, S.-R. Lee, H.-B. Kong, and I. Lee, "3D beamforming designs for single user MISO systems," in *Proc. of IEEE Global Communications Conference (GLOBECOM)*, 2013, pp. 3914–3919.

[17] Y.-H. Nam, M. S. Rahman, Y. Li, G. Xu, E. Onggosanusi, J. Zhang, and J.-Y. Seol, "Full dimension MIMO for LTE-advanced and 5G." Proc. Information Theory and Applications Workshop, San Diego, CA, USA, Feb 2015

[18] M. Shafi, M. Zhang, P. J. Smith, A. L. Moustakas, and A. F. Molisch, "The impact of elevation angle on mimo capacity," in *Proc. of IEEE International Conference on Communications*, vol. 9. IEEE, 2006, pp. 4155–4160.

[19] X. Cheng, B. Yu, L. Yang, J. Zhang, G. Liu, Y. Wu, and L. Wan, "Communicating in the real world: 3D MIMO," *IEEE Wireless Communications magazine*, vol. 21, no. 4, pp. 136–144, 2014.

[20] Y. Li, X. Ji, D. Liang, and Y. Li, "Dynamic beamforming for three-dimensional MIMO technique in LTE-advanced networks," *International Journal of Antennas and Propagation*, vol. 2013, 2013.

[21] 3GPP TR 36.777, "Enhanced LTE support for aerial vehicles (version 15.0.0)," 3GPP, Tech. Rep., December 2017.

[22] M. Mozaffari, W. Saad, M. Bennis, and M. Debbah, "Communications and Control for Wireless Drone-Based Antenna Array," IEEE Transactions on Communications, vol. 67, no. 1, pp. 820 – 834, January 2019. Available: arxiv.org/abs/1712.10291.

[23] N. Rupasinghe, Y. Yapici, I. Guvenc, and Y. Kakishima, "Non-orthogonal multiple access for mmWave drones with multi-antenna transmission," published in Proceedings of 51st Asilomar Conference on Signals, Systems, and Computers, October 2017, Pacific Grove, CA, USA. Available: arxiv.org/abs/1711.10050, 2017.

[24] E. Torkildson, H. Zhang, and U. Madhow, "Channel modeling for millimeter wave MIMO," in *Proc. of Information Theory and Applications Workshop (ITA), 2010*. IEEE, 2010, pp. 1–8.

[25] I. Bor-Yaliniz and H. Yanikomeroglu, "The new frontier in RAN heterogeneity: Multi-tier drone-cells," *IEEE Communications Magazine*, vol. 54, no. 11, pp. 48–55, 2016.

[26] W. Saad, M. Bennis, and M. Chen, "A vision of 6g wireless systems: Applications, trends, technologies, and open research problems," *IEEE Network*, 2019.

[27] J. Gubbi, R. Buyya, S. Marusic, and M. Palaniswami, "Internet of Things (IoT): A vision, architectural elements, and future directions," *Future generation computer systems*, vol. 29, no. 7, pp. 1645–1660, 2013.

[28] Z. Dawy, W. Saad, A. Ghosh, J. G. Andrews, and E. Yaacoub, "Toward massive machine type cellular communications," *IEEE Wireless Communications*, vol. 24, no. 1, pp. 120–128, February 2017.

[29] S.-Y. Lien, K.-C. Chen, and Y. Lin, "Toward ubiquitous massive accesses in 3gpp machine-to-machine communications," *IEEECommunications Magazine*, vol. 49, no. 4, pp. 66–74, April 2011.

[30] A. Zanella, N. Bui, A. Castellani, L. Vangelista, and M. Zorzi, "Internet of Things for smart cities," *IEEE Internet of Things Journal*, vol. 1, no. 1, pp. 22–32, February 2014.

[31] M. Mozaffari, W. Saad, M. Bennis, and M. Debbah, "Unmanned aerial vehicle with underlaid device-to-device communications: Performance and tradeoffs," *IEEE Transactions on Wireless Communications*, vol. 15, no. 6, pp. 3949–3963, June 2016.

[32] J. Chakareski, "Aerial UAV-IoT sensing for ubiquitous immersive communication and virtual human teleportation," in *2017 IEEE Conference on Computer Communications Workshops (INFOCOM WKSHPS)*, Atlanta, GA, USA, May 2017.

[33] M. Chen, W. Saad, and C. Yin, "Echo-liquid state deep learning for 360 content transmission and caching in wireless VR networks with cellular-connected UAVs," *IEEE Trans. Commun.*, vol. 67, no. 9, pp. 6386–6400, September 2019.

[34] N. Bhushan, J. Li, D. Malladi, R. Gilmore, D. Brenner, A. Damnjanovic, R. T. Sukhavasi, C. Patel, and S. Geirhofer, "Network densification: The dominant theme for wireless evolution into 5G," *IEEE Communications Magazine*, vol. 52, no. 2, pp. 82–89, February 2014.

[35] X. Ge, S. Tu, G. Mao, C. X. Wang, and T. Han, "5G ultra-dense cellular networks," *IEEE Wireless Communications*, vol. 23, no. 1, pp. 72–79, February 2016.

[36] Z. Gao, L. Dai, D. Mi, Z. Wang, M. A. Imran, and M. Z. Shakir, "Mmwave massive-MIMO-based wireless backhaul for the 5G ultra-dense network," *IEEE Wireless Communications*, vol. 22, no. 5, pp. 13–21, October 2015.

[37] U. Siddique, H. Tabassum, E. Hossain, and D. I. Kim, "Wireless backhauling of 5G small cells: Challenges and solution approaches," *IEEE Wireless Communications*, vol. 22, no. 5, pp. 22–31, October 2015.

[38] U. Challita and W. Saad, "Network formation in the sky: Unmanned aerial vehicles for multi-hop wireless backhauling," in *Proc. of IEEE Global Telecommunications Conference (GLOBECOM)*, Singapore, December 2017.

[39] Y. Hu, M. Chen, and W. Saad, "Competitive market for joint access and backhaul resource allocation in satellite-drone networks," in *Proc. of 10th IFIP International Conference on New Technologies, Mobility and Security (NTMS), Mobility & Wireless Networks Track*, Canary Islands, Spain, June 2019.

[40] D. Bamburry, "Drones: Designed for product delivery," *Design Management Review*, vol. 26, no. 1, pp. 40–48, 2015.

[41] J. Chen, U. Yatnalli, and D. Gesbert, "Learning radio maps for UAV-aided wireless networks: A segmented regression approach," in *Proc. of IEEE International Conference on Communications (ICC)*, Paris, France, May 2017.

[42] S. Jeong, O. Simeone, and J. Kang, "Mobile edge computing via a UAV-mounted cloudlet: Optimization of bit allocation and path planning," *IEEE Transactions on Vehicular Technology, Early access*, 2017.

[43] G. Lee, W. Saad, and M. Bennis, "Online optimization for UAV-assisted distributed fog computing in smart factories of industry 4.0," in *Proc. of the IEEE Global Communications Conference (GLOBECOM), Selected Areas in Communications Symposiums – Internet of Things Track*, Abu Dhabi, UAE, December 2018.

[44] E. Haas, "Aeronautical channel modeling," *IEEE Transactions on Vehicular Technology*, vol. 51, no. 2, pp. 254–264, March 2002.

[45] H. D. Tu and S. Shimamoto, "A proposal of wide-band air-to-ground communication at airports employing 5-GHz band," in *IEEE Wireless Communications and Networking Conference*, April 2009, pp. 1–6.

[46] Y. S. Meng and Y. H. Lee, "Study of shadowing effect by aircraft maneuvering for air-to-ground communication," *AEU-International Journal of Electronics and Communications*, vol. 66, no. 1, pp. 7–11, January 2012.

[47] R. Sun, D. W. Matolak, and W. Rayess, "Air-ground channel characterization for unmanned aircraft systems – Part IV: Airframe shadowing," *IEEE Transactions on Vehicular Technology*, vol. 66, no. 9, pp. 7643–7652, September 2017.

[48] X. Lin, V. Yajnanarayana, S. D. Muruganathan, S. Gao, H. Asplund, H. L. Maattanen, M. Bergstrom, S. Euler, and Y. P. E. Wang, "The sky is not the limit: LTE for unmanned aerial vehicles," *IEEE Communications Magazine*, vol. 56, no. 4, pp. 204–210, April 2018.

[49] V. Yajnanarayana, Y.-P. E. Wang, S. Gao, S. Muruganathan, and X. Lin, "Interference mitigation methods for unmanned aerial vehicles served by cellular networks," in *IEEE 5G World Forum*, July 2018.

[50] D. W. Matolak and R. Sun, "Air–ground channel characterization for unmanned aircraft systems – Part III: The suburban and near-urban environments," *IEEE Transactions on Vehicular Technology*, vol. 66, no. 8, pp. 6607–6618, August 2017.

[51] K. Takizawa, T. Kagawa, S. Lin, F. Ono, H. Tsuji, and R. Miura, "C-band aircraft-to-ground (A2G) radio channel measurement for unmanned aircraft systems," in *International Symposium on Wireless Personal Multimedia Communications*, 2014, pp. 754–758.

[52] W. Khawaja, I. Guvenc, D. Matolak, U.-C. Fiebig, and N. Schneckenberger, "A survey of air-to-ground propagation channel modeling for unmanned aerial vehicles," *arXiv preprint arXiv:1801.01656*, January 2018.

[53] D. W. Matolak, "Air-ground channels & models: Comprehensive review and considerations for unmanned aircraft systems," in *IEEE Aerospace Conference*, March 2012, pp. 1–17.

[54] M. Walter, S. Gligorević, T. Detert, and M. Schnell, "UHF/VHF air-to-air propagation measurements," in *Proceedings of the Fourth European Conference on Antennas and Propagation*, April 2010, pp. 1–5.

[55] Q. Feng, J. McGeehan, E. K. Tameh, and A. R. Nix, "Path loss models for air-to-ground radio channels in urban environments," in *IEEE Vehicular Technology Conference*, vol. 6, May 2006, pp. 2901–2905.

[56] M. Simunek, P. Pechac, and F. P. Fontan, "Excess loss model for low elevation links in urban areas for UAVs," *Radioengineering*, vol. 20, no. 3, pp. 561–568, September 2011.

[57] A. Al-Hourani, S. Kandeepan, and A. Jamalipour, "Modeling air-to-ground path loss for low altitude platforms in urban environments," in *IEEE Global Communications Conference*, December 2014, pp. 2898–2904.

[58] F. Ono, K. Takizawa, H. Tsuji, and R. Miura, "S-band radio propagation characteristics in urban environment for unmanned aircraft systems," in *International Symposium on Antennas and Propagation*, November 2015, pp. 1–4.

[59] H. T. Friis, "A note on a simple transmission formula," *Proceedings of the IRE*, vol. 34, no. 5, pp. 254–256, May 1946.

[60] A. Goldsmith, *Wireless Communications*. Cambridge, UK: Cambridge University Press, 2005.

[61] D. W. Matolak and R. Sun, "Air–ground channel characterization for unmanned aircraft systems – Part I: Methods, measurements, and models for over-water settings," *IEEE Transactions on Vehicular Technology*, vol. 66, no. 1, pp. 26–44, January 2017.

[62] R. Sun and D. W. Matolak, "Air–ground channel characterization for unmanned aircraft systems – Part II: Hilly and mountainous settings," *IEEE Transactions on Vehicular Technology*, vol. 66, no. 3, pp. 1913–1925, March 2017.

[63] Y. S. Meng and Y. H. Lee, "Measurements and characterizations of air-to-ground channel over sea surface at C-band with low airborne altitudes," *IEEE Transactions on Vehicular Technology*, vol. 60, no. 4, pp. 1943–1948, April 2011.

[64] J. D. Parsons, *The Mobile Radio Propagation Channel*. Hoboken, NJ, USA: Wiley, 2000.

[65] T. S. Rappaport, *Wireless Communications: Principles and Practice*. New Jersey: Prentice Hall PTR, 1996, vol. 2.

[66] F. Ikegami, T. Takeuchi, and S. Yoshida, "Theoretical prediction of mean field strength for urban mobile radio," *IEEE Transactions on Antennas and Propagation*, vol. 39, no. 3, pp. 299–302, March 1991.

[67] K. R. Schaubach, N. Davis, and T. S. Rappaport, "A ray tracing method for predicting path loss and delay spread in microcellular environments," in *Proceedings of IEEE Vehicular Technology Conference*, May 1992, pp. 932–935.

[68] Z. Yun and M. F. Iskander, "Ray tracing for radio propagation modeling: principles and applications," *IEEE Access*, vol. 3, pp. 1089–1100, July 2015.

[69] J. B. Keller, "Geometrical theory of diffraction," *JOSA*, vol. 52, no. 2, pp. 116–130, February 1962.

[70] R. G. Kouyoumjian and P. H. Pathak, "A uniform geometrical theory of diffraction for an edge in a perfectly conducting surface," *Proceedings of the IEEE*, vol. 62, no. 11, pp. 1448–1461, November 1974.

[71] R. Luebbers, "Finite conductivity uniform GTD versus knife edge diffraction in prediction of propagation path loss," *IEEE Transactions on Antennas and Propagation*, vol. 32, no. 1, pp. 70–76, January 1984.

[72] K. A. Remley, H. R. Anderson, and A. Weisshar, "Improving the accuracy of ray-tracing techniques for indoor propagation modeling," *IEEE Transactions on Vehicular Technology*, vol. 49, no. 6, pp. 2350–2358, November 2000.

[73] F. Fuschini, H. El-Sallabi, V. Degli-Esposti, L. Vuokko, D. Guiducci, and P. Vainikainen, "Analysis of multipath propagation in urban environment through multidimensional measurements and advanced ray tracing simulation," *IEEE Transactions on Antennas and Propagation*, vol. 56, no. 3, pp. 848–857, March 2008.

[74] T. S. Rappaport, R. W. Heath Jr, R. C. Daniels, and J. N. Murdock, *Millimeter Wave Wireless Communications*. New York City, NY, USA: Pearson Education, 2014.

[75] P. Pongsilamanee and H. L. Bertoni, "Specular and nonspecular scattering from building facades," *IEEE Transactions on Antennas and Propagation*, vol. 52, no. 7, pp. 1879–1889, July 2004.

[76] V. Degli-Esposti, F. Fuschini, E. M. Vitucci, and G. Falciasecca, "Measurement and modelling of scattering from buildings," *IEEE Transactions on Antennas and Propagation*, vol. 55, no. 1, pp. 143–153, January 2007.

[77] M. Catedra, J. Perez, F. S. De Adana, and O. Gutierrez, "Efficient ray-tracing techniques for three-dimensional analyses of propagation in mobile communications: Application to picocell and microcell scenarios," *IEEE Antennas and Propagation Magazine*, vol. 40, no. 2, pp. 15–28, April 1998.

[78] M. F. Iskander and Z. Yun, "Propagation prediction models for wireless communication systems," *IEEE Transactions on microwave theory and techniques*, vol. 50, no. 3, pp. 662–673, March 2002.

[79] V. Erceg, S. J. Fortune, J. Ling, A. Rustako, and R. A. Valenzuela, "Comparisons of a computer-based propagation prediction tool with experimental data collected in urban microcellular environments," *IEEE Journal on Selected Areas in Communications*, vol. 15, no. 4, pp. 677–684, May 1997.

[80] Q. Feng, E. K. Tameh, A. R. Nix, and J. McGeehan, "WLCp2-06: Modelling the likelihood of line-of-sight for air-to-ground radio propagation in urban environments," in *Proceedings of IEEE Global Telecommunications Conference*, December 2006, pp. 1–5.

[81] I. J. Timmins and S. O'Young, "Marine communications channel modeling using the finite-difference time domain method," *IEEE Transactions on Vehicular Technology*, vol. 58, no. 6, pp. 2626–2637, July 2009.

[82] Y. Wu, Z. Gao, C. Chen, L. Huang, H.-P. Chiang, Y.-M. Huang, and H. Sun, "Ray tracing based wireless channel modeling over the sea surface near Diaoyu islands," in *First International Conference on Computational Intelligence Theory, Systems and Applications*, December 2015, pp. 124–128.

[83] N. Goddemeier, K. Daniel, and C. Wietfeld, "Role-based connectivity management with realistic air-to-ground channels for cooperative UAVs," *IEEE Journal on Selected Areas in Communications*, vol. 30, no. 5, pp. 951–963, June 2012.

[84] W. Khawaja, O. Ozdemir, and I. Guvenc, "UAV air-to-ground channel characterization for mmWave systems," in *Proceedings of IEEE Vehicular Technology Conference*, September 2017, pp. 1–5.

[85] V. Erceg, L. J. Greenstein, S. Y. Tjandra, S. R. Parkoff, A. Gupta, B. Kulic, A. A. Julius, and R. Bianchi, "An empirically based path loss model for wireless channels in suburban environments," *IEEE Journal on Selected Areas in Communications*, vol. 17, no. 7, pp. 1205–1211, July 1999.

[86] R. Amorim, H. Nguyen, P. Mogensen, I. Z. Kovács, J. Wigard, and T. B. Sørensen, "Radio channel modeling for UAV communication over cellular networks," *IEEE Wireless Communications Letters*, vol. 6, no. 4, pp. 514–517, August 2017.

[87] W. Khawaja, I. Guvenc, and D. Matolak, "UWB channel sounding and modeling for UAV air-to-ground propagation channels," in *Proceedings of Global Communications Conference (GLOBECOM)*, December 2016, pp. 1–7.

[88] W. G. Newhall, R. Mostafa, C. Dietrich, C. R. Anderson, K. Dietze, G. Joshi, and J. H. Reed, "Wideband air-to-ground radio channel measurements using an antenna array at 2 GHz for low-altitude operations," in *Proceedings of IEEE Military Communications Conference*, vol. 2, October 2003, pp. 1422–1427.

[89] E. Yanmaz, R. Kuschnig, and C. Bettstetter, "Achieving air-ground communications in 802.11 networks with three-dimensional aerial mobility," in *Proceedings of IEEE INFOCOM*, April 2013, pp. 120–124.

[90] C.-M. Cheng, P.-H. Hsiao, H. Kung, and D. Vlah, "Performance measurement of 802.11a wireless links from UAV to ground nodes with various antenna orientations," in *Proceedings of International Conference on Computer Communications and Networks*, October 2006, pp. 303–308.

[91] J. Allred, A. B. Hasan, S. Panichsakul, W. Pisano, P. Gray, J. Huang, R. Han, D. Lawrence, and K. Mohseni, "Sensorflock: An airborne wireless sensor network of micro-air vehicles," in *Proceedings of the 5th International Conference on Embedded Networked Sensor Systems*, November 2007, pp. 117–129.

[92] E. W. Frew and T. X. Brown, "Airborne communication networks for small unmanned aircraft systems," *Proceedings of the IEEE*, vol. 96, no. 12, December 2008.

[93] M. J. Feuerstein, K. L. Blackard, T. S. Rappaport, S. Y. Seidel, and H. H. Xia, "Path loss, delay spread, and outage models as functions of antenna height for microcellular system design," *IEEE Transactions on Vehicular Technology*, vol. 43, no. 3, pp. 487–498, August 1994.

[94] X. Cai, A. Gonzalez-Plaza, D. Alonso, L. Zhang, C. B. Rodríguez, A. P. Yuste, and X. Yin, "Low altitude UAV propagation channel modelling," in *Proceedings of European Conference on Antennas and Propagation*, March 2017, pp. 1443–1447.

[95] A. Al-Hourani and K. Gomez, "Modeling cellular-to-UAV path-loss for suburban environments," *IEEE Wireless Communications Letters*, vol. 7, no. 1, pp. 82–85, February 2018.

[96] J. Walfisch and H. L. Bertoni, "A theoretical model of UHF propagation in urban environments," *IEEE Transactions on Antennas and Propagation*, vol. 36, no. 12, pp. 1788–1796, December 1988.

[97] F. Ikegami, S. Yoshida, T. Takeuchi, and M. Umehira, "Propagation factors controlling mean field strength on urban streets," *IEEE Transactions on Antennas and Propagation*, vol. 32, no. 8, pp. 822–829, August 1984.

[98] G. L. Turin, F. D. Clapp, T. L. Johnston, S. B. Fine, and D. Lavry, "A statistical model of urban multipath propagation," *IEEE Transactions on Vehicular Technology*, vol. 21, no. 1, pp. 1–9, February 1972.

[99] M. Gudmundson, "Correlation model for shadow fading in mobile radio systems," *Electronics letters*, vol. 27, no. 23, pp. 2145–2146, November 1991.

[100] M. Holzbock and C. Senninger, "An aeronautical multimedia service demonstration at high frequencies," *IEEE Transactions on MultiMedia*, vol. 6, no. 4, pp. 20–29, October 1999.

[101] J. Kunisch, I. De La Torre, A. Winkelmann, M. Eube, and T. Fuss, "Wideband time-variant air-to-ground radio channel measurements at 5 GHz," in *Proceedings of the 5th European Conference on Antennas and Propagation*, April 2011, pp. 1386–1390.

[102] J. Naganawa, J. Honda, T. Otsuyama, H. Tajima, and H. Miyazaki, "Evaluating path loss by extended squitter signals for aeronautical surveillance," *IEEE Antennas and Wireless Propagation Letters*, vol. 16, pp. 1353–1356, 2017.

[103] J. Holis and P. Pechac, "Elevation dependent shadowing model for mobile communications via high altitude platforms in built-up areas," *IEEE Transactions on Antennas and Propagation*, vol. 56, no. 4, pp. 1078–1084, April 2008.

[104] E. Teng, J. D. Falcão, and B. Iannucci, "Holes-in-the-sky: A field study on cellular-connected UAS," in *International Conference on Unmanned Aircraft Systems*, June 2017, pp. 1165–1174.

[105] 3GPP, "Study on channel model for frequencies from 0.5 to 100 GHz," *3GPP TR 38.901, V15.0.0*, June 2018.

[106] ITU-R, "P.1410: Propagation data and prediction methods required for the design of terrestrial broadband radio access systems operating in a frequency range from 3 to 60 GHz," Tech. Rep., February 2012.

[107] A. Al-Hourani, S. Kandeepan, and S. Lardner, "Optimal LAP altitude for maximum coverage," *IEEE Wireless Communications Letters*, vol. 3, no. 6, pp. 569–572, December 2014.

[108] R. I. Bor-Yaliniz, A. El-Keyi, and H. Yanikomeroglu, "Efficient 3-D placement of an aerial base station in next generation cellular networks," in *Proceedings of IEEE International Conference on Communications*, May 2016, pp. 1–5.

[109] M. Mozaffari, W. Saad, M. Bennis, and M. Debbah, "Mobile unmanned aerial vehicles (UAVs) for energy-efficient Internet of Things communications," *IEEE Trans. Wireless Commun.*, vol. 16, no. 11, pp. 7574–7589, November 2017.

[110] M. Alzenad, A. El-Keyi, and H. Yanikomeroglu, "3-D placement of an unmanned aerial vehicle base station for maximum coverage of users with different QoS requirements," *IEEE Wireless Communications Letters*, vol. 7, no. 1, pp. 38–41, February 2018.

[111] H. V. Hitney and L. R. Hitney, "Frequency diversity effects of evaporation duct propagation," *IEEE Transactions on Antennas and Propagation*, vol. 38, no. 10, pp. 1694–1700, October 1990.

[112] H. Heemskerk and R. Boekema, "The influence of evaporation duct on the propagation of electromagnetic waves low above the sea surface at 3-94 GHz," in *International Conference on Antennas and Propagation*, 1993, pp. 348–351.

[113] Z. Xiao, P. Xia, and X.-G. Xia, "Enabling UAV cellular with millimeter-wave communication: Potentials and approaches," *IEEE Communications Magazine*, vol. 54, no. 5, pp. 66–73, May 2016.

[114] ITU-R, "P.838-3: Specific attenuation model for rain for use in prediction methods," Tech. Rep., March 2005.

[115] A. Paier, T. Zemen, L. Bernadó, G. Matz, J. Karedal, N. Czink, C. Dumard, F. Tufvesson, A. F. Molisch, and C. F. Mecklenbrauker, "Non-WSSUS vehicular channel characterization in highway and urban scenarios at 5.2 GHz using the local scattering function," in *International ITG Workshop on Smart Antennas*, February 2008, pp. 9–15.

[116] O. Renaudin, V.-M. Kolmonen, P. Vainikainen, and C. Oestges, "Non-stationary narrowband MIMO inter-vehicle channel characterization in the 5-GHz band," *IEEE Transactions on Vehicular Technology*, vol. 59, no. 4, pp. 2007–2015, May 2010.

[117] P. Bello, "Aeronautical channel characterization," *IEEE Transactions on Communications*, vol. 21, no. 5, pp. 548–563, May 1973.

[118] S. M. Elnoubi, "A simplified stochastic model for the aeronautical mobile radio channel," in *Proceedings of the IEEE Vehicular Technology Conference*, May 1992, pp. 960–963.

[119] M. Walter and M. Schnell, "The Doppler-delay characteristic of the aeronautical scatter channel," in *Proceedings of IEEE Vehicular Technology Conference*, September 2011, pp. 1–5.

[120] M. Walter, D. Shutin, and U.-C. Fiebig, "Joint delay Doppler probability density functions for air-to-air channels," *International Journal of Antennas and Propagation*, vol. 2014, April 2014.

[121] M. Ibrahim and H. Arslan, "Air-ground Doppler-delay spread spectrum for dense scattering environments," in *Proceedings of the EEE Military Communications Conference*, October 2015, pp. 1661–1666.

[122] T. J. Willink, C. C. Squires, G. W. Colman, and M. T. Muccio, "Measurement and characterization of low-altitude air-to-ground MIMO channels," *IEEE Transactions on Vehicular Technology*, vol. 65, no. 4, pp. 2637–2648, April 2016.

[123] R. M. Gutierrez, H. Yu, Y. Rong, and D. W. Bliss, "Time and frequency dispersion characteristics of the UAS wireless channel in residential and mountainous desert terrains," in *IEEE Annual Consumer Communications & Networking Conference*, January 2017, pp. 516–521.

[124] N. Schneckenburger, T. Jost, D. Shutin, M. Walter, T. Thiasiriphet, M. Schnell, and U.-C. Fiebig, "Measurement of the L-band air-to-ground channel for positioning applications," *IEEE Transactions on Aerospace and Electronic Systems*, vol. 52, no. 5, pp. 2281–2297, October 2016.

[125] A. A. Saleh and R. Valenzuela, "A statistical model for indoor multipath propagation," *IEEE Journal on Selected Areas in Communications*, vol. 5, no. 2, pp. 128–137, February 1987.

[126] S. M. Gulfam, S. J. Nawaz, A. Ahmed, and M. N. Patwary, "Analysis on multipath shape factors of air-to-ground radio communication channels," in *Wireless Telecommunications Symposium*, 2016, pp. 1–5.

[127] W. Newhall and J. Reed, "A geometric air-to-ground radio channel model," in *Proceedings of MILCOM*, vol. 1, October 2002, pp. 632–636.

[128] S. Blandino, F. Kaltenberger, and M. Feilen, "Wireless channel simulator testbed for airborne receivers," in *IEEE Globecom Workshops (GC Wkshps)*, December 2015, pp. 1–6.

[129] M. Wentz and M. Stojanovic, "A MIMO radio channel model for low-altitude air-to-ground communication systems," in *IEEE Vehicular Technology Conference*, September 2015, pp. 1–6.

[130] A. Ksendzov, "A geometrical 3D multi-cluster mobile-to-mobile MIMO channel model with Rician correlated fading," in *International Congress on Ultra Modern Telecommunications and Control Systems and Workshops*, 2016, pp. 191–195.

[131] L. Zeng, X. Cheng, C.-X. Wang, and X. Yin, "A 3D geometry-based stochastic channel model for UAV-MIMO channels," in *IEEE Wireless Communications and Networking Conference*, March 2017, pp. 1–5.

[132] P. Chandhar, D. Danev, and E. G. Larsson, "Massive MIMO for communications with drone swarms," *IEEE Transactions on Wireless Communications*, vol. 17, no. 3, pp. 1604–1629, March 2018.

[133] M. Simunek, F. P. Fontán, and P. Pechac, "The UAV low elevation propagation channel in urban areas: Statistical analysis and time-series generator," *IEEE Transactions on Antennas and Propagation*, vol. 61, no. 7, pp. 3850–3858, July 2013.

[134] E. L. Cid, A. V. Alejos, and M. G. Sanchez, "Signaling through scattered vegetation: Empirical loss modeling for low elevation angle satellite paths obstructed by isolated thin trees," *IEEE Vehicular Technology Magazine*, vol. 11, no. 3, pp. 22–28, September 2016.

[135] R. Jain and F. Templin, "Requirements, challenges and analysis of alternatives for wireless datalinks for unmanned aircraft systems," *IEEE Journal on Selected Areas in Communications*, vol. 30, no. 5, pp. 852–860, June 2012.

[136] R. G. Gallager, *Principles of Digital Communication*. Cambridge University Press, 2008, vol. 1.

[137] U. Madhow, *Fundamentals of Digital Communication*. Cambridge University Press, 2008.

[138] M. Marcus, "Spectrum policy challenges of UAV/drones [spectrum policy and regulatory issues]," *IEEE Wireless Communications*, vol. 21, no. 5, pp. 8–9, October 2014.

[139] National Telecommunications and Information Administration, "Aws-3 transition," Tech. Rep.

[140] Z. Wu, H. Kumar, and A. Davari, "Performance evaluation of OFDM transmission in UAV wireless communication," in *Proceedings of the Thirty-Seventh Southeastern Symposium on System Theory*, March 2005, pp. 6–10.

[141] J. Kakar and V. Marojevic, "Waveform and spectrum management for unmanned aerial systems beyond 2025," in *Proceedings of IEEE International Symposium on Personal, Indoor, and Mobile Radio Communications*, October 2017, pp. 1–5.

[142] C. Bluemm, C. Heller, B. Fourestie, and R. Weigel, "Air-to-ground channel characterization for OFDM communication in C-band," in *International Conference on Signal Processing and Communication Systems*, December 2013, pp. 1–8.

[143] Y. Rahmatallah and S. Mohan, "Peak-to-average power ratio reduction in OFDM systems: A survey and taxonomy," *IEEE communications surveys & tutorials*, vol. 15, no. 4, pp. 1567–1592, Fourth quarter 2013.

[144] A. Giorgetti, M. Lucchi, M. Chiani, and M. Z. Win, "Throughput per pass for data aggregation from a wireless sensor network via a UAV," *IEEE Transactions on Aerospace and Electronic Systems*, vol. 47, no. 4, pp. 2610–2626, October 2011.

[145] T. D. Ho, J. Park, and S. Shimamoto, "QoS constraint with prioritized frame selection CDMA MAC protocol for WSN employing UAV," in *Proceedings of IEEE GLOBECOM Workshops*, December 2010, pp. 1826–1830.

[146] J. Li, Y. Zhou, L. Lamont, and M. Déziel, "A token circulation scheme for code assignment and cooperative transmission scheduling in CDMA-based UAV ad hoc networks," *Wireless Networks*, vol. 19, no. 6, pp. 1469–1484, August 2013.

[147] M. Edrich and R. Schmalenberger, "Combined DSSS/FHSS approach to interference rejection and navigation support in UAV communications and control," in *IEEE Seventh International Symposium on Spread Spectrum Techniques and Applications*, vol. 3, September 2002, pp. 687–691.

[148] S. J. Maeng, H.-i. Park, and Y. S. Cho, "Preamble design technique for GMSK-based beamforming system with multiple unmanned aircraft vehicles," *IEEE Transactions on Vehicular Technology*, vol. 66, no. 8, pp. 7098–7113, August 2017.

[149] D. Darsena, G. Gelli, I. Iudice, and F. Verde, "Equalization techniques of control and non-payload communication links for unmanned aerial vehicles," *IEEE Access*, vol. 6, pp. 4485–4496, 2018.

[150] P. G. Sudheesh, M. Mozaffari, M. Magarini, W. Saad, and P. Muthuchidambaranathan, "Sum-rate analysis for high altitude platform (HAP) drones with tethered balloon relay," *IEEE Communications Letters, Early access*, 2017.

[151] M. Haenggi, *Stochastic Geometry for Wireless Networks*. Cambridge, UK: Cambridge University Press, 2012.

[152] F. Baccelli, B. Błaszczyszyn et al., "Stochastic geometry and wireless networks: Volume ii applications," *Foundations and Trends® in Networking*, vol. 4, no. 1–2, pp. 1–312, 2010.

[153] V. V. Chetlur and H. S. Dhillon, "Downlink coverage analysis for a finite 3-D wireless network of unmanned aerial vehicles," *IEEE Transactions on Communications*, vol. 65, no. 10, pp. 4543–4558, October 2017.

[154] N. Lee, X. Lin, J. G. Andrews, and R. Heath, "Power control for D2D underlaid cellular networks: Modeling, algorithms, and analysis," *IEEE Journal on Selected Areas in Communications*, vol. 33, no. 1, pp. 1–13, February 2015.

[155] X. Lin, R. Heath, and J. Andrews, "The interplay between massive MIMO and underlaid D2D networking," *IEEE Transactions on Wireless Communications*, June 2015.

[156] M. Afshang, H. S. Dhillon, and P. H. J. Chong, "Modeling and performance analysis of clustered device-to-device networks," *available online: arxiv.org/abs/1508.02668*, 2015.

[157] A. Hourani, S. Kandeepan, and A. Jamalipour, "Modeling air-to-ground path loss for low altitude platforms in urban environments," in *Proc. of IEEE Global Telecommunications Conference (GLOBECOM)*, Austin, TX, USA, December 2014.

[158] A. Hourani, K. Sithamparanathan, and S. Lardner, "Optimal LAP altitude for maximum coverage," *IEEE Wireless Communication Letters*, vol. 3, no. 6, pp. 569–572, December 2014.

[159] F. Baccelli and B. Blaszczyszyn, "Stochastic geometry and wireless networks, volume ii-applications," Foundations and Trends in Networking, vol. 4, no.1-2, 2009.

[160] E. Artin, *The Gamma Function*. Mineola, NY, USA: Courier Dover Publications, 2015.

[161] R. K. Ganti, "A stochastic geometry approach to the interference and outage characterization of large wireless networks," Ph.D. dissertation, University of Notre Dame, 2009.

[162] S. P. Weber, X. Yang, J. G. Andrews, and G. De Veciana, "Transmission capacity of wireless ad hoc networks with outage constraints," *IEEE Transactions on Information Theory*, vol. 51, no. 12, pp. 4091–4102, November 2005.

[163] M. Haenggi and R. K. Ganti, *Interference in Large Wireless Networks*. Hanover, MA, USA: Foundations and Trends in Networking, 2009.

[164] S. Shalmashi, E. Björnson, M. Kountouris, K. W. Sung, and M. Debbah, "Energy efficiency and sum rate tradeoffs for massive MIMO systems with underlaid device-to-device communications," *available online: arxiv.org/abs/1506.00598.*, 2015.

[165] R. Kershner, "The number of circles covering a set," *American Journal of Mathematics*, pp. 665–671, 1939.

[166] G. F. Tóth, "Thinnest covering of a circle by eight, nine, or ten congruent circles," *Combinatorial and Computational Geometry*, vol. 52, no. 361, p. 59, 2005.

[167] M. Mozaffari, W. Saad, M. Bennis, and M. Debbah, "Drone small cells in the clouds: Design, deployment and performance analysis," in *Proceedings of IEEE Global Communications Conference (GLOBECOM)*, San Diego, CA, USA, December 2015.

[168] M. Mozaffari, W. Saad, M. Bennis, and M. Debbah, "Efficient deployment of multiple unmanned aerial vehicles for optimal wireless coverage," *IEEE Communications Letters*, vol. 20, no. 8, pp. 1647–1650, August 2016.

[169] E. Kalantari, H. Yanikomeroglu, and A. Yongacoglu, "On the number and 3D placement of drone base stations in wireless cellular networks," in *Proc. of IEEE Vehicular Technology Conference*, 2016.

[170] R. Yaliniz, A. El-Keyi, and H. Yanikomeroglu, "Efficient 3-D placement of an aerial base station in next generation cellular networks," in *Proc. of IEEE International Conference on Communications (ICC)*, Kuala Lumpur, Malaysia, May. 2016.

[171] A. M. Hayajneh, S. A. R. Zaidi, D. C. McLernon, and M. Ghogho, "Drone empowered small cellular disaster recovery networks for resilient smart cities," in *Proc. of IEEE International Conference on Sensing, Communication and Networking (SECON Workshops)*, June 2016.

[172] J. Kosmerl and A. Vilhar, "Base stations placement optimization in wireless networks for emergency communications," in *Proc. of IEEE International Conference on Communications (ICC)*, Sydney, Australia, June. 2014.

[173] M. Alzenad, A. El-Keyi, F. Lagum, and H. Yanikomeroglu, "3-D placement of an unmanned aerial vehicle base station (UAV-BS) for energy-efficient maximal coverage," *IEEE Wireless Communications Letters*, vol. 6, no. 4, pp. 434–437, August 2017.

[174] M. Alzenad, A. El-Keyi, and H. Yanikomeroglu, "3-D placement of an unmanned aerial vehicle base station for maximum coverage of users with different QoS requirements," *IEEE Wireless Communications Letters*, vol. 7, no. 1, pp. 38–41, February 2018.

[175] E. Kalantari, M. Z. Shakir, H. Yanikomeroglu, and A. Yongacoglu, "Backhaul-aware robust 3D drone placement in 5G+ wireless networks," in *Proc. of IEEE International Conference on Communications Workshops (ICC Workshops)*, May 2017, pp. 109–114.

[176] H. A. Eiselt and V. Marianov, *Foundations of Location Analysis*. Berlin, Germany: Springer Science & Business Media, 2011, vol. 155.

[177] H. M. Farahani, R.Z., *Facility Location: Concepts, Models, Algorithms and Case Studies*. Physica-Verlag, Heidelberg, 2009.

[178] S. Ahmadian, Z. Friggstad, and C. Swamy, "Local-search based approximation algorithms for mobile facility location problems," in *Proc. of the Twenty-Fourth Annual ACM-SIAM Symposium on Discrete Algorithms*. SIAM, 2013, pp. 1607–1621.

[179] R. L. Graham, B. D. Lubachevsky, K. J. Nurmela, and P. R. Östergård, "Dense packings of congruent circles in a circle," *Discrete Mathematics*, vol. 181, no. 1-3, pp. 139–154, 1998.

[180] Z. Gáspár and T. Tarnai, "Upper bound of density for packing of equal circles in special domains in the plane," *Periodica Polytechnica Civil Engineering*, vol. 44, no. 1, pp. 13–32, 2000.

[181] K. Venugopal, M. C. Valenti, and R. W. Heath Jr, "Device-to-device millimeter wave communications: Interference, coverage, rate, and finite topologies," *available online: arxiv.org/abs/1506.07158*, 2015.

[182] C. A. Balanis, *Antenna Theory: Analysis and Design*. Hoboken, NJ, USA: John Wiley & Sons, 2016.

[183] K.-C. Chen and S.-Y. Lien, "Machine-to-machine communications: Technologies and challenges," *Ad Hoc Networks*, vol. 18, pp. 3–23, July. 2014.

[184] 3GPP, "Study on RAN improvements for machine type communication," *TR 37.868*, Sept. 2011.

[185] X. Jian, X. Zeng, Y. Jia, L. Zhang, and Y. He, "Beta/M/1 model for machine type communication," *IEEE Communications Letters*, vol. 17, no. 3, pp. 584–587, March 2013.

[186] M. Tavana, V. Shah-Mansouri, and V. W. S. Wong, "Congestion control for bursty M2M traffic in LTE networks," in *Proc. of IEEE International Conference on Communications (ICC)*, London, UK, June 2015.

[187] A. K. Gupta and S. Nadarajah, *Handbook of Beta Distribution and Its Applications*. Boca Raton, FL,USA : CRC Press, 2004.

[188] R. D. Yates, "A framework for uplink power control in cellular radio systems," *IEEE Journal on Selected Areas in Communications*, vol. 13, no. 7, pp. 1341–1347, September 1995.

[189] R. Sun, M. Hong, and Z. Q. Luo, "Joint downlink base station association and power control for max-min fairness: Computation and complexity," *IEEE Journal on Selected Areas in Communications*, vol. 33, no. 6, pp. 1040–1054, June. 2015.

[190] P. T. Boggs and J. W. Tolle, "Sequential quadratic programming," *Acta Numerica*, vol. 4, pp. 1–51, 1995.

[191] M. Peng, Y. Sun, X. Li, Z. Mao, and C. Wang, "Recent advances in cloud radio access networks: System architectures, key techniques, and open issues," *IEEE Communications Surveys and Tutorials*, vol. 18, no. 3, pp. 2282–2308, Thirdquarter 2016.

[192] M. Chen, M. Mozaffari, W. Saad, C. Yin, M. Debbah, and C. S. Hong, "Caching in the sky: Proactive deployment of cache-enabled unmanned aerial vehicles for optimized quality-of-experience," *IEEE J. Select. Areas Commun.*, vol. 35, no. 5, pp. 1046 – 1061, May 2017.

[193] T. S. Rappaport, F. Gutierrez, E. Ben-Dor, J. N. Murdock, Y. Qiao, and J. I. Tamir, "Broadband millimeter-wave propagation measurements and models using adaptive-beam antennas for outdoor urban cellular communications," *IEEE Transactions on Antennas and Propagation*, vol. 61, no. 4, pp. 1850–1859, April. 2013.

[194] A. Al-Hourani, S. Kandeepan, and A. Jamalipour, "Modeling air-to-ground path loss for low altitude platforms in urban environments," in *Proc. of IEEE Global Communications Conference (GLOBECOM)*, Austin, TX, USA, December 2014.

[195] O. Somekh, O. Simeone, Y. Bar-Ness, A. M. Haimovich, and S. Shamai, "Cooperative multicell zero-forcing beamforming in cellular downlink channels," *IEEE Transactions on Information Theory*, vol. 55, no. 7, pp. 3206–3219, June. 2009.

[196] F. Hoppner and F. Klawonn, *Clustering with Size Constraints*. Berlin, Germany: Springer Berlin Heidelberg, 2008.

[197] M. Bennis, S. Perlaza, P. Blasco, Z. Han, and H. Poor, "Self-organization in small cell networks: A reinforcement learning approach," *IEEE Transactions on Wireless Communications*, vol. 12, no. 7, pp. 3202–3212, June. 2013.

[198] M. Chen, W. Saad, and C. Yin, "Echo state networks for self-organizing resource allocation in LTE-U with uplink-downlink decoupling," *IEEE Transactions on Wireless Communications*, vol. 1, no. 1, January 2017.

[199] M. Chen, W. Saad, C. Yin, and M. Debbah, "Echo State Networks for Proactive Caching in Cloud-Based Radio Access Networks with Mobile Users," IEEE Transactions on Wireless Communications, vol. 16, no. 6, pp. 3520–3535, June 2017.

[200] M. V. Menshikov, "Estimates for percolation thresholds for lattices in R^n," *Dokl. Akad. Nauk SSSR*, vol. 284, pp. 36–39, 1985.

[201] H. Kesten, "Asymptotics in high dimensions for percolation," in *Disorder in Physical Systems: A Volume in Honour of John Hammersley*, G. R. Grimmett and D. J. A. Welsh, Eds. Oxford, UK: Oxford University Press, 1990, pp. 219–240.

[202] D. Reimer, "Proof of the van den Berg–Kesten conjecture," *Combin. Probab. Comput.*, vol. 9, pp. 27–32, 2000.

[203] J. M. Hammersley and G. Mazzarino, "Properties of large Eden clusters in the plane," *Combin. Probab. Comput.*, vol. 3, pp. 471–505, 1994.

[204] J. M. Hammersley, "Percolation processes: Lower bounds for the critical probability," *Ann. Math. Statist.*, vol. 28, pp. 790–795, 1957.

[205] M. Aizenman and D. J. Barsky, "Sharpness of the phase transition in percolation models," *Comm. Math. Phys.*, vol. 108, pp. 489–526, 1987.

[206] M. V. Menshikov, S. A. Molchanov, and A. F. Sidorenko, "Percolation theory and some applications," in *Probability theory. Mathematical statistics. Theoretical cybernetics, Vol. 24 (Russian)*. Akad. Nauk SSSR Vsesoyuz. Inst. Nauchn. i Tekhn. Inform., 1986, pp. 53–110, translated in *J. Soviet Math.* **42** (1988), no. 4, 1766–1810.

[207] J. M. Hammersley, "Comparison of atom and bond percolation processes," *J. Mathematical Phys.*, vol. 2, pp. 728–733, 1961.

[208] J. M. Hammersley and G. Mazzarino, "Markov fields, correlated percolation, and the Ising model," in *The Mathematics and Physics of Disordered Media (Minneapolis, Minn., 1983)*, ser. Lecture Notes in Math. Springer, 1983, vol. 1035, pp. 201–245.

[209] J. M. Hammersley and D. J. A. Welsh, "First-passage percolation, subadditive processes, stochastic networks, and generalized renewal theory," in *Proc. Internat. Research Seminar, Statist. Lab., Univ. California, Berkeley, Calif.* New York City, Ny, USA: Springer, 1965, pp. 61–110.

[210] J. Yoon, Y. Jin, N. Batsoyol, and H. Lee, "Adaptive path planning of UAVs for delivering delay-sensitive information to ad-hoc nodes," in *Proc. IEEE Wireless Communications and Networking Conference (WCNC)*, March 2017, pp. 1–6.

[211] Y. Zeng and R. Zhang, "Energy-efficient UAV communication with trajectory optimization," *IEEE Transactions on Wireless Communications*, vol. 16, no. 6, pp. 3747–3760, June 2017.

[212] M. Messous, S. Senouci, and H. Sedjelmaci, "Network connectivity and area coverage for UAV fleet mobility model with energy constraint," in *Proc. IEEE Wireless Communications and Networking Conference*, April 2016, pp. 1–6.

[213] X. Wang, A. Chowdhery, and M. Chiang, "Networked drone cameras for sports streaming," in *Proc. IEEE International Conference on Distributed Computing Systems (ICDCS)*, June 2017, pp. 308–318.

[214] A. Al-Hourani, S. Kandeepan, and A. Jamalipour, "Modeling air-to-ground path loss for low altitude platforms in urban environments," in *Proc. IEEE Global Communications Conference*, December 2014, pp. 2898–2904.

[215] 3GPP TR 25.942 v2.1.3, "3rd generation partnership project; technical specification group (TSG) RAN WG4; RF system scenarios," Tech. Rep., 2000.

[216] D. Bertsekas and R. Gallager, *Data Networks*. Upper Saddle River, NJ, USA: Prentice Hall, March 1992.

[217] Z. Han, D. Niyato, W. Saad, T. Başar, and A. Hjorungnes, *Game Theory in Wireless and Communication Networks: Theory, Models, and Applications*. Cambridge, UK: Cambridge University Press, 2012.

[218] W. Kwon, I. Suh, S. Lee, and Y. Cho, "Fast reinforcement learning using stochastic shortest paths for a mobile robot," in *Proc. IEEE/RSJ International Conference on Intelligent Robots and Systems*, October 2007, pp. 82–87.

[219] U. Challita, W. Saad, and C. Bettstetter, "Interference management for cellular-connected UAVs: A deep reinforcement learning approach," *IEEE Trans. Wireless Commun.*, vol. 18, no. 4, pp. 2125–2140, April 2019.

[220] M. Osborne, *An Introduction to Game Theory*. Oxford, UK: Oxford University Press, 2004.

[221] M. Chen, U. Challita, W. Saad, C. Yin, and M. Debbah, "Artificial neural networks-based machine learning for wireless networks: A tutorial," *IEEE Communications Surveys and Tutorials*, 2019.

[222] M. Chen, U. Challita, W. Saad, C. Yin, and M. Debbah, "Machine learning for wireless networks with artificial intelligence: A tutorial on neural networks," *CoRR*, vol. abs/1710.02913, 2017.

[223] C. Gallicchio and A. Micheli, "Echo state property of deep reservoir computing networks," *Cognitive Computation*, vol. 9, pp. 337–350, May 2017.

[224] H. Jaeger, M. Lukosevicius, D. Popovici, and U. Siewert, "Optimization and applications of echo state networks with leaky-integrator neurons," *Neural Networks*, vol. 20, no. 3, pp. 335–352, 2007.

[225] I. Szita and A. L. V. Gyenes, *Reinforcement Learning with Echo State Networks*. Germany: Springer, Berlin, Heidelberg, 2006, vol. 4131.

[226] R. Sutton and A. Barto, *Introduction to Reinforcement Learning*. Cambridge, MA, USA: MIT Press, 1998.

[227] A. Ghaffarkhah and Y. Mostofi, "Path planning for networked robotic surveillance," *IEEE Transactions on Signal Processing*, vol. 60, no. 7, pp. 3560–3575, July 2012.

[228] M. Mozaffari, W. Saad, M. Bennis, and M. Debbah, "Wireless communication using unmanned aerial vehicles (UAVs): Optimal transport theory for hover time optimization," *IEEE Trans. Wireless Commun.*, vol. 16, no. 12, pp. 8052–8066, December 2017.

[229] V. Sharma, M. Bennis, and R. Kumar, "UAV-assisted heterogeneous networks for capacity enhancement," *IEEE Communications Letters*, vol. 20, no. 6, pp. 1207–1210, June 2016.

[230] M. Mozaffari, W. Saad, M. Bennis, and M. Debbah, "Optimal transport theory for power-efficient deployment of unmanned aerial vehicles," in *Proc. of IEEE International Conference on Communications (ICC)*, May 2016.

[231] S. Niu, J. Zhang, F. Zhang, and H. Li, "A method of UAVs route optimization based on the structure of the highway network," *International Journal of Distributed Sensor Networks*, December 2015.

[232] K. Dorling, J. Heinrichs, G. G. Messier, and S. Magierowski, "Vehicle routing problems for drone delivery," *IEEE Transactions on Systems, Man, and Cybernetics: Systems*, vol. 47, no. 1, pp. 70–85, January 2017.

[233] Y. Zeng and R. Zhang, "Energy-efficient UAV communication with trajectory optimization," *IEEE Transactions on Wireless Communications*, vol. 16, no. 6, pp. 3747–3760, June 2017.

[234] V. V. Chetlur and H. S. Dhillon, "Downlink coverage analysis for a finite 3D wireless network of unmanned aerial vehicles," *IEEE Transactions on Communications, Early access*, 2017.

[235] F. Jiang and A. L. Swindlehurst, "Optimization of UAV heading for the ground-to-air uplink," *IEEE Journal on Selected Areas in Communications*, vol. 30, no. 5, pp. 993–1005, June 2012.

[236] D. Orfanus, E. P. de Freitas, and F. Eliassen, "Self-organization as a supporting paradigm for military UAV relay networks," *IEEE Communications Letters*, vol. 20, no. 4, pp. 804–807, 2016.

[237] Q. Wu, Y. Zeng, and R. Zhang, "Joint trajectory and communication design for multi-UAV enabled wireless networks," *IEEE Transactions on Wireless Communications*, vol. 17, no. 3, pp. 2109–2121, March 2018.

[238] S. Chandrasekharan, K. Gomez, A. Al-Hourani, S. Kandeepan, T. Rasheed, L. Goratti, L. Reynaud, D. Grace, I. Bucaille, T. Wirth, and S. Allsopp, "Designing and implementing future aerial communication networks," *IEEE Communications Magazine*, vol. 54, no. 5, pp. 26–34, May 2016.

[239] ITU-R, "Rec. p.1410-2 propagation data and prediction methods for the design of terrestrial broadband millimetric radio access systems," *Series, Radiowave propagation*, 2003.

[240] F. Aurenhammer, "Voronoi diagrams: A survey of a fundamental geometric data structure," *ACM Computing Surveys (CSUR)*, vol. 23, no. 3, pp. 345–405, 1991.

[241] A. Silva, H. Tembine, E. Altman, and M. Debbah, "Optimum and equilibrium in assignment problems with congestion: Mobile terminals association to base stations," *IEEE Transactions on Automatic Control*, vol. 58, no. 8, pp. 2018–2031, August 2013.

[242] C. Villani, *Topics in optimal transportation*. American Mathematical Soc., 2003, no. 58.

[243] G. Crippa, C. Jimenez, and A. Pratelli, "Optimum and equilibrium in a transport problem with queue penalization effect," *Advances in Calculus of Variations*, vol. 2, no. 3, pp. 207–246, 2009.

[244] M. C. Achtelik, J. Stumpf, D. Gurdan, and K. M. Doth, "Design of a flexible high performance quadcopter platform breaking the MAV endurance record with laser power beaming," in *Proc. of IEEE International Conference on Intelligent Robots and Systems*, September 2011.

[245] M. Mozaffari, A. T. Z. Kasgari, W. Saad, M. Bennis, and M. Debbah, "Beyond 5G with UAVs: Foundations of a 3D wireless cellular network," *IEEE Trans. Wireless Commun.*, vol. 18, no. 1, p. 357–372, January 2019.

[246] F. Lagum, I. Bor-Yaliniz, and H. Yanikomeroglu, "Strategic densification with UAV-BSs in cellular networks," *IEEE Wireless Communications Letters, Early access*, 2017.

[247] M. Mozaffari, W. Saad, M. Bennis, and M. Debbah, "Optimal transport theory for cell association in UAV-enabled cellular networks," *IEEE Communications Letters*, vol. 21, no. 9, pp. 2053–2056, September 2017.

[248] J. Lyu, Y. Zeng, and R. Zhang, "UAV-aided offloading for cellular hotspot," *IEEE Transactions on Wireless Communications*, vol. 17, no. 6, pp. 3988–4001, June 2018.

[249] E. Kalantari, I. Bor-Yaliniz, A. Yongacoglu, and H. Yanikomeroglu, "User association and bandwidth allocation for terrestrial and aerial base stations with backhaul considerations,"

in *Proc. IEEE Annual International Symposium on Personal, Indoor, and Mobile Radio Communications (PIMRC)*, Montreal, QC, Canada, October 2017.

[250] M. M. Azari, F. Rosas, A. Chiumento, and S. Pollin, "Coexistence of terrestrial and aerial users in cellular networks," in *Proc. of IEEE Global Telecommunications Conference (GLOBECOM) Workshops*, Singapore, December 2017.

[251] S. Zhang, Y. Zeng, and R. Zhang, "Cellular-enabled UAV communication: Trajectory optimization under connectivity constraint," in *Proc. of IEEE International Conference on Communications (ICC), to appear*, Kansas City, USA, May. 2018.

[252] J. Horwath, N. Perlot, M. Knapek, and F. Moll, "Experimental verification of optical backhaul links for high-altitude platform networks: Atmospheric turbulence and downlink availability," *International Journal of Satellite Communications and Networking*, vol. 25, no. 5, pp. 501–528, 2007.

[253] B. Galkin, J. Kibilda, and L. A. DaSilva, "Backhaul for low-altitude UAVs in urban environments," in *Proc. of IEEE International Conference on Communications (ICC)*, May 2018, pp. 1–6.

[254] S. Alam and Z. J. Haas, "Coverage and connectivity in three-dimensional networks," in *Proceedings of Annual International Conference on Mobile Computing and Networking*, Los Angeles, CA, USA, September 2006.

[255] H. S. M. Coxeter, *Regular Polytopes*. North Chelmsford, MA, USA: Courier Corporation, 1973.

[256] J. Liu, M. Sheng, L. Liu, and J. Li, "Effect of densification on cellular network performance with bounded pathloss model," *IEEE Communications Letters*, vol. 21, no. 2, pp. 346–349, 2017.

[257] D. Athukoralage, I. Guvenc, W. Saad, and M. Bennis, "Regret based learning for UAV assisted LTE-U/WiFi public safety networks," in *IEEE Global Communications Conference*, 2016, pp. 1–7.

[258] H. Wang, G. Ding, F. Gao, J. Chen, J. Wang, and L. Wang, "Power control in UAV-supported ultra dense networks: Communications, caching, and energy transfer," *IEEE Communications Magazine*, vol. 56, no. 6, pp. 28 – 34, June 2018.

[259] A. Azizi, N. Mokari, and M. R. Javan, "Joint radio resource allocation, 3D placement and user association of aerial base stations in IoT networks," *arXiv:1710.05315*, October 2017. [Online]. Available: https://arxiv.org/abs/1710.05315

[260] Y. Zeng, R. Zhang, and T. J. Lim, "Throughput maximization for UAV-enabled mobile relaying systems," *IEEE Trans. Commun.*, vol. 64, no. 12, pp. 4983–4996, December 2016.

[261] M. Liu, J. Yang, and G. Gui, "DSF-NOMA: UAV-assisted emergency communication technology in a heterogeneous internet of things," *IEEE Internet of Things Journal*, vol. 6, no. 3, pp. 5508 – 5519, June 2019.

[262] J. Lyu, Y. Zeng, and R. Zhang, "Spectrum sharing and cyclical multiple access in UAV-aided cellular offloading," in *Proc. IEEE Global Communication Conference*, Singapore, December 2017.

[263] J. Zhang, Y. Zeng, and R. Zhang, "Spectrum and energy efficiency maximization in UAV-enabled mobile relaying," in *Proc. Int. Conf. on Communications*, Paris, France, May 2017.

[264] R. Zhang, M. Wang, L. X. Cai, Z. Zheng, X. Shen, and L. L. Xie, "LTE-unlicensed: The future of spectrum aggregation for cellular networks," *IEEE Wireless Comm.*, vol. 22, no. 3, pp. 150–159, June 2015.

[265] Q. Chen, G. Yu, H. Shan, A. Maaref, G. Y. Li, and A. Huang, "Cellular meets WiFi: Traffic offloading or resource sharing?" *IEEE Trans. Wireless Commun.*, vol. 15, no. 5, p. 3354–3367, May 2016.

[266] G. Bianchi, "Performance analysis of IEEE 802.11 distributed coordination function," *IEEE Journal on Selected Areas in Communications*, vol. 18, no. 3, pp. 535–547, March 2000.

[267] M. J. Neely, "Stochastic network optimization with application to communication and queueing systems," *Synthesis Lectures on Communication Networks*, vol. 3, no. 1, pp. 1–211, 2010.

[268] M. Chen, W. Saad, and C. Yin, "Liquid state machine learning for resource and cache management in LTE-U unmanned aerial vehicle (UAV) networks," *IEEE Trans. Commun.*, vol. 18, no. 3, pp. 1504 – 1517, March 2019.

[269] W. Maas, "Liquid state machines: Motivation, theory, and applications," *Computability in Context: Computation and Logic in the Real World*, p. 275–296, 2011.

[270] R. Amer, W. Saad, H. Elsawy, M. Butt, and N. Marchetti, "Caching to the sky: Performance analysis of cache-assisted CoMP for cellular-connected UAVs," in *Proc. IEEE Wireless Communications and Networking Conf.*, Marrakech, Morocco, April 2019.

[271] C. Zhu and W. Yu, "Stochastic modeling and analysis of user-centric network MIMO systems," *IEEE Transactions on Communications*, vol. 66, no. 12, pp. 6176–6189, December 2018.

[272] P. Series, "Propagation data and prediction methods required for the design of terrestrial broadband radio access systems operating in a frequency range from 3 to 60 GHz," ITU-R Report, 2013.

[273] R. W. Heath Jr, T. Wu, Y. H. Kwon, and A. C. Soong, "Multiuser MIMO in distributed antenna systems with out-of-cell interference," *IEEE Transactions on Signal Processing*, vol. 59, no. 10, pp. 4885–4899, October 2011.

[274] D. K. Cheng, "Optimization techniques for antenna arrays," *Proceedings of the IEEE*, vol. 59, no. 12, pp. 1664–1674, December 1971.

[275] W. L. Stutzman and G. A. Thiele, *Antenna Theory and Design*. Hoboken, NJ, USA: John Wiley & Sons, 2012.

[276] S. Vaidyanathan and C.-H. Lien, *Applications of Sliding Mode Control in Science and Engineering*.Hoboken, NJ, USA Springer, 2017, vol. 709.

[277] L. C. Evans, "An introduction to mathematical optimal control theory," *Lecture Notes, University of California, Department of Mathematics, Berkeley*, 2005.

[278] Y. Mutoh and S. Kuribara, "Control of quadrotor unmanned aerial vehicles using exact linearization technique with the static state feedback," *Journal of Automation and Control Engineering*, vol. 4, no. 5, pp. 340–346, October 2016.

[279] R. Amer, W. Saad, and N. Marchetti, "Towards a connected sky: Performance of beamforming with down-tilted antennas for ground and UAV user co-existence," *IEEE Comm. Letters*, 2019.

[280] T. Zeng, M. Mozaffari, O. Semiari, W. Saad, M. Bennis, and M. Debbah, "Wireless communications and control for swarms of cellular-connected UAVs," in *Proc. Asilomar Conf. on Signals, Systems, and Computers*, Pacific Grove, CA, USA, November 2018.

[281] G. Yang, X. Lin, Y. Li, H. Cui, M. Xu, D. Wu, H. Rydén, and S. B. Redhwan, "A telecom perspective on the internet of drones: From LTE-Advanced to 5G," *arXiv preprint arXiv:1803.11048*, March 2018.

[282] Qualcomm, "Lte unmanned aircraft systems; trial report (version 1.0.1)," Tech. Rep., May 2017.

[283] X. Lin, J. Li, R. Baldemair, T. Cheng, S. Parkvall, D. Larsson, H. Koorapaty, M. Frenne, S. Falahati, A. Grövlen et al., "5G new radio: Unveiling the essentials of the next generation wireless access technology," *arXiv preprint arXiv:1806.06898*, June 2018.

[284] X. Lin, J. Andrews, A. Ghosh, and R. Ratasuk, "An overview of 3GPP device-to-device proximity services," *IEEE Communications Magazine*, vol. 52, no. 4, pp. 40–48, April 2014.

[285] 3GPP, "New WID on enhanced LTE support for aerial vehicles," *RP-172826*, January 2018.

[286] X. Lin, J. Bergman, F. Gunnarsson, O. Liberg, S. M. Razavi, H. S. Razaghi, H. Rydn, and Y. Sui, "Positioning for the Internet of Things: A 3GPP perspective," *IEEE Communications Magazine*, vol. 55, no. 12, pp. 179–185, December 2017.

[287] UAS Identification and Tracking (UAS ID) Aviation Rulemaking Committee (ARC), "ARC recommendations final report," Tech. Rep., September 2017.

[288] 3GPP, "New study on remote identification of unmanned aerial systems," *SP-180172*, March 2018.

[289] E. Dahlman, S. Parkvall, and J. Skold, *4G: LTE/LTE-Advanced for Mobile Broadband*. Cambridge, MA, USA: Academic Press, 2013.

[290] FAA "Federal aviation administration reports," Available: www.faa.gov/about/plans-reports.

[291] J. G. Andrews, "Seven ways that HetNets are a cellular paradigm shift," *IEEE Communications Magazine*, vol. 51, no. 3, pp. 136–144, March 2013.

[292] X. Lin, R. K. Ganti, P. J. Fleming, and J. G. Andrews, "Towards understanding the fundamentals of mobility in cellular networks," *IEEE Transactions on Wireless Communications*, vol. 12, no. 4, pp. 1686–1698, April 2013.

[293] 3GPP, "Mobility enhancements in heterogeneous networks," *3GPP TR 38.839, V11.1.0*, January 2013.

[294] ——, "Radio resource control (RRC); protocol specification," *3GPP TS 36.331, V15.2.2*, July 2018.

[295] S. Euler, H.-L. Maattanen, X. Lin, Z. Zou, M. Bergström, and J. Sedin, "Mobility support for cellular connected unmanned aerial vehicles: Performance and analysis," *arXiv preprint arXiv:1804.04523*, 2018.

[296] 3GPP, "Study on scenarios and requirements for next generation access technologies," *3GPP TR 38.913, V15.0.0*, July 2018.

[297] A. Kumbhar, I. Güvenç, S. Singh, and A. Tuncer, "Exploiting LTE-advanced HetNets and FeICIC for UAV-assisted public safety communications," *IEEE Access*, vol. 6, pp. 783–796, 2018.

[298] A. Merwaday and I. Guvenc, "UAV assisted heterogeneous networks for public safety communications," in *IEEE Wireless Communications and Networking Conference Workshops (WCNCW)*, 2015, pp. 329–334.

[299] M. Moradi, K. Sundaresan, E. Chai, S. Rangarajan, and Z. M. Mao, "SkyCore: Moving core to the edge for untethered and reliable UAV-based LTE networks," in *Proceedings of the 24th Annual International Conference on Mobile Computing and Networking*. ACM, 2018, pp. 35–49.

[300] S. Rohde and C. Wietfeld, "Interference aware positioning of aerial relays for cell overload and outage compensation," in *IEEE Vehicular Technology Conference*, September 2012, pp. 1–5.

[301] R. Gangula, O. Esrafilian, D. Gesbert, C. Roux, F. Kaltenberger, and R. Knopp, "Flying rebots: First results on an autonomous UAV-based LTE relay using open airinterface," in *IEEE 19th International Workshop on Signal Processing Advances in Wireless Communications (SPAWC)*, 2018, pp. 1–5.

[302] 3GPP, "Remote identification of unmanned aerial systems; stage 1," *3GPP TS 22.825, V16.0.0*, September 2018.

[303] E. Dahlman, S. Parkvall, and J. Skold, *5G NR: The Next Generation Wireless Access Technology*. Cambridge, MA, USA: Academic Press, 2018.

[304] ITU-R SG05, "Draft new report ITU-R M.[IMT-2020.TECH PERF REQ] - minimum requirements related to technical performance for IMT-2020 radio interface(s)," February 2017.

[305] 3GPP, "NR; multi-connectivity; overall description; stage-2," *version 15.2.0*, June 2018.

[306] NGMN, "Description of network slicing concept," *V1.0.8*, September 2016.

[307] P. Rost, C. Mannweiler, D. S. Michalopoulos, C. Sartori, V. Sciancalepore, N. Sastry, O. Holland, S. Tayade, B. Han, D. Bega et al., "Network slicing to enable scalability and flexibility in 5G mobile networks," *IEEE Communications Magazine*, vol. 55, no. 5, pp. 72–79, May 2017.

[308] 3GPP, "NR; overall description; stage-2," *3GPP TS 38.300, V15.2.0*, June 2018.

[309] ——, "System architecture for the 5G system," *3GPP TS 23.501, V15.2.0*, June 2018.

[310] ——, "5G system; network data analytics services; stage 3," *3GPP TS 29.520, V15.0.0*, June 2018.

[311] ——, "Study of enablers for network automation for 5G," *3GPP TR 23.791, V0.5.0*, July 2018.

[312] ——, "Study on RAN-centric data collection and utilization for LTE and NR," *3GPP TR 37.816, V0.1.0*, October 2018.

[313] ITU-T Focus Group, "Machine learning for future networks including 5G," November 2017.

[314] N. M. Rodday, R. d. O. Schmidt, and A. Pras, "Exploring security vulnerabilities of unmanned aerial vehicles," in *NOMS 2016-2016 IEEE/IFIP Network Operations and Management Symposium*. IEEE, 2016, pp. 993–994.

[315] K. Mansfield, T. Eveleigh, T. H. Holzer, and S. Sarkani, "Unmanned aerial vehicle smart device ground control station cyber security threat model," in *2013 IEEE International Conference on Technologies for Homeland Security (HST)*. IEEE, 2013, pp. 722–728.

[316] A. Y. Javaid, W. Sun, V. K. Devabhaktuni, and M. Alam, "Cyber security threat analysis and modeling of an unmanned aerial vehicle system," in *2012 IEEE Conference on Technologies for Homeland Security (HST)*. IEEE, 2012, pp. 585–590.

[317] A. J. Kerns, D. P. Shepard, J. A. Bhatti, and T. E. Humphreys, "Unmanned aircraft capture and control via GPS spoofing," *Journal of Field Robotics*, vol. 31, no. 4, pp. 617–636, 2014.

[318] D. He, S. Chan, and M. Guizani, "Communication security of unmanned aerial vehicles," *IEEE Wireless Communications*, vol. 24, no. 4, pp. 134–139, 2016.

[319] J. Valente and A. A. Cardenas, "Understanding security threats in consumer drones through the lens of the discovery quadcopter family," in *Proceedings of the 2017 Workshop on Internet of Things Security and Privacy*. ACM, 2017, pp. 31–36.

[320] Q. Wu, W. Mei, and R. Zhang, "Safeguarding wireless network with UAVs: A physical layer security perspective," *arXiv:1902.02472*, February 2019.

[321] M. Cui, G. Zhang, Q. Wu, and D. W. K. Ng, "Robust trajectory and transmit power design for secure UAV communications," *IEEE Trans. Veh. Technol.*, vol. 67, no. 9, pp. 9042–9046, September 2018.

[322] A. Hussain, J. Heidemann, J. Heidemann, and C. Papadopoulos, "A framework for classifying denial of service attacks," in *Proceedings of the 2003 conference on Applications,*

technologies, architectures, and protocols for computer communications. ACM, 2003, pp. 99–110.

[323] M. Mozaffari, W. Saad, M. Bennis, Y.-H. Nam, and M. Debbah, "A tutorial on UAVs for wireless networks: Applications, challenges, and open problems," *IEEE Communications Surveys & Tutorials*, 2019.

[324] M. Hooper, Y. Tian, R. Zhou, B. Cao, A. P. Lauf, L. Watkins, W. H. Robinson, and W. Alexis, "Securing commercial WiFi-based UAVs from common security attacks," in *MILCOM 2016-2016 IEEE Military Communications Conference*. IEEE, 2016, pp. 1213–1218.

[325] A. Perkins, L. Dressel, S. Lo, and P. Enge, "Antenna characterization for UAV-based GPS jammer localization," in *Proceedings of the 28th International Technical Meeting of The Satellite Division of the Institute of Navigation (ION GNSS+ 2015)*, 2015.

[326] K. C. Zeng, Y. Shu, S. Liu, Y. Dou, and Y. Yang, "A practical GPS location spoofing attack in road navigation scenario," in *Proceedings of the 18th International Workshop on Mobile Computing Systems and Applications*. ACM, 2017, pp. 85–90.

[327] A. Eldosouky, A. Ferdowsi, and W. Saad, "Drones in distress: A game-theoretic countermeasure for protecting UAVs against GPS spoofing," April 2019. [Online]. Available: https://arxiv.org/abs/1904.11568

[328] A. El-Dosouky, W. Saad, and D. Niyato, "Single controller stochastic games for optimized moving target defense," in *Proc. Int. Conf. on Communications*, Kuala lumpur, Malaysia, May 2016.

[329] S. Čapkun, M. Hamdi, and J.-P. Hubaux, "GPS-free positioning in mobile ad hoc networks," *Cluster Computing*, vol. 5, no. 2, pp. 157–167, 2002.

[330] M. Baza, M. Nabil, N. Bewermeier, K. Fidan, M. Mahmoud, and M. Abdallah, "Detecting sybil attacks using proofs of work and location in vanets," *arXiv preprint arXiv:1904.05845*, 2019.

[331] M. McFarland, "Google drones will deliver chipotle burritos at Virginia Tech," *CNN Money*, September 2016.

[332] Amazon, "Amazon prime air," 2016.

[333] A. Y. Javaid, W. Sun, V. K. Devabhaktuni, and M. Alam, "Cyber security threat analysis and modeling of an unmanned aerial vehicle system," in *IEEE Conference on Technologies for Homeland Security (HST)*, November 2012, pp. 585–590.

[334] K. Mansfield, T. Eveleigh, T. H. Holzer, and S. Sarkani, "Unmanned aerial vehicle smart device ground control station cyber security threat model," in *IEEE International Conference on Technologies for Homeland Security (HST)*, November 2013, pp. 722–728.

[335] N. M. Rodday, R. d. O. Schmidt, and A. Pras, "Exploring security vulnerabilities of unmanned aerial vehicles," in *IEEE/IFIP Network Operations and Management Symposium (NOMS)*, April 2016, pp. 993–994.

[336] J. Pagliery, "Sniper attack on California power grid may have been an insider, dhs says," *CNN. com*, October 2015.

[337] G. Xiang, A. Hardy, M. Rajeh, and L. Venuthurupalli, "Design of the life-ring drone delivery system for rip current rescue," in *IEEE Systems and Information Engineering Design Symposium (SIEDS)*, April 2016, pp. 181–186.

[338] V. Gatteschi, F. Lamberti, G. Paravati, A. Sanna, C. Demartini, A. Lisanti, and G. Venezia, "New frontiers of delivery services using drones: A prototype system exploiting a quadcopter for autonomous drug shipments," in *39th IEEE Annual Computer Software and Applications Conference (COMPSAC)*, vol. 2, July 2015, pp. 920–927.

[339] A. Sanjab, W. Saad, and T. Başar, "Prospect theory for enhanced cyber-physical security of drone delivery systems: A network interdiction game," in *Proc. Int. Conf. on Communications*, Paris, France, May 2017.

[340] A. Sanjab, W. Saad, and T. Başar, "A game of drones: Cyber-physical security of time-critical UAV applications with cumulative prospect theory perceptions and valuations," *arXiv:1902.03506*, February 2019. [Online]. Available: https://arxiv.org/abs/1902.03506

[341] T. Başar and G. J. Olsder, *Dynamic Noncooperative Game Theory*. Philadelphia, PA, USA: SIAM Series in Classics in Applied Mathematics, January 1999.

[342] R. K. Wood, "Deterministic network interdiction," *Mathematical and Computer Modeling*, vol. 17, no. 2, pp. 1–18, 1993.

[343] T. Başar and G. J. Olsder, *Dynamic Noncooperative Game Theory*. Philadelphia, PA: SIAM Series in Classics in Applied Mathematics, January 1999.

[344] D. Kahneman and A. Tversky, "Prospect theory: An analysis of decision under risk," *Econometrica*, vol. 47, pp. 263–291, 1979.

Index

3D cellular network, 159, 161
3D deployment, 161
3D wireless network, 159, 168
3GPP
  Release 15, 209, 231
  Release 16, 232
3GPP standardization, 208, 227
3GPP standards, 209
4G, 212
5G, 15, 208, 234, 237
5G new radio, 207, 208, 234
6G, 207

Aerial wireless channels, 27, 37, 40, 51, 56, 67
  characteristics, 27
  height, 28
Air traffic management, 209
Airframe shadowing, 28, 43
Angel of arrival, 35
Angel of departure, 35
Angular spread, 57
Antenna array, 15, 181, 192–195, 198, 203
Antenna configuration, 29
Antenna tilt, 216
Artificial intelligence, 238

Backhaul, 18, 159, 169
Bang-bang solution, 201
Beamforming, 15, 115, 116, 181, 192, 193

Caching, 112–114, 116, 117, 120, 169, 174, 175, 183, 189, 191
Cauchy-Schwarz inequality, 186
Cell association, 101, 103, 117, 126, 133, 134, 145, 146, 149, 152, 159, 161, 165
Cellular technologies, 208
Channel modeling, 36, 40, 229
  weather effects, 50
Circle packing, 93, 97, 122
Cloud radio access network, 112, 118
Coherence time, 52
CoMP, 183, 185, 189, 191, 205

Complex baseband signal, 60
Continuous phase modulation, 65
Control time, 193–195, 198–200, 202
Cooperation, 181
Cooperative communications, 181, 191
Coordinated multi-point transmission, 181, 218
Coverage probability, 73, 74, 78, 79, 83, 86, 96, 97, 181, 186, 187, 189, 191
Covering problem, 93

D2D, 70, 71, 77, 82, 84
Data acquisition, 233
Deep echo state network, 131, 132, 134
Deep reinforcement learning, 131, 134
Delay spread, 54
Device-to-device communication, 13, 70, 209
Diffraction, 35, 40, 42, 46
Direct sequence spread spectrum, 22, 64
  Rake receiver, 65
  spreading waveform, 64
Disk covering problem, 79
Doppler effects, 53
Doppler shift, 25, 26, 29, 63
Doppler spread, 29, 52
Down-tilt, 184
Duty cycle, 170

Echo state networks, 116, 131
Empirical path loss model, 40
Excess path loss model, 42

Facility location, 91, 92
Fading, 27, 35, 52, 188
Flight time, 68, 81, 146
Floating intercept model, 39
Flying mode detection, 231
Flying taxi, 21
Frequency division duplex, 211
Frequency hopping spread spectrum, 64
Frequency planning, 162, 163
Frequency reuse factor, 163
Frequency selectivity, 54

Index

Fresnel zone, 45
Fronthaul, 113, 117, 118, 169, 173

Game theory, 123, 128, 244
   behavioral strategy, 130
   dynamic noncooperative game, 128
   Nash equilibrium, 130, 249
   subgame perfect Nash equilibrium, 130, 132, 135
General ray tracing, 35
Global navigation satellite system, 210
GPRS, 212
GPS spoofing, 240, 242

Handover, 221, 230
Handover failure, 220, 222
High-altitude platforms, 3, 6, 10, 18, 44, 159
Hover time, 145, 148, 150, 155

Information dissemination, 13
Interference, 8, 70, 73, 82, 99, 109, 113, 123, 126, 136, 181, 183, 185, 218
Interference detection, 229, 232
Interference mitigation, 230
Internet of Things, 16, 90, 100, 101, 107, 108, 111

Jain fairness index, 154

Large-scale propagation effects, 24, 30
Latency, 68, 81, 125, 138, 141, 160, 167, 223
Licensed band, 169, 172, 174
Licensed-assisted access, 226
Line-of-sight probability, 44, 47, 50, 73, 85, 97, 102, 148
Line-of-sight propagation, 23, 28, 45
Location-based services, 209
Log-distance path loss, 37
   basic model, 37
   dual-slope, 40
   modified model, 39
   multi-slope, 40
Loon, 3
Low altitude platforms, 3, 6, 159
LTE, 207, 209, 210, 216
   evolved packet core, 212
   introduction, 210
   radio access network, 212
   radio interface, 213
   RAN protocol stack, 213
   system architecture, 212
LTE-U, 169, 171, 226

Machine intelligence, 238
Machine learning, 90, 112, 116, 128, 160, 175, 179
   k-mean clustering, 117
   liquid state machines, 175
   recurrent neural networks, 112, 116, 131

   spiking neural networks, 175
Millimeter wave, 14
Mobile broadband, 207
Mobility management, 9, 220
Mobility models, 223
Multi-antenna techniques, 211
Multipath, 26, 29, 35, 51
Multiple access, 81, 146

Nakagami fading, 59, 185
Network interdiction game, 246
Network slicing, 237

OFDM, 211, 215, 235
   cyclic prefix, 215
Optimal control, 195, 200, 201
Optimal transport theory, 151, 165
   Monge-Kantorovich problem, 152
Orthogonal frequency division multiplexing, 62
   cyclic prefix, 63
   waveform, 63
Outage probability, 82, 87

Path loss, 25, 30, 31, 33, 34, 37, 39, 40
Path loss exponent, 37, 38
Path planning, 123, 124, 128, 134
Performance analysis, 68, 72, 80
Perturbation technique, 196
Power control, 103–105, 126, 131, 232
Propagation modeling, 7, 22, 23, 27, 36
Prospect theory, 244
   framing, 251
   rationality parameter, 252
   weighting effect, 251
Public safety, 12, 13

Q-learning, 176
Quadrotor, 193, 199

Radio wave propagation
   absorption, 24
   diffraction, 23
   fundamentals, 23
   reflection, 23
   refraction, 23
   scattering, 24
Ray tracing, 24, 31, 47
Reflection, 40
Reinforcement learning, 123, 131, 134
Reliability, 107, 109, 224, 225
Remote radio heads, 112, 115
Reservoir computing, 131
Resource management, 7, 145, 159, 169, 174, 180
Resource planning, 159
Rician channel model, 58, 125

Scattering, 35, 40, 47
Security strategy, 248
Shadowing, 30, 42
Small-scale propagation effects, 25, 51
Smart city, 20
Spatial selectivity, 56
Spectrum allocation, 174, 175
Spectrum management, 145, 169
Stochastic geometry, 68, 181
    binomial point process, 184
    Poisson point process, 69, 183
Stop points, 80, 83

Thrust, 199
Time division duplex, 211
Time selectivity, 52
Trajectory optimization, 123
Truncated octahedron, 161
Two-ray model, 31, 33, 34, 37
    height dependent, 41

UAV, 1
    applications, 12
    classification, 3
    command and control, 223, 237
    definition, 1
    deployment, 90, 100, 105, 116, 117, 161
    history, 1
    identification, 209, 230–232
    interference, 217
    localization, 210
    mission time, 124
    mobility, 9, 79, 106, 125, 220, 230
    positioning, 210
    regulation criteria, 5
    regulations, 4
    security, 240
    trajectory, 123
    wireless communications and networking, 5
    wireless networking scenarios, 12
UAV base station, 6, 12, 14–16, 18, 68, 70, 73, 79, 97–101, 104, 145, 147, 149, 159, 166, 170, 181, 192, 203, 208, 226
    challenges, 7
UAV BS, 94
UAV relays, 9, 14, 18, 227
    research challenges, 10
UAV security
    communication channel attacks, 240
    delivery systems, 243
    denial-of-service attacks, 241
    eavesdropping, 241
    false data injection, 241
    fly-away attack, 241
    GPS attack, 240
    GPS attacks, 241
    information attacks, 241
    man-in-the-middle attack, 241
    physical attacks, 244
UAV user equipment, 8, 13, 18, 19, 41, 123, 124, 128, 134, 159, 166, 181, 183, 184, 188, 208, 216, 224, 228, 229
    identification, 9
    research challenges, 8
Unlicensed band, 169, 174

Virtual reality, 18
Virtual reality, ECHOTL,MING01, 16
Virtual reality, MING01, 237
Voronoi diagram, 149, 153, 166

Waveform, 22
Waveform design, 22, 60
Wind dynamics, 195, 199
Wireless research challenges, 6

Zephyr, 3